# Fault Diagnosis and Sustainable Control of Wind Turbines

Robust Data-Driven and Model-Based Strategies

# Fault Diagnosis and Sustainable Control of Wind Turbines

## Robust Data-Driven and Model-Based Strategies

Silvio Simani
Saverio Farsoni

Butterworth-Heinemann
An imprint of Elsevier

Butterworth-Heinemann is an imprint of Elsevier
The Boulevard, Langford Lane, Kidlington, Oxford OX5 1GB, United Kingdom
50 Hampshire Street, 5th Floor, Cambridge, MA 02139, United States

**Library of Congress Cataloging-in-Publication Data**
A catalog record for this book is available from the Library of Congress

**British Library Cataloguing-in-Publication Data**
A catalogue record for this book is available from the British Library

ISBN: 978-0-12-812984-5

For information on all Butterworth-Heinemann publications
visit our website at https://www.elsevier.com/books-and-journals

Working together
to grow libraries in
developing countries

www.elsevier.com • www.bookaid.org

*Publisher:* Mara Conner
*Acquisition Editor:* Sonnini R. Yura
*Editorial Project Manager:* Katie Chan
*Production Project Manager:* Surya Narayanan Jayachandran
*Designer:* Miles Hitchen

Typeset by VTeX

# Dedication

*To our families*

# Contents

# Foreword

This monograph aims to report and encourage technology transfer in control engineering applied to wind turbines. The rapid development of control technology has an impact on all areas of the control discipline. In the recent years, new theoretical and technological developments have widen the scope of applications and challenges that can be addressed. Much of this development work resides in industrial reports, feasibility study papers and the reports of advanced collaborative projects. This monograph offers an opportunity for researchers, practitioners, and students to have an extended and clear exposition of new investigations in applied industrial control and diagnosis for wider and rapid dissemination.

As many technological systems become more complex, widespread, and integrated, the effects of system faults can be simply devastating to the infrastructure of modern society. Feedback control is just one important component of total system supervision. Fault diagnosis and sustainable (fault tolerant) control represents further components with extensive commercial, industrial, and societal implications if only we could work out how to do it in a reliable and inexpensive manner. Hardware or physical redundancy is the usual solution of the practical fault diagnosis and sustainable control problems, but as Farsoni and Simani note in this monograph, it is capital and maintenance costly. The search for reliable, robust, and inexpensive fault diagnosis and sustainable control methods has been active since the early 1980s. Since 1991, the International Federation of Automatic Control (IFAC) has created the SAFEPROCESS Steering Committee to promote research, developments, and applications in the fault diagnosis and fault tolerant control fields. The last decade has seen the formalization of several theoretical approaches accompanied by some attempts to standardize nomenclature in the field.

The related literature does not have many entries from this important research area applied to wind turbines, even if several monographs can represent interesting contributions on control and condition monitoring for such systems, even if they use quite different ideas and principles. To the existing references, we can now add this monograph by Farsoni and Simani. Key features of this text include useful survey material, new approaches based on data-driven and model-based methodologies, as well as an extended application study using a well-established wind turbine benchmark. Different groups of readers ranging from industrial engineers wishing to gain insight into the applications potential of new fault diagnosis and sustainable control methods, to the academic control community looking for new problems to tackle will find much to learn from this monograph.

**Vicenc Puig**
Barcelona, Spain
August 2017

# Preface

Control devices, which are nowadays exploited to improve the overall performance of industrial processes, involve both sophisticated digital system design techniques and complex hardware (input–output sensors, actuators, components, and processing units). Such complexity results in an increased probability of failure. As a direct consequence of this, control systems must include automatic supervision of the closed-loop operation to detect, isolate, and accommodate malfunctions as early as possible.

Since the early 1980s the problem of fault diagnosis and fault tolerant control in dynamic processes has received great attention, and a large variety of methodologies have been studied and developed based upon both physical and analytical redundancy. In the first case, the system is equipped with redundant physical devices, like sensors and actuators, so that if a fault occurs, the redundant device replaces the functionality of the faulty one.

The analytical redundancy approach is based on a completely different principle. The basic idea consists of using an accurate model of the system to mimic the real process behavior. If a fault occurs, the residual signal (i.e., the difference between real system and model behaviors) can be used to diagnose and isolate the malfunction. This approach has some advantages with respect to physical (hardware/software) redundancy, mainly in economical and practical aspects. The analytical redundancy approach does not require additional equipment, but also suffers from some potential disadvantages, which are principally related to the need of an accurate model of the real system. Model-based method reliability, which also includes false alarm rejection, is strictly related to the "quality" of the model and measurements exploited for fault diagnosis, as model uncertainty and noisy data can prevent an effective application of analytical redundancy methods.

On the other hand, modern technological systems rely on sophisticated control systems suitable to meet increased performance and safety requirements. A conventional feedback control design for a complex system may result in an unsatisfactory performance, or even instability, in the event of malfunctions in actuators, sensors, or other system components. To overcome such weaknesses, new approaches to control system design have been developed in order to tolerate component malfunctions, while maintaining desirable stability and performance properties. Such control systems are often known as sustainable (fault tolerant control) systems, which possess the ability to accommodate the effects of faults automatically. For this reason, the closed-loop control system can usually handle malfunctions, while maintaining desirable performance. A fault tolerant control system can be also based on the fault diagnosis scheme. In general, fault tolerant control methods are classified into two types, i.e., passive fault tolerant control schemes and active fault tolerant control schemes. In the passive solutions, controllers are fixed and are designed to be robust against a class of presumed faults. This approach needs neither fault diagnosis schemes nor controller reconfiguration, but it has limited fault-tolerant capabilities. In contrast to passive schemes, active methodologies react to the system component failures actively by reconfiguring control actions so that the stability and acceptable performance of the entire system can be maintained. A successful active fault tolerant control scheme can rely heavily on real-time fault diagnosis methods to provide the most up-to-date information about the true status of the system.

This is not a simple problem. As model-based methods are designed to detect any discrepancy between real system and model behaviors, it is assumed that a discrepancy signal is related to (has a response from) a fault. However, the same difference signal can respond to model mismatch or noise in real measurements, which can be (erroneously) detected as a fault, giving rise to a "false alarm" in detection. These considerations have led to research in the field of "robust" methods, in which particular attention is paid to the discrimination between actual faults and errors due to model mismatch. On the other hand, the availability of a "good" model of the monitored system can significantly improve the performance of diagnostic tools, minimizing the probability of false alarms.

This monograph focuses on the explanation of what is a "good" model suitable for robust diagnosis and control of system performance and operation. The book also carefully describes how "accurate models" can be obtained from real data. A large amount of attention is paid to the "real system modeling problem", with reference to either linear–nonlinear model structures. Special treatment is given to the case in which noise and disturbance affect the acquired data. The mathematical description of the monitored system can be obtained by means of both model-based and data-driven approaches. Comparisons between these strategies are also addressed.

After an overview of the state-of-the-art of the related literature, the discussion of the mathematical description of the system under diagnosis, followed by a review of data-driven and model-based procedures for fault diagnosis and fault tolerant control, the monograph focuses on the fault diagnosis and fault tolerant control problems for several fault situations affecting wind turbine systems.

The purpose of the monograph is to provide guidelines for the modeling of wind turbine systems oriented to robust and reliable fault diagnosis and sustainable control. Hence significant attention is paid to practical application of the methods described to realistic wind turbine installations, as reported in the last chapter.

This monograph aims also to report and encourage technology transfer in different control engineering fields. In fact, the advanced development of control technology and its related fields have impacts on all areas of the control discipline. New modeling theory, novel control strategies, actuator and sensor devices, computer applications represent new challenges regarding the proposed issues. Most of these investigations can be found in technical reports and research papers, as well as reports of collaborative projects. Therefore this monograph offers an opportunity for researchers, practitioners, and students to have an extended and clear exposition of new investigations in modeling and advanced control of wind turbine installations for wider and rapid dissemination.

The related literature does not have many entries from this important research area, but previous monographs by the authors represent interesting contributions even if similar ideas and principles are applied to different industrial processes. The key features of this monograph include useful survey material, new approaches relying on data-driven and model-based methodologies for robust and reliable fault diagnosis and fault tolerant control, as well as extended application studies using different wind turbine benchmarks and installations.

Both theoretical and practical arguments have been presented and discussed in a homogeneous manner and the book targets both professional engineers working in industry and researchers in academic and scientific institutions. In this way different groups of readers ranging from industrial engineers wishing to gain insight into the applications potential of new fault diagnosis and sustainable control methods, to the academic control community looking for new problems to tackle will find much to learn from this work.

**Silvio Simani**
**Saverio Farsoni**
Ferrara
February 2017

# Glossary

The symbols and abbreviations listed here are used unless otherwise stated.

**AF**      Adaptive Filter
**AFTC**    Active Fault Tolerant Control
**ANFIS**   Adaptive Neuro-Fuzzy Inference System
**ARMAX**   AutoRegressive Moving Average eXogenous
**ARX**     AutoRegressive eXogenous
**CART**    Classification and Regression Trees
**DOS**     Dedicated Observer Scheme
**EA**      Evolutionary Algorithms
**EE**      Equation Error
**EIV**     Errors-In-Variables
**EKF**     Extended Kalman Filter
**FAST**    Fatigue, Aerodynamics, Structures, and Turbulence
**FDD**     Fault Detection & Diagnosis
**FMEA**    Failure Mode & Effect Analysis
**FDI**     Fault Detection & Isolation
**FTC**     Fault Tolerant Control
**GK**      Gustafson–Kessel
**GOS**     Generalized Observer Scheme
**HAWT**    Horizontal-Axis Wind Turbine
**KF**      Kalman Filter
**LS**      Least-Squares
**MIMO**    Multiple-Input Multiple-Output
**MISO**    Multiple-Input Single-Output
**MF**      Membership Function
**MLP**     Multi-Layer Perceptron
**NF**      Neuro-Fuzzy
**NLGA**    NonLinear Geometric Approach
**NARX**    Nonlinear AutoRegressive eXogenous
**NN**      Neural Network
**NREL**    National Renewable Energy Laboratory
**OE**      Output Error
**OO**      Output Observer
**OLS**     Ordinary Least-Squares
**PFTC**    Passive Fault Tolerant Control
**RBF**     Radial Basis Function
**RLS**     Recursive Least-Squares
**SISO**    Single-Input Single-Output
**SOI**     Severity Occurrence Index
**TS**      Takagi–Sugeno
**UIKF**    Unknown Input Kalman Filter
**UIO**     Unknown Input Observer

Chapter 1

# Introduction

## 1.1 INTRODUCTION

With increasing concerns about climate change and the depletion of fossil fuels, renewable energy has become a topical area for research. To date, significant penetration of, for example, wind and solar electricity generation systems has taken place in an effort to supplement thermal power and other traditional modes of electricity generation. However, the costs associated with these developing technologies are high, and favorable feed-in tariffs are usually employed to encourage development. In addition, the variability of the production from such intermittent renewables incurs extra grid integration costs. Nevertheless, wind and other less mature renewable technologies, such as wave energy, are required to satisfy increasing world energy demand in a climate-neutral and cost-effective way. There is therefore an imperative to reduce the cost of renewable energy, in order to compete with, and eventually displace, significant number of conventional electricity generation plants.

With this economic goal in mind, it is important that renewable energy converters incorporate sufficient intelligence to allow them to convert the available energy as efficiently as possible for a given capital investment, while prolonging the life of the conversion systems and their components. It is clear that control systems technology has a strong role to play in optimizing the operation of these conversions systems. In particular, this monograph presents a selection of modeling, control system, supervision approaches, and methodologies in the wind application area as a sample of the spectrum of possibilities. In addition, since this application area and the application of control technology in it have grown up relatively independently, it gives an opportunity to examine some key aspects of both the modeling and control challenges of this application area and the nature of the control systems being developed. In addition, there is an opportunity for cross-pollination between domains, especially in view of the fact that wind energy is now relatively mature from the modeling point of view, while presenting challenging aspects in the fields of control, supervision, fault diagnosis, and fault tolerance.

The monograph covers different aspects of the wind conversion systems, spanning also the application area of wind energy.

First, the monograph attempts to provide an introduction to wind conversion systems, presenting an overview of the conversion systems and principles, the mathematical models that are used to describe them and the wind resources that drive them, and some control possibilities. An important feature of this monograph is the comparisons and contrasts made between the different techniques and technologies in this application area.

Second, the monograph is focused on both model-based and data-driven approaches to the modeling and control of wind turbines. Different data-driven schemes for monitoring and fault detection in wind turbines are considered, which use the measurements already available in wind turbines, thus avoiding the introduction of expensive auxiliary condition monitoring modules. The main advantage with respect to the model-based approaches is that only the data of the system behavior are exploited. On the other hand, model-based strategies that may require accurate analytical descriptions of the systems under investigation can provide more efficient regulation and quicker reaction and recovery against disturbance and uncertainty affecting the system, with respect to data-driven schemes available in the literature.

The monograph covers modeling, supervision, and control approaches ranging from the more traditional techniques typically found in feedback regulatory systems to more advanced methods, also appearing in the solutions of optimal and nonlinear control. Optimal and suboptimal methodologies are considered as viable approaches to maximize the energy capture for wind energy devices, but also focused on the important issue of the system availability.

Finally, on the one hand, the simulations considered in the monograph adopt a standard statistics approach, where the wind excitation and the uncertainty affecting the system are modeled as a Gaussian stochastic disturbance. On the other hand, the experimental results are obtained by considering high-fidelity simulators or hardware-in-the-loop tools which exhibit realistic complex and nonlinear behaviors of wind turbine systems. Key considerations will concern the calculation of optimal solutions, which take into account of nonideal effects and model-reality features of the energy conversion systems.

In summary, the topics addressed in this monograph, including several overview studies, help to show the relative extent to which control technology has penetrated wind energy. It could reasonably be argued that the greater advancement of

**Fault Diagnosis and Sustainable Control of Wind Turbines.** DOI: 10.1016/B978-0-12-812984-5.00001-8

control technology in wind energy reflects the greater commercial development of wind energy. Aside from maximizing converted power in view of the physical constraints of the system and a need to maximize the lifetime of device components, there is clear value in advanced modeling, monitoring, and control strategies.

## 1.2 MOTIVATIONS

The increased level of wind-generated energy in power grids worldwide raises the levels of reliability and sustainability required from wind turbines. Wind farms should have the capability to generate the desired value of electrical power continuously, depending on the current wind speed level and on the grid demand.

As a consequence, the possible faults affecting the system have to be properly identified and treated, before they endanger the correct functioning of the turbines or become critical faults. Wind turbines in the megawatt range are extremely expensive systems, therefore their availability and reliability must be high, in order to assure the maximization of the generated power while minimizing the operation and maintenance (O&M) services. Alongside the fixed costs of the produced energy, mainly due to the installation and the foundation of the wind turbine, the O&M costs could increase the total energy cost up to about the 30%, particularly considering the offshore installation (Odgaard and Patton, 2012).

These considerations motivate the introduction of fault diagnosis system coupled with fault tolerant controllers. Currently, most of the turbines feature a simple conservative approach against faults that consists in the shutdown of the system to wait for maintenance service. Hence effective strategies coping with faults have to be studied and developed to improve the turbine performance, particularly in faulty working conditions. Their benefits would concern the prevention of critical failures that jeopardize wind turbine components, thus avoiding unplanned replacement of functional parts, as well as the reduction of the O&M costs and the increment of the energy production. The advent of computerized control, communication networks, and information techniques brings interesting challenges concerning the development of novel real-time monitoring and fault tolerant control design strategies for industrial processes.

Indeed, in the recent years, many contributions have been proposed related to the topics of fault diagnosis of wind turbines; see, e.g., Chen et al. (2011), Gong and Qiao (2013b). Some of them highlight the difficulties to achieve the diagnosis of particular faults, e.g., those affecting the drive-train, at wind turbine level. However, these faults are better dealt with at wind farm level, when the wind turbine is considered in comparison to other wind turbines of the wind farm (Odgaard and Stoustrup, 2013a). Moreover, fault tolerant control of wind turbines has been investigated, e.g., in Odgaard and Stoustrup (2015a), Parker et al. (2011) and international competitions on these issues arose (Odgaard and Stoustrup, 2012c; Odgaard and Shafiei, 2015b).

Hence the modeling, fault diagnosis, and *sustainable* control issues for wind turbine systems have been proven to be a challenging task and motivate the research activities carried out through this monograph.

## 1.3 NOMENCLATURE

Since 1991 the Symposium on Fault Detection Supervision and Safety for Technical Processes (SAFEPROCESS), organized by the International Federation of Automatic Control (IFAC), has represented one of the most important international gatherings of academia and industry experts, focused on the topics of diagnosis of dynamic systems, fault tolerant control, process supervision, system reliability and safety, etc. Because of the past inconsistency of the related literature in terms of nomenclature, the SAFEPROCESS committee suggested a set of definitions that represent a good introduction to the issue of fault diagnosis and fault tolerant control. The main definitions, taken from Isermann and Balle (1998) and listed in the following with a brief description, regard the system states and signals, the functions and tasks, the system properties, and the fault characteristics.

- **System states and signals**:
  - **Fault** is an unpermitted deviation of at least one characteristic property or parameter of the system from the acceptable, usual, or standard conditions.
  - **Failure** is a interruption of a system's ability to perform a required function under specified operating conditions.
  - **Malfunction** is an intermittent irregularity in the fulfilment of a system's desired function.
  - **Error** is a deviation between a measured or computed value of an output variable and its true or theoretically correct one.
  - **Disturbance** is an unknown and uncontrolled input acting on a system.
  - **Residual** is a fault indicator, based on a deviation between measurements and model-equation-based computations.
  - **Symptom** is a change of an observable quantity from normal behavior.

- **Functions and tasks**:
  - **Fault detection** is determination of the presence of faults and the time of occurrence.
  - **Fault isolation** is determination of the kind, location, and time of detection of a fault. Follows fault detection.
  - **Fault identification** is determination of the size and time-variant behavior of a fault. Follows fault isolation.
  - **Fault diagnosis** is determination of the kind, size, location, and time of detection of a fault. Follows fault detection. Includes fault detection and identification.
  - **FDI** is an acronym of Fault Detection and Isolation.
  - **FDD** is an acronym of Fault Detection and Identification (Diagnosis).
  - **FTC** is an acronym of Fault Tolerant Control.
  - **AFTC** is an acronym of Active Fault Tolerant Control (an FTC approach).
  - **PFTC** is an acronym of Passive Fault Tolerant Control (an FTC approach).
  - **Fault accommodation** is the task of using the knowledge of the fault function in order to compensate its effect.
  - **Control reconfiguration** represents an active approach in control theory to achieve the FTC for dynamic systems.
- **Models**:
  - **Quantitative model** is using static and dynamic relations among system variables and parameters in order to describe the system behavior in quantitative mathematical terms.
  - **Qualitative model** is using static and dynamic relations among system variables in order to describe the system behavior in qualitative terms, such as causalities.
  - **Analytical model** describes the mathematical relations among the system variables which are based on physical laws derived from the knowledge of the system behavior.
  - **Data-driven model** gives the mathematical relations among the system variables which are inferred on the basis of a set data coming from the system itself.
  - **Model-based model** gives the mathematical relations among the system variables which are derived on the basis of the physical laws describing the system behavior itself.
- **System properties**:
  - **Reliability** is the ability of a system to perform a required function under stated conditions, within a given scope, during a given period of time.
  - **Safety** is the ability of a system not to cause danger to persons or equipment or the environment.
  - **Availability** is the probability that a system or equipment will operate satisfactorily and effectively at any point of time.
- **Faults, time dependency**:
  - **Abrupt fault** is modeled as a stepwise function. It represents bias in the monitored signal.
  - **Incipient fault** is modeled by using ramp signals. It represents drift of the monitored signal.
  - **Intermittent fault** is a combination of pulses with different amplitudes and lengths.
- **Faults, typology**:
  - **Additive fault** influences a variable by an addition of the fault itself. It may represent, e.g., offset of a sensor.
  - **Multiplicative fault** is represented by the product of a variable with the fault itself. Such faults can appear as parameter changes within a process.

## 1.4   INTRODUCTION TO WIND TURBINE MODELING

Modern wind turbines have large flexible structures operating in uncertain environments, thus representing interesting cases for advanced control solutions (Munteanu and Bratcu, 2008). Advanced controllers can help achieve the desirable goal of decreasing the wind energy cost by increasing the efficiency, and thus the energy capture, or by reducing structural loading and increasing the lifetimes of the components and turbine structures (Bianchi et al., 2007; Rivkin et al., 2012).

Horizontal-Axis Wind Turbines (HAWT) have the advantage that the rotor is placed atop a tall tower, where it can take advantage of larger wind speeds higher above the ground. Moreover, HAWT used for utility-scale installations include pitchable blades, improved power capture and structural performance, as well as no need for tensioned cables used to add structural stability (Bianchi et al., 2007; Pramod, 2010; Garcia-Sanz and Houpis, 2012). Vertical-axis (VAWT) solutions are more common for smaller turbines, where these disadvantages become less important and the benefits of reduced noise and omnidirectionality become more pronounced (Burton et al., 2011; Garcia-Sanz and Houpis, 2012). Note finally that

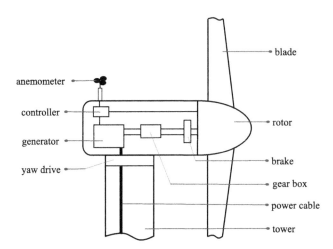

**FIGURE 1.1** Main wind turbine components.

the generating capacity of modern and commercial turbines ranges from less than 1 kW to several MW. The proper wind turbine system modeling oriented to the design of a suitable control strategy is more cost-effective for large wind turbines.

Another important issue derives from the steadily increasing sizes and a growing complexity of wind turbines, thus giving rise to more severe requirements regarding the system safety, reliability, and availability (Kusiak and Li, 2011; Rivkin and Silk, 2012; Odgaard et al., 2013; Ningsu et al., 2014). The safety demand can be commonly achieved by introducing redundancy in the system architecture, like additional sensors, which become vital for a safer operation of wind turbines. The classic example regards the pitch system for adjusting the angles of a rotor blade. For each of the three blades, one totally independent pitch system is used, such that in the worst case of a malfunction in one or two pitch systems, the remaining one or two would still be able to bring the turbine to a standstill. This solution improves the system safety, but it generates additional costs and possibly additional turbine downtimes due to faults in the redundant system parts. The enhanced safety may lead to a reduction of the system availability. This motivates the use of functional redundancies, as addressed, e.g., in Echavarria et al. (2012).

The main components of an HAWT that are visible from the ground are its tower, nacelle, and rotor, as can be seen in Fig. 1.1. The nacelle houses the generator, which is driven by a high-speed shaft. The high-speed shaft is in turn usually driven by a gear-box, which steps up the rotational speed from a low-speed shaft. The low-speed shaft is connected to the rotor, which includes the airfoil-shaped blades. These blades capture the kinetic energy in the wind and transform it into the rotational kinetic energy of the wind turbine.

The main wind turbine components are briefly described in the following (Darling, 2008). More details will be provided in Chapter 2.

- **Blades** interact with the wind, producing the movement of the rotor shaft. The profile of the blade is designed in order to obtain a good value of the aerodynamic lift with respect to the aerodynamic resistance and, at the same time, to oppose the proper stiffness to the applied variable mechanical loads that determine the wear and tear effect along time. The construction materials should be light, such as plastic materials reinforced with glass, aluminium or carbon, depending on the blade size.
- **Hub** constrains the blades to the rotor shaft, transmitting the extracted wind power. It can contain the pitch actuator that forces the blade to have a certain orientation relative to the wind main direction for control purposes.
- **Brakes** can be different (mechanical, electrical, hydraulic), and they are used as parking brakes to avoid the rotor movement when the turbine has to be kept in nonoperating conditions.
- **Gear-box** (having one or more stages) is used to adapt the mechanical power of the rotor shaft to the generator shaft, by increasing the rotational speed and by decreasing the torque, in order to permit an efficient conversion of energy. Often the gear box ratio can be greater than 1:100. The design of the gear box involves epicycloidal or parallel axis gears.
- **Generator** converts the mechanical energy of the connected shaft to electrical energy. It can be an asynchronous or a synchronous machine, the former consists of an induction three-phase motor that works as a generator providing energy to the grid, as its speed is higher than the synchronous speed and the applied torque is motive. The usual structure of an asynchronous generator involves a squirrel cage rotor, which implies a low difference between the synchronous speed and the actual speed in the nominal working region, so that it can be considered a constant speed machine. The configuration,

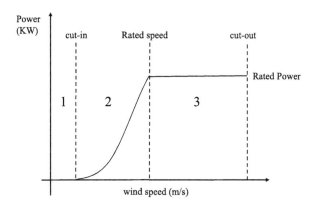

**FIGURE 1.2** Example of wind turbine working conditions.

called doubly-fed, with a power converter located between the rotor and the grid allows the functioning of the system at a variable rotor speed. Otherwise, the synchronous generator provides a voltage with frequency proportional to its rotational speed. In the configuration called *full-converter*, similarly to the doubly-fed generators, a power converter has to be interposed between the generator and the grid, in order to permit a variable speed functioning. The output power of a modern turbine can be up to five megawatts.

- **Nacelle** is the shell containing the shafts, the brakes, the gear box, the generator, and the control equipment. It is located at the top of the tower.
- **Anemometer** is the sensor that provides the current wind speed. Its measurements are exploited in the control system.
- **Yaw mechanism** contains several electrical motors for the heading orientation of the turbine parallel to the main wind direction.
- **Tower** supports the nacelle, hub, and blades. The height of the tower determines the height of the hub, which is a prominent value in the generation of power, since the wind speed increases with the distance from the earth surface.

Wind turbine control goals and strategies are affected by turbine configuration. HAWTs may be "upwind", with the rotor on the upwind side of the tower, or "downwind." The choice of upwind versus downwind configuration affects the choice of yaw controller and the turbine dynamics, and thus the structural design. Wind turbines may also be variable pitch or fixed pitch, meaning that the blades may or may not be able to rotate along their longitudinal axes. Although fixed-pitch machines are less expensive initially, the reduced ability to control loads and change the aerodynamic torque means that they are becoming less common within the realm of large wind turbines. Variable-pitch turbines may allow all or part of their blades to rotate along the pitch axis.

Moreover, wind turbines can be variable speed or fixed speed. Variable-speed turbines tend to operate closer to their maximum aerodynamic efficiency for a higher percentage of the time, but require electrical power processing so that the generated electricity can be fed into the electrical grid at the proper frequency. As generator and power electronics technologies improve and costs decrease, variable-speed turbines are becoming more popular than constant-speed turbines at the utility scale.

Fig. 1.2 shows an example power curve for a variable-speed wind turbine. When the wind speed is low (usually below 6 m/s), the power available in the wind is low compared to losses in the turbine system, so the turbine is not running. This operational region is sometimes known as Region 1. When the wind speed is high, Region 3 (above 11.7 m/s), power is limited to avoid exceeding safe electrical and mechanical load limits.

The main difference between fixed-speed and variable speed wind turbines appears for mid-range wind speeds, the Region 2 in Fig. 1.2, which normally encompasses wind speeds between 6 and 11.7 m/s. Except for one design operating point (10 m/s), a variable-speed turbine captures more power than a fixed-speed turbine. The reason for the discrepancy is that variable-speed turbines can operate at maximum aerodynamic efficiency over a wider range of wind speeds than fixed-speed turbines.

Even a perfect wind turbine cannot fully capture the power available in the wind. In fact, actuator disc theory shows that the theoretical maximum aerodynamic efficiency, which is called the Betz Limit, is approximately 60% of the wind power (Betz and Randall, 1966). The reason that an efficiency of 100% cannot be achieved is that the wind must have some kinetic energy remaining after passing through the rotor disc. If it did not, the wind would by definition be stopped and no more wind would be able to pass through the rotor to provide energy to the turbine.

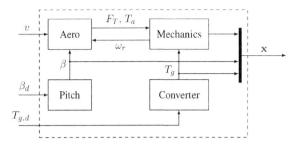

**FIGURE 1.3** Block diagram of the complete wind turbine model.

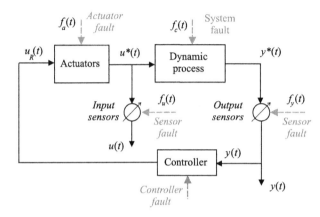

**FIGURE 1.4** Possible fault locations in the system under diagnosis.

The complete wind turbine model consists of several submodels for the mechanical structure, the aerodynamics, as well as the dynamics of the pitch system and the generator/converter system, as sketched in Fig. 1.3. The generator/converter dynamics are usually described as a first order delay system. However, when the delay time constant is very small, an ideal converter can be assumed, such that the reference generator torque signal is equal to the actual generator torque. In this situation, the generator torque can be considered as a system input.

Fig. 1.3 reports also the wind turbine inputs and outputs. In particular, $v$ is wind speed, $F_T$ and $T_a$ correspond to the rotor thrust force and rotor torque, respectively; $\omega_r$ is the rotor angular velocity, $x$ the state vector, $T_g$ the generator torque, and $T_{g,d}$ the demanded generator torque. $\beta$ is the pitch angle, whilst $\beta_d$ its demanded value.

Wind turbine high-fidelity simulators, which were described, for example, in Odgaard and Johnson (2013), consider white noise added to all measurements. This relies on the assumption that noisy sensor signals should represent more realistic scenarios. However, this is not the case, as a realistic simulation would require an accurate knowledge of each sensor and its measurement reliability. To the best of the authors' knowledge and from their experience with wind turbine systems, all main measurements acquired from the wind turbine process (rotor and generator speed, pitch angle, generator torque), are virtually noise-free or affected by very weak noise.

A more accurate and detailed overview of the state-of-the-art of FDI/FDD techniques is provided in Chapter 3.

## 1.5 INTRODUCTION TO FAULT DIAGNOSIS METHODS

There exist several ways in which faults can affect the system. According to Fig. 1.4, a closed-loop system can be viewed as the union of interconnected elements, namely the main process, actuators, controller, input sensors, and output sensors. Each of these components can be associated to a fault, so that process faults, actuator faults, controller faults, and sensor faults can be considered.

In several application domains a typical approach to fault diagnosis involves the so-called *hardware redundancy* (Isermann, 1997b). It consists in the usage of multiple sensors, actuators or components to measure, or control, a particular signal. The related diagnosis is based on the comparison among the different redundant hardware information, hence a voting technique is adopted to decide if and where a fault occurs. Although the hardware redundancy can be very effective, it involves extra cost concerning the equipment and the maintenance operations that could be in conflict with the sustainability requirements.

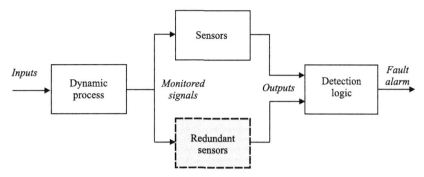

**FIGURE 1.5**   The hardware redundancy scheme.

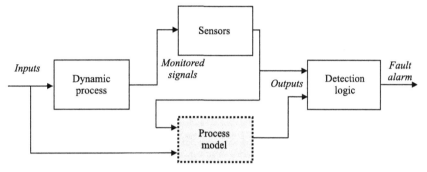

**FIGURE 1.6**   The analytical (software) redundancy principle.

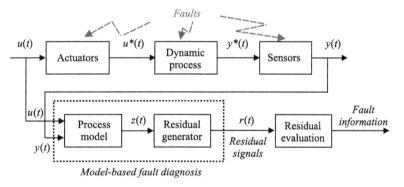

**FIGURE 1.7**   The fault diagnosis logic relying on the model-based residual generation and evaluation.

On the other hand, the so-called *analytical redundancy* (Patton et al., 1989b) does not need additional hardware components, but it exploits a model of the system under investigation for estimating the value of a particular variable on the basis of the preexistent sensors, so that a residual signal can be generated and a diagnosing logic can be inferred. In particular, the residual signal, already defined in Section 1.3, should be close to zero in normal operating condition and significantly different from zero when a fault occurs. The analytical redundancy is often referred to as *model-based* approach, because of its dependence on the system model. A model-based module can be implemented via software on a process control computer, without additional costs on the equipment, as already remarked. Some drawbacks of the model-based approach concern the accuracy of the generated estimations, as well as the disturbances and the noise affecting the process, that can lead to false alarm or missed fault.

The block diagram of Figs. 1.5 and 1.6 illustrate possible implementations of hardware and analytical (software) redundancy schemes.

In more details, the detection of a fault by means of the model-based approach relies on the generation and the evaluation of the residual, as depicted in the block diagram of Fig. 1.7.

The residual is firstly generated as the error between the measured and the estimated variable. The latter comes from the model elaboration of the input–output measurements. The residual, at this stage, should be independent from system input and output, as it is ideally zero for every input–output conditions in fault-free case. Then the residual has to be evaluated, in terms of fault likelihood, and a decision rule has to be applied to determine the fault occurrence. At this stage the residual can be simply compared to a fixed threshold (also called geometrical methods), or preprocessed through a proper function, e.g., a moving average or more complex statistical methods, ahead of the threshold test.

A particular class of the model-based methods, called *signal-based* technique, occurs when only output measurements are available. It includes the case of vibration analysis (e.g., related to rotating machinery) that are performed by means of band-pass filters or spectral analysis.

Among the basic model-based FDI techniques (Simani et al., 2003b), we can mention the output-observer, parity equation, and parameter estimation. They require an adequate knowledge of the state-space or input–output behavior of the system under investigation that is expressed in terms of analytical relationships. However, most of the real industrial processes have a strongly nonlinear behavior that cannot be modeled using a single model for all the operating conditions. Indeed, the parameters of the system may vary with time and the unavoidable disturbance and noise effects have unknown characteristics. The result is a discrepancy between the plant behavior and its mathematical description, even in fault-free condition, that can lead to the impossibility of generating a residual, as the fault may be hidden by the modeling errors. The Unknown Input Observer (UIO), eigenstructure assignment, and parity relation methods are FDI strategies that take into account this key issue (Simani et al., 2003b).

Therefore, the proper mathematical description of the system under investigation, required by classical model-based FDI strategy, is very difficult to derive in practice, sometimes even impossible. Because of these assumptions, data-driven modeling approaches offer a natural tool to handle a poor knowledge of the system, together with disturbances and noises. Indeed, their implementations exploit input–output data directly acquired from the system, for deriving the relations among those variables. For example, fuzzy logic theory (Babuška, 2012) allows the representation of the process under investigation by means of a collection of local affine fuzzy models, among which the transition are handled by identified fuzzy parameters. Furthermore, also neural networks can handle complex nonlinear behaviors (Liu, 2012), as they have the capability of learning the system functioning on the basis of the information provided by the training data.

These considerations motivate the studies on data-driven strategies applied to the wind turbine context, because their behavior is characterized by very complex and nonlinear dynamics, as described in Chapter 2.

## 1.6 INTRODUCTION TO FAULT TOLERANT CONTROL METHODS

Control solutions that are able to cope with possible malfunctions and fault situations are usually referred to as *Fault Tolerant* Control system (FTC), as they possess the capability to automatically manage the faults affecting the components (Patton, 2015). Fault tolerant control methods can be classified into two main categories, namely Passive Fault Tolerant Control (PFTC) and Active Fault Tolerant Control (AFTC) schemes. PFTC techniques require neither fault diagnosis nor controller reconfiguration, as they are designed to be robust against a set of possible faults. However, although they do not involve the problem related to fault diagnosis, as false alarms or missed fault, they have limited fault tolerant capabilities. They make use of robust control techniques to ensure that the considered closed-loop system remains insensitive to faults, using fixed controller parameter, without needing information regarding the fault occurrence. This is accomplished by designing a controller that is optimized for fault-free situations, while satisfying some graceful degradation requirements in the faulty case.

On the other hand, AFTC reacts to the faults actively by reconfiguring the control actions so that the performance of the fault-free system can be maintained, even in presence of faults. In particular, AFTC is mainly based on a fault diagnosis block, that provides the real-time information about the faulty, or fault-free, status of the system under monitoring (Mahmoud et al., 2004). The controller exploits a further control loop aimed at the compensation of the faulty signals (fault *accommodation*). The main advantage of this approach is that the controller can be designed considering only the nominal operating conditions. The structures of the considered AFTC schemes are addressed in Chapter 4.

As already remarked, the need of advanced control solutions for these very demanding systems motivated also the requirement of reliability, availability, maintainability, and safety over power conversion efficiency. These issues have begun to stimulate research and development of FTC systems.

In general, wind turbines in the megawatt range are expensive, and hence their availability and reliability must be high in order to maximize the energy production. This issue could be particularly important for offshore installations, where O&M services have to be minimized, since they represent one of the main factors of the energy cost. The capital cost, as well as the wind turbine foundation and installation, determines the basic term in the cost of the produced energy, which constitute the energy "fixed cost." The O&M represents a "variable cost" that can increase the energy cost up to about the 30%. At the same time, industrial systems have become more complex and expensive, with less tolerance for performance degradation, productivity decrease, and safety hazards. This leads also to an ever increasing requirement on reliability and safety of control systems subjected to process abnormalities and component faults. As a result, it is extremely important the FDD or the FDI tasks, as well as the achievement of fault-tolerant features for minimizing possible performance degradation and avoiding dangerous situations. With the advent of computerized control, communication networks, and information techniques, it becomes possible to develop novel real-time monitoring and fault-tolerant design techniques for industrial processes, but it also brings challenges.

In the last years, many works have been proposed on wind turbine FDI/FDD, and the most relevant are, e.g., in Gong and Qiao (2013a), Freire et al. (2013). On the other hand, regarding the FTC problem for wind turbines, it was recently analyzed with reference to an offshore wind turbine benchmark, e.g., in Odgaard et al. (2013). As already remarked, FTC methods are classified into two types, PFTC and AFTC schemes (Mahmoud et al., 2003). In particular for wind turbines, some FTC designs were considered and compared in Odgaard et al. (2013). These processes are nonlinear dynamic systems, whose aerodynamics are nonlinear and unsteady, whilst their rotors are subject to complicated turbulent wind inflow fields driving fatigue loading. Therefore, the so-called wind turbine *sustainable* control represents a complex and challenging task (Odgaard and Stoustrup, 2013b).

Therefore the purpose of this monograph is to revise the basic solutions to sustainable control design, which are capable of handling faults affecting the controlled wind turbine. For example, changing dynamics of the pitch system due to a fault cannot be accommodated by signal correction. Therefore, it should be considered in the controller design, to guarantee stability and a satisfactory performance. Among the possible causes for changed dynamics of the pitch system, they can be due to a change in the air content of the hydraulic system oil. This fault is considered since it is the most likely to occur, and since the reference controller becomes unstable when the hydraulic oil has a high air content. Another issue arises when the generator speed measurement is unavailable, and the controller should rely on the measurement of the rotor speed, which is contaminated with much more noise than the generator speed measurement. This makes it necessary to reconfigure the controller to obtain a reasonable performance of the control system.

In order to outline and compare the controllers developed using AFTC and PFTC design approaches, they should be derived using the same procedures in the fault-free case. In this way, any differences in their performance or design complexity would be caused only by the fault tolerance approach, rather than the underlying controller solutions. Furthermore, the controllers should manage the parameter-varying nature of the wind turbine along its nominal operating trajectory caused by the aerodynamic nonlinearities. Usually, in order to comply with these requirements, the controllers are usually designed for example using Linear Parameter-Varying (LPV) modeling or fuzzy descriptions (Bianchi et al., 2007; Galdi et al., 2008).

The two FTC solutions have different structures as shown in Fig. 1.8. Note that only the AFTC system can rely on a fault diagnosis algorithm (FDD). This represents the main difference between the two control schemes.

The main point between AFTC and PFTC schemes is that an active fault-tolerant controller relies on a fault diagnosis system, which provides information about the faults $f$ to the controller. In the considered case the fault diagnosis system FDD contains the estimation of the unknown input (fault) affecting the system under control. The knowledge of the fault $f$ allows the AFTC to reconfigure the current state of the system. On the other hand, the FDD is able to improve the controller performance in fault-free conditions, since it can compensate, e.g., the modeling errors, uncertainty, and disturbances. On the other hand, the PFTC scheme does not rely on a fault diagnosis algorithm, but is designed to be robust towards any possible faults. This is accomplished by designing a controller that is optimized for the fault-free situation, while satisfying some graceful degradation requirements in the faulty cases. However, with respect to the robust control design, the PFTC strategy provides reliable controllers that guarantee the same performance with no risk of false FDI or reconfigurations.

In general, the methods used in the fault-tolerant controller designs should rely on output feedback, since only part of the state vector is measured. Additionally, they should take the measurement noise into account. Moreover, the design methods should be suited for nonlinear systems or linear systems with varying parameters. The latest proposed solutions for the derivation of both active and the passive fault-tolerant controllers rely on LPV and fuzzy descriptions, to which the fault-tolerance properties are added, since these frameworks methods are able to provide stability and guaranteed performance with respect to parameter variations, uncertainty and disturbance. Additionally, LPV and fuzzy controller design methods are well-established in multiple applications including wind turbines (Bianchi et al., 2007; Galdi et al., 2008).

To add fault-tolerance to the common LPV and fuzzy controller formulation, different approaches can be exploited. For example, the AFTC scheme can use the parameters of both the LPV and fuzzy structures estimated by the FDD module for scheduling the controllers (Sloth et al., 2010; Sami and Patton, 2012; Simani and Castaldi, 2013). On the other hand, different approaches can be used to obtain fault-tolerance in the PFTC methods. For this purpose, the design methods described in Bianchi et al. (2007), Galdi et al. (2008) can be modified to cope with parametric uncertainties, as addressed, e.g., in Puig (2010), Rodrigues et al. (2007).

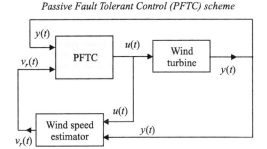

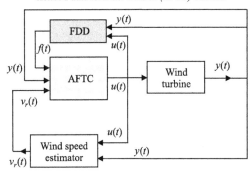

**FIGURE 1.8** Structures of the *active* and *passive* FTC systems.

Alternatively, other methods could have been used such as Niemann and Stoustrup (2005), which preserves the nominal performance. Generally, these approaches rely on solving some optimization problems where a controller is calculated subject to maximizing the disturbance attenuation. These problems are formulated as Linear Matrix Inequalities (LMI) (Chen et al., 1997).

A more accurate and detailed overview of the state-of-the-art of FTC approaches is provided in Chapter 4.

## 1.7 MODELING AND ADVANCED CONTROL BENCHMARKING

Prior to designing and applying a new control strategy on a real wind turbine, the efficacy of the control scheme has to be tested in detailed aero-elastic simulation model. Several simulation packages exist that are commonly used in academia and industry for wind turbine load simulation. One of the most used simulation packages is the Fatigue, Aerodynamics, Structures, and Turbulence (FAST) code (Jonkman and Buhl, 2005) provided by the National Renewable Energy Laboratory (NREL), since it represents a reference simulation environment for the development of high-fidelity wind turbine prototypes that are taken as a reference test-cases for many practical studies (Jonkman et al., 2009). FAST provides a high-fidelity wind turbine model with 24 degrees of freedom, which is appropriate for testing the developed control algorithms but not for control design. For the latter purpose, a reduced-order dynamic wind turbine model, which captures only dynamic effects directly influenced by the control, is recalled in this section and it can be used for model-based control design (Bianchi et al., 2007).

Control systems for wind turbines are now well developed (Bianchi et al., 2007; Pao and Johnson, 2011), and many of these schemes can be successfully exploited in wind energy devices. In this case, the general problem is to maximize energy capture, subject to grid and environmental constraints. However, the objective of energy capture maximization might be modified to that of maximization of economic return (Bianchi et al., 2007), which requires a balance to be achieved between maximizing energy capture and minimizing wear on components. For the current analysis, in order to retain a focus on the fundamental control issues, the general problem of energy capture maximization will be considered. There are two broad approaches which may be taken to solve the energy maximization problem:

1. Overall extremum seeking control (Pao and Johnson, 2011), with little use of a detailed model of the system;
2. Determination of an optimal setpoint for the system, which gives maximum energy capture, followed by a regulatory to make sure this setpoint is achieved (Johnson et al., 2006a).

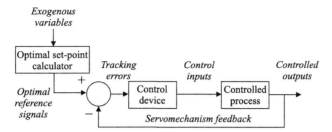

**FIGURE 1.9**  Diagram highlighting the optimal setpoint (feed-forward) selector and the control section.

Item 1 is attractive from the point of view of the lack of requirement for a detailed model, but may have dynamic performance limitations in convergence rates, and may have difficulty finding a global maximum over a nonconvex performance surface. On the other hand, a different framework may be adopted for item 2, which requires an accurate mathematical description of the system under diagnosis.

In general, control attempts to devise algorithms that force a system to follow a desired path, objective, or behavior modality. Traditionally, the control problem is defined by a tracking problem, where the objective is for the system output to follow the reference input. While problems of these type do occur in energy conversion applications, for example, speed control of wind turbines, it is more useful to broaden the set of problem descriptions and potential solutions a little, in order to assess the potential of control engineering in the general energy conversion context. A generic problem framework, as shown in Fig. 1.9, consisting of an upper (optimal) setpoint generation stage and a lower control loop to ensure tracking of the setpoint. Both sets of control calculations must be mindful of physical constraints in the system. In the wind energy case, for variable speed turbines, an optimal rotational speed is first calculated, and torque and/or blade pitch control used to achieve the required rotational speed.

In general, the control problem definition requires the maximization or the minimization if prescribed performance objective (such as the max. energy, min. error) subject to proper system constraints (see, e.g., amplitudes, rates, forces, etc.), i.e., a constrained optimization problem. The definition considered here is not inconsistent with the purpose of a classic controlled system with a feedback loop, where the objective function is usually some measure (e.g., a quadratic measure) of the difference between the controller output and its desired value, i.e., the tracking error, with respect to the reference or the setpoint. In this way, the desired performance of the tracking system in closed-loop can be specified in a variety of ways (Kuo, 1995; Cheng, 1984; Lewis, 1986; Zhou et al., 1996b):

**1.** Desired transient response;
**2.** Desired steady-state response;
**3.** Desired closed-loop poles (roots of the closed-loop transfer function);
**4.** Trade-off between control energy and tracking error;
**5.** Minimization of the sensitivity of the closed-loop system to variations in the system description;
**6.** Minimization of the sensitivity of the closed-loop system to external disturbances.

Items 5 and 6 in the list above relate to the system robustness, and specific control methodologies to address these objectives have been developed since the late 1970s. In most cases, control design methods provide an explicit solution for the feedback controllers, while some methods solve the more general optimization problem defined at each time step. In the monograph, specific or general solutions can be useful in the FDI/FDD and FTC of wind turbines will be recalled and analyzed.

Note finally that some control methods require a mathematical model of the system, in order to determine the control algorithm, and such methods are termed model-based. The requirement for an accurate mathematical system model often involves considerably more work than the calculation of the controller itself, though system identification techniques (Simani et al., 2003a) can be employed to determine a black-box model, i.e., a model which has no structural relationship to the physical system. The combination of data-driven techniques with a mathematical procedure for controller determination can be used to develop adaptive controllers, which have the capability to adapt to unknown (in "self-tuning mode") or time-varying systems. Adaptive control schemes based on linear system models also have the capability to track variations in a linear model due to the presence of nonlinearity, though nonlinear systems are best controlled with a dedicated fixed-parameter nonlinear controller. Significant care and attention must also be paid, e.g., to adaptive schemes to ensure stability and convergence over all operating regimes (Ioannou and Sun, 2012).

## 1.8 OUTLINE OF THE MONOGRAPH

The contents of the chapters of the monograph are briefly summarized in the following, in order to provide an overview of the whole book.

- **Chapter 2** introduces the systems under investigation (i.e., the wind turbines), describing their characteristics, categories, and components, together with some global statistics that highlight the importance of the discussed topics. Then two realistic benchmark systems are detailed presented, which represent high-fidelity simulators of a single wind turbine and of a wind farm, respectively. They rely on the analytical description of the system behaviors and provide also the models of typical fault cases. The chapter describes also the faults affecting the wind turbine systems and determine their effect on the system behavior. Redundancies in the system are also identified in order to define the detectable faults and the remedial actions that must be conducted to stop the propagation of the fault are established.
- **Chapter 3** describes the main fault diagnosis algorithms applied to wind turbine systems. These techniques can be used for both condition monitoring purposes and fault diagnosis, in order to provide the information required by the fault tolerant control module. Model-based and data-driven strategies are summarized, which rely on measured as well as estimated variables.
- **Chapter 4** presents the fault tolerant control algorithms applied to wind turbine systems. They are based on the signal correction principle, which means that the control system is not modified since the inputs and outputs of the baseline controller are compensated according to the estimated faults. The fault tolerant control algorithms recalled in this chapter rely on the fault diagnosis design addressed in Chapter 3. Passive and active fault tolerant control systems are also discussed and compared, in order to highlight the achievable performances and the complexity of their design procedures. Controller reconfiguration mechanisms are also considered, which are able to guarantee the system stability and satisfactory performance.
- **Chapter 5** addresses the performances achieved by developed fault diagnosis and fault tolerant control systems, analyzed and applied to both the single wind turbine and the wind farm benchmark models. The chapter reports also the result obtained in comparison with the different fault diagnosis and fault tolerant control schemes proposed in Chapters 3 and 4. The robustness and reliability features of the proposed techniques are analyzed by means of the Monte Carlo tool that is able to take into account possible disturbance and uncertainties affecting the systems under diagnosis.
- **Chapter 6** describes the Matlab and Simulink implementations of the proposed data-driven and model-based approaches to the fault diagnosis and fault tolerant control of wind turbine and wind farm benchmark models. The main Simulink modules implementing the suggested strategies are briefly discussed and analyzed. These simulation tools are available from the link of the book's authors homepage. Finally, Hardware-In-the-Loop tests are carried out in order to highlight the controller performances in more realistic and effective frameworks.
- **Chapter 7** summarizes and discusses the obtained results, and suggests some future investigations about the discussed topic.

## 1.9 SUMMARY

This chapter summarized the most important modeling and control issues for wind turbines from a systems and control engineering point of view that will be discussed in detail in the next chapters. A walk around the wind turbine control loops discussed the goals of the most common solutions and reviewed the typical actuation and sensing available on commercial turbines. The chapter intended to provide also an updated and broader perspective by covering not only the modeling and control of individual wind turbines, but also outlining a number of areas for further research, and anticipating new issues that can open up new paradigms for advanced control approaches. In summary, wind energy is a fast growing industry, and this growth has led to a large demand for better modeling and control of wind turbines. Uncertainty, disturbance, and other deviations from normal working conditions of the wind turbines make the control challenging, thus motivating the need for advanced modeling and a number of so-called sustainable control approaches that should be explored to reduce the cost of wind energy. By enabling this clean renewable energy source to provide and reliably meet the world's electricity needs, the tremendous challenge of solving the world's energy requirements in the future will be enhanced. The wind resource available worldwide is large, and much of the world's future electrical energy needs can be provided by wind energy alone if the technological barriers are overcome. The application of sustainable controls for wind energy systems is still in its infancy, and there are many fundamental and applied issues that can be addressed by the systems and control community to significantly improve the efficiency, operation, and lifetimes of wind turbines.

Chapter 2

# System and Fault Modeling

## 2.1 INTRODUCTION

The motivation for this chapter comes from a real need to have an overview about the main challenges of modeling for very demanding systems, such as wind turbine systems, which require reliability, availability, maintainability, and safety over power conversion efficiency. These issues have begun to stimulate research and development in the wide control community, particularly for these installations that need a high degree of "sustainability." Note that this represents a key point for offshore wind turbines and wind park installations, since they are characterized by expensive and/or safety critical maintenance work. In this case, a clear conflict exists between ensuring a high degree of availability and reducing maintenance times, which affect the final energy cost. On the other hand, wind turbines have highly nonlinear dynamics, with a stochastic and uncontrollable driving force as input in the form of wind speed, thus representing an interesting challenge also from the modeling point of view. As shown in Chapters 3 and 4, suitable fault diagnosis and fault tolerant control methods can provide a sustainable optimization of the energy conversion efficiency over wider than normally expected working conditions. Moreover, a proper mathematical description of the wind turbine system should be able to capture the complete behavior of the process under monitoring, thus providing an important impact on the control design itself. In this way, the fault diagnosis and control schemes could guarantee prescribed performance, whilst also giving a degree of "tolerance" to possible deviation of characteristic properties or system parameters from standard conditions, if properly included in the wind turbine model itself.

## 2.2 SYSTEM DESCRIPTION

Wind energy represents a fast-developing interdisciplinary field comprising many different branches of engineering and science. According to the National Renewable Energy Laboratory (NREL), the installed capacity of wind grew at a rate of about 30% from 2002 to 2007 (Pao and Johnson, 2009; Laks et al., 2009; Johnson and Fleming, 2011). According to the report describing the installed wind energy capacity (Global Wind Energy Council, 2014), it can be seen how global wind power installations increased by 35,467 MW in 2013, bringing total installed capacity up to 318,137 MW. During 2010–2011 more than half of all new wind power was added outside of the traditional markets of Europe and North America, mainly driven by the continuing boom in China which accounted for nearly half of all of the installations at 18,000 MW in 2011. China now has 91,424 MW of wind power installed. Several countries have achieved relatively high levels of wind power penetration, such as 21% of stationary electricity production in Denmark, 18% in Portugal, 16% in Spain, 14% in Ireland, and 9% in Germany in 2010 (Global Wind Energy Council, 2014). As of 2011, 83 countries around the world are using wind power on a commercial basis. It is clear why wind power is recognized as an effective and green solution for energy harvesting. However, even if the US receives less than 2% of its electrical energy from wind, NREL report lays the framework for achieving 20% of the domestic electrical energy generation from wind in the US by the year 2030 (Pao and Johnson, 2009; Laks et al., 2009; Johnson and Fleming, 2011).

Despite the expected growth in the installed capacity of wind turbines in recent years, engineering and science challenges still exist (Pramod, 2010; Burton et al., 2011). Since wind turbine installations must guarantee both power capture and economical advantages, also the size of wind turbines has grown dramatically from 1980, as reported in Intergovernmental Panel on Climate Change IPCC (2011).

### 2.2.1 Wind Turbine Categories

Modern wind turbines have large, flexible structures operating in uncertain environments, thus representing interesting cases for advanced control solutions (Munteanu and Bratcu, 2008), as it will be addressed in Chapters 3 and 4. In particular, Chapter 4 will show that advanced controller solutions can help achieve the desirable goal of decreasing the wind energy cost by increasing the efficiency, and thus the energy capture, or by reducing structural loading and increasing the lifetimes of the components and turbine structures (Bianchi et al., 2007; Rivkin et al., 2012). This monograph aims also at sketching

Fault Diagnosis and Sustainable Control of Wind Turbines. DOI: 10.1016/B978-0-12-812984-5.00002-X

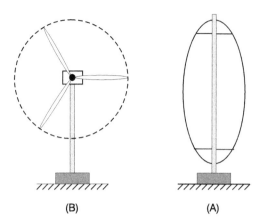

**FIGURE 2.1** (A) Vertical-axis and (B) horizontal-axis wind turbine examples.

the main challenges that exist in the wind industry and at stimulating new research topics in this area. Although wind turbines come in both vertical- and horizontal-axis configurations, as sketched in Fig. 2.1, this book will focus only on Horizontal-Axis Wind Turbines (HAWTs), since they represent the most commonly produced large-scale installations today.

HAWTs have the advantage that the rotor is placed atop a tall tower, where it can take advantage of larger wind speeds higher above the ground. Moreover, HAWTs used for utility-scale installations include pitchable blades, improved power capture and structural performance, as well as no need for tensioned cables used to add structural stability (Bianchi et al., 2007; Pramod, 2010; Garcia-Sanz and Houpis, 2012).

Vertical-Axis Wind Turbine (VAWT) installations are more common for smaller turbines, where these disadvantages become less important and the benefits of reduced noise and omnidirectionality become more pronounced (Burton et al., 2011; Garcia-Sanz and Houpis, 2012). Note finally that the generating capacity of modern and commercial turbines ranges from less than 1 kW to several MW. The proper wind turbine system modeling oriented to the design of a suitable control strategy is more cost-effective for large wind turbines, and therefore this monograph will refer to wind turbines with capacities of several MW.

Another important issue that will be discussed in Chapter 4 derives from the steadily increasing sizes and a growing complexity of wind turbines, thus giving rise to more severe requirements regarding the system safety, reliability, and availability (Kusiak and Li, 2011; Rivkin and Silk, 2012; Odgaard et al., 2013; Ningsu et al., 2014). The safety demand can be commonly achieved by introducing redundancy in the system architecture, like additional sensors, which become vital for a safer operation of wind turbines. The classic example regards the pitch system for adjusting the angles of a rotor blade, as shown in Fig. 2.1. For each of the three blades, one totally independent pitch system is used, so that in the worst case of a malfunction in one or two pitch systems, the remaining one or two would still be able to bring the turbine to a standstill. This scheme improves the system safety, but it generates additional costs and possibly additional turbine downtimes due to faults in the redundant system parts. The enhanced safety may lead to reducing the system availability. This motivates the use of functional redundancies, which will be remarked in Chapter 3, as addressed, e.g., in Echavarria et al. (2012).

Even when reducing hardware redundancies, large wind turbines are prone to unexpected malfunctions or alterations of the nominal working conditions. Many of these anomalies, even if not critical, often lead to turbine shutdowns, again for safety reasons. Especially in offshore wind turbines, this may result in a substantially reduced availability, because rough weather conditions may prevent the prompt replacement of the damaged system parts. The need for reliability and availability that guarantees the continuous energy production requires the so-called sustainable control solutions, as addressed in Section 4. These schemes enable keeping the turbine in operation in the presence of anomalous situations, perhaps with reduced performance, while managing the maintenance operations.

Apart from increasing availability and reducing turbine downtimes, sustainable control schemes might also obviate the need for more hardware redundancy, if virtual sensors could replace redundant hardware sensors (Blanke et al., 2006). The schemes discussed in Chapter 3 and currently employed in wind turbines are typically on the level of the supervisory control, where commonly used strategies include sensor comparison, model comparison, and thresholding tests (Johnson and Fleming, 2011). These strategies enable safe turbine operations, which involve shutdowns in case of critical situations, but they are not able to actively counteract anomalous working conditions. Therefore, the goal of this monograph is also to revise these so-called sustainable control strategies, which allow obtaining a system behavior that is close to the nominal

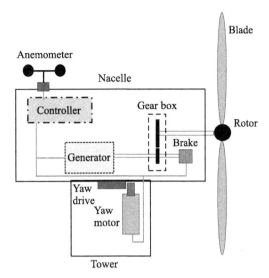

**FIGURE 2.2** Main components of a wind turbine system.

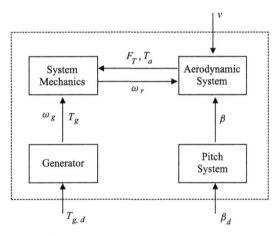

**FIGURE 2.3** Blocks comprising the overall wind turbine system.

situation in the presence of unpermitted deviations of any characteristic properties or system parameters from standard conditions (i.e., a fault) (Isermann, 2005; Ningsu et al., 2014). Moreover, these schemes should provide the reconstruction of the equivalent unknown input that represents the effect of a fault, thus achieving the so-called fault diagnosis task (Chen and Patton, 2012; Blanke et al., 2006; Ding, 2008).

## 2.3 WIND TURBINE MAIN COMPONENTS

This section sketches the main components of a HAWT, which are visible from the ground. They consist of its tower (i.e., the load-carrying structure), the nacelle, and the rotor, as depicted in Fig. 2.2. The nacelle comprises the generator, which is driven by a high-speed shaft. The high-speed shaft is in turn usually driven by a gear-box, which steps up the rotational speed from a low-speed shaft. The low-speed shaft is connected to the rotor, which includes airfoil-shaped blades. These blades capture the kinetic energy in the wind and transform it into the rotational kinetic energy of the wind turbine. These elements are represented in Fig. 2.2.

The complete description of the wind turbine system consists of several submodels for the mechanical structure, the aerodynamics, as well as the dynamics of the pitch system and the generator/converter system, as sketched in Fig. 2.3. The generator/converter dynamics are usually described as a first order delay system. However, when the delay time constant is very small, an ideal converter can be assumed, such that the reference generator torque signal is equal to the actual generator torque. In this situation the generator torque can be considered as a system input.

Fig. 2.3 reports also the main input and output signals that drive the wind turbine system. In particular, $v$ is wind speed, $F_T$ and $T_a$ correspond to the rotor thrust force and rotor torque, respectively. The signal $\omega_r$ is the rotor angular velocity, $T_g$ the generator torque, $\omega_g$ the generator angular velocity, and $T_{g,d}$ is the demanded generator torque. Finally, $\beta$ is the pitch angle, whilst $\beta_d$ corresponds to its demanded value.

### 2.3.1 Aerodynamic System

The aerodynamic submodel consists of the expressions for the thrust force $F_T$ acting on the rotor and the aerodynamic rotor torque $T_a$. They are determined by the reference force $F_{st}$ and by the aerodynamic rotor thrust and torque coefficients $C_T$ and $C_Q$ (Gasch and Twele, 2012):

$$\begin{cases} F_T = F_{st} \, C_T \, (\lambda, \, \beta) \,, \\ T_a = F_{st} \, R \, C_Q \, (\lambda, \, \beta) \,. \end{cases} \tag{2.1}$$

The reference force $F_{st}$ is defined from the impact pressure $\frac{1}{2} \rho \, v^2$ and the rotor swept area $\pi \, R^2$ (with rotor radius $R$), where $\rho$ denotes the air density,

$$F_{st} = \frac{1}{2} \rho \, \pi \, R^2 \, v^2. \tag{2.2}$$

It is worth noting that for simulation purposes, the static wind speed $v$ is used. However, a more accurate model should exploit the effective wind speed $v_e = v - (\dot{y}_T + \dot{y}_B)$, i.e., the static wind speed corrected by the tower and blade motion effects, as described in Section 2.3.3. However, the aerodynamic maps used for the calculation of the rotor thrust and torque are usually represented as static two-dimensional tables, which already take into account the dynamic contributions of both the tower and blade motions.

As highlighted in the expressions of Eq. (2.1), the rotor thrust and torque coefficients $\left(C_T, \, C_Q\right)$ depend on the tip-speed ratio $\lambda = \frac{\omega_r \, R}{v}$ and the pitch angle $\beta$. Therefore, the rotor thrust $F_T$ and torque $T_a$ assume the following expressions:

$$\begin{cases} F_T = \frac{1}{2} \rho \, \pi \, R^2 \, C_T \, (\lambda, \, \beta) \, v^2, \\ T_a = \frac{1}{2} \rho \, \pi \, R^3 \, C_Q \, (\lambda, \, \beta) \, v^2. \end{cases} \tag{2.3}$$

The expressions (2.3) highlight that the rotor thrust $F_T$ and torque $T_a$ are nonlinear functions depending on the wind speed $v$, the rotor speed $\omega_r$, and the pitch angle $\beta$. These functions are usually expressed as two-dimensional maps, which must be known for the whole range of variation of both the pitch angles and tip-speed ratios. These maps are usually a static approximation of more detailed aerodynamic computations that can be obtained using, for example, the Blade Element Momentum (BEM) method. In this case the aerodynamic lift and drag forces at each blade section are calculated and integrated in order to obtain the rotor thrust and torque (Gasch and Twele, 2012). More accurate maps can be obtained by exploiting the calculations implemented via the AeroDyn module of the FAST code, where the maps are extracted from several simulation runs (Laino and Hansen, 2002).

It is worth noting that for simulation purposes, the tabulated versions of the aerodynamic maps $C_Q$ and $C_T$ are sufficient. On the other hand, for control design, the derivatives of the rotor torque (and thrust) are needed, thus requiring a description of the aerodynamic maps as analytical functions. Therefore, these maps can be approximated using combinations of polynomial and exponential functions, whose powers and coefficients can be estimated via, e.g., modeling (Heier, 2014) or data-driven identification (Simani and Castaldi, 2011, 2014) approaches.

### 2.3.2 Drive-Train Model

The drive-train consisting of a rotor, shaft, and generator is modeled as a two-mass inertia system, including shaft torsion, where the two inertias are connected with a torsional spring with spring constant $k_S$ and a torsional damper with damping constant $d_S$, as illustrated in Fig. 2.4.

With reference to Fig. 2.4, the angular velocities $\omega_r$ and $\omega_g$ are the time derivatives of the rotation angles $\theta_r$ and $\theta_g$. In this case, the rotor torque $T_a$ is generated by the lift forces on the individual blade elements, whilst $T_g$ represents the generator torque. The ideal gear-box effect can be simply included in the generator model by multiplying the generator inertia $J_g$ by the square of the gear-box ratio $n_g$.

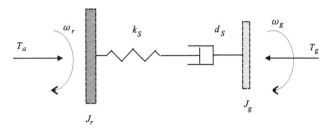

**FIGURE 2.4**  Diagram of the drive-train system.

The motion equations are derived by means of Lagrangian dynamics, which first requires to define the generalized coordinates and generalized external forces. In this way the energy terms of the system are derived, as well as the motion equations. The vector of generalized coordinates is given by $\mathbf{q} = \left[ y_T, \, y_B, \, \theta_r, \, \theta_g \right]^T$, whilst the vector of external forces is defined as $\mathbf{f} = \left[ F_T, \, F_T, \, T_a, \, -T_g \right]$.

The generalized force $F_T$ represents the rotor thrust force, which can be computed from the wind speed at the blade and from the aerodynamic map of the thrust coefficient. On the other hand, the generalized force $T_a$ is given by the aerodynamic rotor torque, which can be calculated from the wind speed and from the aerodynamic map of the torque coefficient described in Section 2.3.1. By considering the tower dynamics, the complete blade tip displacement is given by $y_T + y_B$, and the kinetic energy has the following form:

$$E_K = \frac{1}{2} m_T \, \dot{y}_T^2 + \frac{1}{2} N \, m_B \, (\dot{y}_T + \dot{y}_B)^2 + \frac{1}{2} J_r \dot{\theta}_r^2 + \frac{1}{2} J_g \, \dot{\theta}_g^2. \tag{2.4}$$

In the same way the potential energy has the form

$$E_P = \frac{1}{2} k_T \, y_T^2 + \frac{1}{2} N \, k_B \, y_B^2 + \frac{1}{2} k_S \left( \theta_r - \frac{1}{n_g} \theta_g \right)^2, \tag{2.5}$$

with $n_g$ being the gear-box ratio. The dampings in the system produce generalized friction forces, which can be written as derivatives of a quadratic form, e.g., the dissipation function (Landau and Lifshitz, 1976). In this case it assumes the form

$$P_D = \frac{1}{2} d_T \, \dot{y}_T^2 + \frac{1}{2} N \, d_B \, \dot{y}_B^2 + \frac{1}{2} d_S \left( \dot{\theta}_r - \frac{1}{n_g} \dot{\theta}_g \right)^2. \tag{2.6}$$

The Lagrangian equations of second order including the dissipation term are given by (Landau and Lifshitz, 1976)

$$\frac{d}{dt} \left( \frac{\partial L}{\partial \dot{q}_i} \right) - \frac{\partial L}{\partial q_i} = f_i - \frac{\partial P_D}{\partial \dot{q}_i}, \tag{2.7}$$

where the Lagrangian function $L$ denotes the difference between kinetic and potential energy. As the kinetic energy in Eq. (2.4) does not depend on the generalized coordinates and the potential energy in Eq. (2.5) does not depend on the generalized velocities, the motion equations in the following form are obtained:

$$\begin{cases} (m_T + N \, m_B) \, \ddot{y}_T + N \, m_B \, \ddot{y}_B + d_T \, \dot{y}_T + k_T \, y_T = F_T, \\ N \, m_B \, \ddot{y}_T + N \, m_B \, \ddot{y}_B + N \, d_B \, \dot{y}_B + N \, k_B \, y_B = F_T, \\ J_r \ddot{\theta}_r + d_S \left( \omega_r - \frac{1}{n_g} \omega_g \right) + k_S \left( \theta_r - \frac{1}{n_g} \theta_g \right) = T_a, \\ J_r \frac{1}{n_g} \ddot{\theta}_g - d_S \left( \omega_r - \frac{1}{n_g} \omega_g \right) - k_S \left( \theta_r - \frac{1}{n_g} \theta_g \right) = -T_g. \end{cases} \tag{2.8}$$

System (2.8) can be rewritten in matrix form as

$$\mathbf{M} \ddot{q} + \mathbf{D} \dot{q} + \mathbf{K} q = \mathbf{f}, \tag{2.9}$$

where the mass matrix $\mathbf{M}$, damping matrix $\mathbf{D}$, and stiffness matrix $\mathbf{K}$ have the form:

$$\mathbf{M} = \begin{bmatrix} m_T + N\,m_B & N\,m_B & 0 & 0 \\ N\,m_B & N\,m_B & 0 & 0 \\ 0 & 0 & J_r & 0 \\ 0 & 0 & 0 & J_g \end{bmatrix},$$

$$\mathbf{D} = \begin{bmatrix} d_T & 0 & 0 & 0 \\ 0 & N\,d_B & 0 & 0 \\ 0 & 0 & d_S & -\frac{d_S}{n_g} \\ 0 & 0 & -\frac{d_S}{n_g} & \frac{d_S}{n_g^2} \end{bmatrix}, \tag{2.10}$$

$$\mathbf{K} = \begin{bmatrix} k_T & 0 & 0 & 0 \\ 0 & N\,k_B & 0 & 0 \\ 0 & 0 & k_S & -\frac{k_S}{n_g} \\ 0 & 0 & -\frac{k_S}{n_g} & \frac{k_S}{n_g^2} \end{bmatrix}.$$

The second order system of differential equations (2.9) can be transformed into a first order state-space model by introducing the state vector $x = \left[ q, \dot{q} \right]^T$. To this aim, the expression (2.9) is solved with respect to the second time derivative of the coordinate vector $\mathbf{q}$. The equivalent state-space model is thus obtained in the form

$$\begin{cases} \dot{\mathbf{x}} = \mathbf{A}_m\,\mathbf{x} + \mathbf{B}_m\,\mathbf{u}_m, \\ \mathbf{y} = \mathbf{C}_m\,\mathbf{x}, \end{cases} \tag{2.11}$$

where the state vector is given by $\mathbf{x} = \left[ y_T, y_B, \theta_r, \theta_g, \dot{y}_T, \dot{y}_B, \dot{\theta}_r, \dot{\theta}_g \right]^T$, the input vector is $\mathbf{u}_m = \left[ F_T, T_a, T_g \right]^T$, whilst the system matrices have the form:

$$\mathbf{A}_m = \begin{bmatrix} 0_{4\times 4} & I_{4\times 4} \\ -\mathbf{M}^{-1}\mathbf{K} & -\mathbf{M}^{-1}\mathbf{D} \end{bmatrix}, \quad \mathbf{B}_m = \begin{bmatrix} 0_{4\times 3} \\ \mathbf{M}^{-1}\mathbf{Q} \end{bmatrix},$$

$$\mathbf{C}_m = \mathbf{I}_{8\times 8}, \qquad \text{with } \mathbf{Q} = \begin{bmatrix} 1 & 0 & 0 \\ 1 & 0 & 0 \\ 0 & 1 & 0 \\ 0 & 0 & -1 \end{bmatrix}. \tag{2.12}$$

### 2.3.3 Load Carrying Structure and Blade Models

As an example this section considers a mechanical wind turbine model with four degrees of freedom, since they are the most strongly affected by the wind turbine control. In particular, they represent the fore-aft tower bending, flap-wise blade bending, rotor rotation, and generator rotation (Gasch and Twele, 2012).

Both the tower and blade bending are not modeled by means of bending beam models, but only the translational displacement of the tower top and blade tip are considered, where the bending stiffness parameters are transformed into equivalent translational stiffness parameters, as depicted in Figs. 2.5 and 2.6.

For the tower the equivalent translational stiffness parameter is derived by means of a direct stiffness method common in structural mechanics calculations (Gasch and Twele, 2012). In fact, since the blades move with the tower, the blade tip displacement is considered in the moving tower coordinate system and the tower motion must be taken into account for the derivation of the kinetic energy of the blade. The force $F_T$ acts both on the tower and on $N$ blades. Only one collective blade degree of freedom is considered.

Note that the $N$ blade degrees of freedom would have to be considered individually if control strategies for load reduction involving individual blade pitch control were designed. The assumption that the same external force $F_T$ acts on both the tower and blade degrees of freedom (with $N$ blades) is a simplification. It is reasonable, however, because the

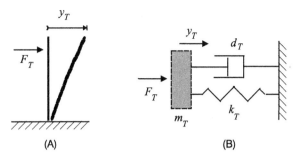

FIGURE 2.5   (A) Tower bending and (B) tower mechanical model with spring and damper.

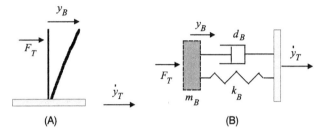

FIGURE 2.6   (A) Blade bending and (B) blade mechanical model with spring and damper.

rotor thrust force, which is caused by the aerodynamic lift forces acting on the blade elements, acts on the tower top, thus causing a distributed load on each blade. This distributed load generates a bending of the blade, which could be modeled as a bending beam. As shown in Figs. 2.5 and 2.6, a beam subjected not to a distributed load but to a concentrated load at the upper point must have a higher bending beam stiffness, in order for the same displacement to result at the upper point. However, a reduced-order wind turbine model considers only the blade tip displacement, which requires the assumption of a translational stiffness. To obtain an adequate translational stiffness constant, the bending stiffness of the bending beam must thus be larger than the case of a distributed load.

### 2.3.4   Power System Model

An explicit model for the generator/converter dynamics can be included into the complete wind turbine system model. Note that for mere simulation purposes, this is not necessary, since the generator/converter dynamics are relatively fast. However, when advanced control designs are considered, an explicit generator/converter model might be required in order to take into account generator torque fast dynamics. In this case, a simple first order delay model can be sufficient, as described, e.g., in Odgaard et al. (2013):

$$\dot{T}_g = -\frac{1}{\tau_g} T_g + \frac{1}{\tau_g} T_{g,d},$$ (2.13)

where $T_{g,d}$ represents the demanded generator torque, whilst $\tau_g$ is the delay time constant.

### 2.3.5   Pitch System Model

In pitch-regulated wind turbines, the pitch angle of the blades is controlled only in the full load region to reduce the aerodynamic rotor torque, thus maintaining the turbine at the desired rotor speed. Moreover, the pitching of the blades to feather position (i.e., 90 degrees) is used as main braking system to bring the turbine to standstill in critical situations. Two different types of pitch technologies are usually exploited in wind turbines, i.e., hydraulic and electromechanical pitch systems. For hydraulic pitch systems the dynamics can be modeled by means of a second-order delay model (Odgaard et al., 2013), which is able to display oscillatory behavior.

For electromechanical pitch systems, which are more commonly used, a first-order delay model is sufficient. In this monograph, the first-order delay model is considered, namely

$$\dot{\beta} = -\frac{1}{\tau} \beta + \frac{1}{\tau} \beta_d,$$ (2.14)

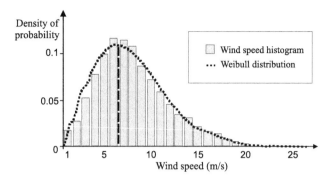

**FIGURE 2.7** Example histogram of wind speed and Weibull function.

where $\beta$ and $\beta_d$ are the physical and demanded pitch angles, respectively, as already highlighted in Fig. 2.3. The parameter $\tau$ denotes the delay time constant.

## 2.3.6 Wind Model

The differential heating of the earth's atmosphere is the driving mechanism for wind. Various atmospheric phenomena, such as the nocturnal low-level jet, sea breezes, frontal passages, and mountain and valley flows, affect the wind inflow across a wind turbine rotor plane (Manwell et al., 2002), which spans from 60 m to 180 m above the ground for megawatt utility-scale wind turbines. Given the large rotor plane and the variability of the wind, hundreds of sensors would be required to characterize the spatial variation of the wind speed encountered over the entire span of each blade.

The available wind resource can be characterized by the spatial or temporal average of the wind speed; the frequency distribution of wind speeds; the temporal and spatial variation in wind speed; the most frequent wind direction, also known as the prevailing wind direction; and the frequency of the remaining wind directions (Manwell et al., 2002). The probability of the wind speed being above a given turbine rated wind speed can be used to predict how often the turbine operates in Region 3 at its maximum, that is, rated power capacity. The capacity factor $CF$ is defined by the ratio

$$CF = \frac{E_{\text{out}}}{E_{\text{cap}}}, \tag{2.15}$$

where $E_{\text{out}}$ is a wind turbine energy output over a period of time and $E_{\text{cap}}$ is the energy the turbine would have produced if it had run at rated power for the same amount of time.

To predict the capacity factor and maintenance requirements for a wind turbine, it is useful to understand wind characteristics over both long and short time scales, ranging from multiyear to subsecond. Determining whether a location is suitable and economically advantageous for setting a wind turbine depends on the ability to measure and predict the available wind resource at that site. Significant variations in seasonal average wind speeds affect a local area's available wind resource over the course of each year. Wind speed and direction variations caused by the differential heating of the earth's surface during the daily solar radiation cycle occur on a diurnal, that is, daily time scale. The ability to predict hourly wind speed variations can help utilities plan their energy resource portfolio mix of wind energy and additional sources of energy. Finally, knowledge of short-term wind speed variations, such as gusts and turbulence, is used in both turbine and control design processes so that structural loading can be mitigated during these events.

Therefore it is very important for the wind industry to be able to describe the variation of wind speeds. Turbine designers need the information to optimize the design of their turbines, so as to minimize generating costs. Turbine investors need the information to estimate their income from electricity generation.

If wind speeds are measured throughout a year, in most areas strong gale force winds will be rare, while moderate and fresh winds are quite common. The wind variation for a typical site is usually described using the Weibull distribution, as represented in Fig. 2.7. This particular example site is characterized by a mean wind speed of 7 m/s, and the shape of the curve is determined by a so-called shape parameter of 2.

The example of Fig. 2.7 shows that 6.6 m/s is the median of the distribution, which is skewed, i.e., it is not symmetrical. Sometimes very high wind speeds occur, but they are very rare. Wind speeds of 5.5 m/s, on the other hand, are the most common ones; 5.5 m/s is called the modal value of the distribution. The probability distribution function has the form

$$p(v) = \frac{k}{A} \left(\frac{v}{A}\right)^{k-1} e^{-\left(\frac{v}{A}\right)^k},$$ (2.16)

where $A > 0$ and $k > 0$ are the scale and shape parameters, respectively, which determine the function form. In particular, $k$ determines the decrease rate of the function, whilst $A$ represents the function skewness. Properly chosen parameters and a value for $k$ indicate that the average speed and wind energy calculated from the gross Weibull distribution will be equal to that calculated from the histogram of the example depicted in Fig. 2.7.

The statistical distribution of wind speeds varies from place to place around the globe, depending upon local climate conditions, the landscape, and its surface. The Weibull distribution may thus vary, both in its shape, and in its mean value. If the shape parameter is exactly 2, as in Fig. 2.7, the distribution is known as a Rayleigh distribution. Wind turbine manufacturers often give standard performance figures for their machines using the Rayleigh distribution.

It is worth noting that more detailed models of the wind are not usually exploited in the related literature, as shown, for example, in Odgaard et al. (2013), Odgaard and Stoustrup (2013b, 2015b). However, in the following, a typical wind description is briefly outlined (Burton et al., 2011). Wind can be modeled as the sum of a steady state mean wind and a perturbation wind, accounting for turbulence and/or gusts. The deterministic component of the wind field implements the transients specified by IEC 61400-1 (Bottasso et al., 2007), the exponential and logarithmic wind shear models, and the tower shadow effects, which include the potential flow model for a conical tower, the downwind empirical model (Bottasso et al., 2007), or an interpolation of these two models. Their expressions will be omitted for brevity. The stochastic component of the wind field can be described according to the von Karman or Kaimal turbulence models.

In this way, the wind model generates, from a scalar mean wind speed at hub height, a time-varying matrix that contains the wind speed for each point in the wind field, namely

$$V_{\text{field}}(t, R, \theta) = v_{\text{mean}}(t) + V_{ws}(t, R, \theta) + V_{ts}(t, R, \theta) + V_{wk}(t, R, \theta),$$ (2.17)

where $V_{\text{field}}$ is the total wind speed field, $v_{\text{mean}}$ is the mean wind speed, $V_{ws}$ is the wind shear component, $V_{ts}$ is the tower shadow component, and $V_{wk}$ is the far wake component of one preceding wind turbine (relevant for the case of wind farms). Notice the dependence on the rotor radius $R$ and rotor azimuth angle $\theta$. When required, the simplified wake model is represented as a part of the wind field (i.e., a circle) with a lower wind speed (Friis et al., 2011). The wake is centered around a point $(R_0, \varphi)$ placed on the rim of the wind field, and with the form

$$R^2 - 2 R R_0 \cos(\theta - \varphi) + R_0^2 = W^2,$$ (2.18)

where $R_0$ is the radial coordinate for the center of the wake, $\varphi$ is the angular coordinate of the center of the wake, and $W$ is the radius of the wake.

Finally, stochastic variables can be added to the wind components except tower shadow, giving a closer to reality parameterization of the wind speeds throughout the rotor plane. In this way, the wind field is converted to equivalent winds signals that acts on two distinct parts of the blades, namely the tip and root sections, in order to obtain a linearizable model description.

### 2.3.7 Model–Reality Mismatch

Prior to the design and application of new control strategies on real wind turbines, the efficacy of the control scheme has to be tested in detailed aero-elastic simulation model. Several simulation packages exist that are commonly used in academia and industry for wind turbine load simulation. One of the most used simulation packages is the Fatigue, Aerodynamics, Structures, and Turbulence (FAST) code (Jonkman and Buhl, 2005) provided by the National Renewable Energy Laboratory (NREL) in Golden (Colorado, USA), since it represents a reference simulation environment for the development of high-fidelity wind turbine prototypes that are taken as a reference test-cases for many practical studies (Jonkman et al., 2009). FAST provides a high-fidelity wind turbine model with 24 degrees of freedom, which is appropriate for testing the developed control algorithms but not for control design. For the latter purpose, a reduced-order dynamic wind turbine model, which captures only dynamic effects directly influenced by the control, is recalled in this section and it can be used for model-based control design (Bianchi et al., 2007).

Due to the competitive nature of the wind turbine industry and possible confidentiality issues, the modeling available in the wind turbine literature is usually kept at a conceptual level. For more detailed modeling of pitch regulated wind turbines, see, e.g., Burton et al. (2011), Muljadi and Butterfield (1999), Knudsen et al. (2011). It is worth noting also that, in the wind turbine area, there have been a number of IFAC and IEEE publications with sessions and special issues starting from 2009, based also on competition studies, e.g., Ostergaard et al. (2009), Pao and Johnson (2011), Odgaard and Stoustrup (2012d). These sessions and special issues have led to important results and publications that will be recalled in this monograph, in order to give readers a basic research review.

Previous studies have shown that linear aero-elastic models used for the analysis of wind turbines are commonly of very high order. Multibody dynamics coupled with unsteady aerodynamics (e.g., dynamic stall) are among the recently developments in wind turbine aero-elasticity (Rasmussen et al., 2003; Bianchi et al., 2007; Hansen, 2011). The resulting models contain hundreds or even thousands of flexible modes and aerodynamic delays. In order to synthesize wind turbine controllers, a common practice is to obtain linear time-invariant (LTI) models from a nonlinear model for different operating points. Modern control analysis and synthesis tools are inefficient for such high-order dynamical systems; reducing the model size is crucial to analyze and synthesize model-based controllers. The most interesting modeling solution available in the literature relies on the Linear Parameter Varying (LPV) framework, as it has shown to be suitable to cope, in a systematic manner, with the inherent varying dynamics of a wind turbine over the operating envelope (Bianchi et al., 2007; Ostergaard et al., 2009; Adegas et al., 2012, 2013).

Wind turbine LPV models are usually simple, first-principles based, often neglecting dynamics related to aerodynamic phenomena and some structural modes. This in turn restricted LPV control of wind turbines to the academic environment only. A procedure to encapsulate high-fidelity dynamics of wind turbines as an LPV system would be beneficial to facilitate industrial use of LPV control.

Other modeling approached that one may find in the literature are based on some type of simplified wind turbine descriptions (Pedersen and Fossen, 2012). These may have the form of lookup tables as in Bianchi et al. (2007) or linear models obtained from complex numerical simulation tools (Namik and Stol, 2010). Hybrid models blending lookup tables with mechanical models have also been used (Bottasso et al., 2007). These and even simpler approaches dominate. Linear models can be valid in a small envelope around the linearization point, which requires several individual models to cover the operational domain of the turbine (Pintea et al., 2010).

However, most of the control algorithms for modern variable-pitch wind turbines that one may find in the literature are usually based on some type of simplified wind turbine linear model. Therefore, after these considerations, this chapter addresses the most important components of a HAWT used for the linear modeling of a wind turbine installation. As already remarked, they consist of the wind turbine tower, its nacelle, and the rotor, visible from the ground, as depicted in the example of Fig. 2.2. Moreover, as reported in Fig. 2.3, the complete wind turbine model consists of several submodels for the mechanical structure, aerodynamic system, as well as the dynamics of the pitch system, and the generator/converter system. The generator/converter dynamics are usually described as a first order delay system. However, when the delay time constant is very small, an ideal converter can be assumed, such that the reference generator torque signal is equal to the actual generator torque. In this situation, the generator torque can be considered as a system input, whilst the generator is the device that converts mechanical energy from the aerodynamic torque to electrical energy.

The purpose of this section is also to outline simplifications, which are introduced in the modeling of the wind turbine benchmark considered in Chapter 5. These simplifications are made since only a reduced set of model parameters are available, and to limit the extent of the modeling effort. Generally, two simplifications are introduced: modeling only the dominant modes of the system, while neglecting all others and assuming that the parameters of the wind turbine are constant, although they depend on the operating conditions.

In addition to the general assumptions, the following significant simplifications are introduced:

**Perfect yaw alignment.** The wind is assumed to be perpendicular to the rotor plane at all times. This simplification eliminates some periodic fluctuations caused by yaw misalignment of the wind turbine.

**Static aerodynamic model.** The aerodynamics is assumed to possess static properties, thus neglecting the dynamical properties caused by changes in the wind, rotor speed, and pitch angles. This assumption is made since only a $C_p$-table is available for the considered wind turbine.

**Stiff blades.** The blades are assumed to be stiff, but are in fact flexible, especially on large wind turbines similar to the considered one. This simplification eliminates all bending modes of the blades and transfers all forces acting on the blades directly to the tower.

The presented simplifications should be taken into account when assessing the results of this monograph, since they may affect the performance of the designed algorithms when applied to a real wind turbine.

In order to obtain the nonlinear simulation model, all submodels derived in the previous sections are combined directly. However, in some parts of the design procedure it is favorable to exploit a linearized model, as shown in Chapter 5.

### 2.3.8    Actuator and Sensor Models

Wind turbine high-fidelity simulators, which were described, for example, in Odgaard et al. (2013), Odgaard and Stoustrup (2013b), Odgaard and Johnson (2013), consider white noise added to all measurements achieved from the wind turbine actuator and sensor signals. The rationale relies on the assumption that noisy sensor signals should represent more realistic scenarios. However, this is not the case, as a realistic simulation would require accurate knowledge of each sensor and its measurement reliability. To the best of the authors' knowledge and from their experience with wind turbine systems, all main measurements acquired from the wind turbine process (rotor and generator speed, pitch angle, generator torque) are virtually noise-free or affected by very weak noise.

On the other hand, with reference to real wind turbines, one characteristic frequency is usually affecting the generator speed signal due to the periodic rotor excitation of the drive-train. When using the generator speed as a controlled variable, a notch filter is thus applied to smooth the generator speed signal. Such a notch filter is usually applied to industrial wind turbine control, as described, e.g., in Bossanyi and Hassan (2000). On the other hand, the rotor speed measurement is normally assumed to be a continuous signal. However, in many wind turbines, the rotor speed signal is discretized, due to the limited number of metal pieces on the main shaft that are scanned by a magnetic sensor, although better sensing technologies, which could yield nearly continuous signals, are already available, like optical scanning of densely spaced barcodes.

### 2.3.9    Overall Model Structure

By replacing the expressions of Eq. (2.3) for the rotor thrust and torque into the mechanical model of Eq. (2.11) and adding the models of Eqs. (2.14) and (2.13) for the pitch and the generator/converter dynamics, a nonlinear state-space description is obtained:

$$\begin{cases} \dot{\mathbf{x}} = \mathbf{A}\mathbf{x} + \mathbf{B}\mathbf{u} + g\,(\mathbf{x},\,v)\,, \\ \mathbf{y} = \mathbf{C}\mathbf{x}, \end{cases} \tag{2.19}$$

with a state vector that now includes the pitch angle and the generator torque $\mathbf{x} = \left[ y_T,\ y_B,\ \theta_r - \theta_g,\ \dot{y}_T,\ \dot{y}_B,\ \dot{\theta}_r,\ \dot{\theta}_g,\ \beta \right]^T$.

Since the rotor thrust force and the rotor torque have been used as inputs for the vector $\mathbf{u}_m$ in the mechanical submodel of Eq. (2.11), a new input vector is defined for the complete state-space model (2.19), i.e., $\mathbf{u} = \left[ \beta_d,\ T_g \right]^T$ whose components are the demanded pitch angle and the generator torque, respectively. The wind speed is normally considered as a disturbance input. The linear part of the state-space model of Eq. (2.19) is defined by the matrices:

$$\mathbf{A} = \begin{bmatrix} 0_{3\times3} & \tilde{\mathbf{L}} & 0_{3\times1} & 0_{3\times1} \\ \mathbf{M}^{-1}\tilde{\mathbf{K}} & \mathbf{M}^{-1}\mathbf{D} & 0_{4\times1} & \begin{bmatrix} 0_{3\times1} \\ -\frac{1}{J_g} \end{bmatrix} \\ 0_{1\times3} & 0_{1\times4} & -\frac{1}{\tau} & 0 \\ 0_{1\times3} & 0_{1\times4} & 0 & -\frac{1}{\tau_g} \end{bmatrix}, \quad \mathbf{B} = \begin{bmatrix} 0_{7\times1} & 0_{7\times1} \\ \frac{1}{\tau} & 0 \\ 0 & -\frac{1}{\tau_g} \end{bmatrix}, \tag{2.20}$$

with

$$\tilde{\mathbf{L}} = \begin{bmatrix} 1 & 0 & 0 & 0 \\ 0 & 1 & 0 & 0 \\ 0 & 0 & 1 & -1 \end{bmatrix} \quad \text{and} \quad \tilde{\mathbf{K}} = \begin{bmatrix} k_T & 0 & 0 \\ 0 & Nk_B & 0 \\ 0 & 0 & k_S \\ 0 & 0 & -k_S \end{bmatrix}. \tag{2.21}$$

Moreover, the system vector in Eq. (2.19) nonlinearly depends on the state and input vector

$$g\,(\mathbf{x},\,v) = \begin{bmatrix} 0_{4\times1} \\ \frac{1}{m_B} F_T\,(\mathbf{x},\,v) \\ \frac{1}{J_r} T_a\,(\mathbf{x},\,v) \\ 0_{3\times1} \end{bmatrix}. \tag{2.22}$$

Here the rotor thrust and torque expressions are given in Eq. (2.3), whilst the mass and damping matrices are defined in Eq. (2.10).

It is worth noting that in a real wind turbine, the centrifugal forces acting on the rotating rotor blades lead to a stiffening of the blades. As a consequence, the bending behavior of the rotor blades depends on the rotor speed itself. By considering again the translational spring–mass system of the blade-tip displacement, this second-order effect can be included in the model of Eq. (2.19) by introducing a translational blade stiffness parameter $k_B$ dependent on the rotor speed, i.e., $k_B(\omega_r) = \alpha m_B r_B \omega_r^2$, where $r_B$ denotes the distance from the blade root to the blade center of mass and $\alpha$ is a tuning parameter. In this way, by including the centrifugal stiffening correction, the nonlinear system vector $g(\mathbf{x}, v)$ in Eq. (2.22) has the form

$$g(\mathbf{x}, v) = \begin{bmatrix} 0_{3 \times 1} \\ \frac{N}{m_T} k_B(\omega_r) y_B \\ \frac{1}{N m_B} F_T(\mathbf{x}, v) + \frac{m_T + N m_B}{m_T m_B} k_B(\omega_r) y_B \\ \frac{1}{J_r} T_a(\mathbf{x}, v) \\ 0_{3 \times 1} \end{bmatrix}. \tag{2.23}$$

The inclusion of the centrifugal term is inspired from the FAST code, in order to obtain a high-fidelity wind turbine simulation model. For example, the translational blade bending model could be required when overspeed scenarios shall be taken into account. However, for the usual operating regimes of a wind turbine, the corrections induced by the centrifugal blade stiffening have only minor effects on the final results. Therefore the centrifugal correction has been recalled here for the sake of completeness, but it has limited interest in real cases.

## 2.4 WIND TURBINE CONTROL ISSUES

One of the main goals of this chapter is to introduce control engineers to the technical challenges that exist in the energy conversion industry and to encourage new modeling strategies and control systems research in this area. In fact, wind turbines are complex structures operating in uncertain environments and lend themselves nicely to advanced control solutions. Advanced controllers can help achieve the overall goal of decreasing the cost of wind energy by increasing the efficiency, and thus energy capture, or by reducing structural loading and increasing the lifetimes of the components and turbine structures (Bossanyi, 2003).

Although wind turbines come in both vertical- and horizontal-axis configurations, as already remarked, the monograph will focus on HAWTs, which have an advantage over VAWTs in that the entire rotor can be placed atop a tall tower, where they can take advantage of larger wind speeds higher above the ground. Some of the other advantages of HAWTs over VAWTs for utility-scale turbines include pitchable blades, improved power capture, and structural performance. VAWTs are much more common as smaller turbines, where these disadvantages become less important and the benefits of reduced noise and omnidirectionality become more pronounced. Active control is most cost-effective on larger wind turbines, and therefore this chapter will refer to wind turbines with relatively large capacities. As remarked in Pao and Johnson (2009), active control refers to those active actions allowing conversion energy systems to achieve optimal power capture and structural performance, such as the use of pitchable blades, power and torque control techniques. On the other hand, the term "active" has been extended to fault diagnosis and fault tolerant control fields (Chen and Patton, 2012; Mahmoud et al., 2003; Zhang and Jiang, 2008; Ding, 2008), as outlined in more detail in Chapters 3 and 4.

It is worth also noting that the mathematical description used for wind turbine modeling and control is quite basic, as the monograph focuses on the related fundamental aspects. On the other hand, real system cases require much more complex modeling and control considerations, which have been highlighted through proper bibliographical references.

### 2.4.1 Advanced Control Solutions

In the case of a wind turbine, optimal blade pitch, $\beta$, and rotor velocity (via the tip/speed ratio, $\lambda$) are set based on the incident wind flow velocity, in order to maximize the power coefficient, $C_Q$. The manipulated variable for the pitch control is the power to the pitch actuators (voltage and/or current). For torque control, the generator excitation is used as a control actuator. It is worth noting that the relationship between $\beta$, $\lambda$, and $C_Q$ is specific to each wind turbine, and must be determined for each particular case. However, this relationship is then fixed, though some slight variation may occur due, for example, to component wear or installation errors. Note also that when a wind turbine reaches its rated power (i.e., above the rated wind speed), the turbine needs to be "depowered" in order to avoid exceeding any rated specifications.

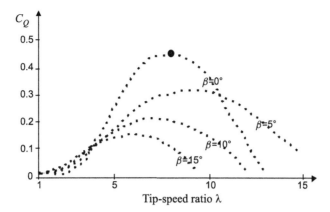

**FIGURE 2.8**  Power coefficient curve example.

In this situation, it is not required to maximize power conversion (i.e., the wind power that can be converted into electric energy) and, for variable pitch turbines, blade pitch can be adjusted in order to limit power converted.

As already remarked in Section 2.3, in the wind area there have been a number of IFAC and IEEE publications, sessions, and special issues starting from 2011, based also on competition studies, addressing basic and advanced wind turbine control issues, e.g., Odgaard and Stoustrup (2011), Diaz-Guerra et al. (2012), Biegel et al. (2013), Pao and Johnson (2011), Adegas and Stoustrup (2012), Odgaard and Stoustrup (2012d).

On the other hand, previous investigations, e.g., Muljadi and Butterfield (1999), Leithead and Connor (2000), Bossanyi and Hassan (2000), Bianchi et al. (2007) have shown that linear, time-invariant methods provide good closed-loop results when observing local behavior. A natural choice for controller design covering the entire operating envelope is therefore to design linear controllers along a chosen operating trajectory and then to interconnect them in an appropriate way in order to get a control formulation for the entire operating region. This approach is denoted as gain scheduling, and in Cutululis et al. (2006) this is done by interpolating the outputs of a set of local controllers (either by linear interpolation or by switching). Alternatively, parameters of the controller are updated according to a prespecified function of a measured/estimated variable (Leithead and Connor, 2000). A systematic way of designing such parameter-dependent controllers is within the framework of LPV systems, already recalled in Section 2.3. In this case the model is represented by a linear model at all operating conditions, and a controller with similar parameter dependency is synthesized to guarantee a certain performance specification for all possible parameter values within a specified set. A major difference to classical gain scheduling is that it is possible to take into account that the scheduling parameters can vary in time (Ostergaard et al., 2009). Other controllers with different structures, e.g., linear quadratic, and repetitive model predictive, to mention a few more (Adegas and Stoustrup, 2012; Diaz-Guerra et al., 2012; Adegas and Stoustrup, 2011), were also designed and applied to wind turbine systems.

After these considerations control systems for wind turbines seem now well developed (Bianchi et al., 2007) and the fundamental control strategies are sketched below, in order to provide the readers a basic research review.

The primary Region 2 control objective for a variable-speed wind turbine is to maximize the power coefficient, and in particular the $C_Q$ map in Eq. (2.1). The relationship between $C_Q$ and the tip-speed ratio $\lambda$ is a turbine-specific nonlinear function. $C_Q$ also depends on the blade pitch angle in a nonlinear way, and these relationships have the same basic shape for most modern wind turbines. An example of $C_Q$ surface is shown in Fig. 2.8 for a generic wind turbine.

As shown in Fig. 2.8, the turbine will operate at its highest aerodynamic efficiency point, $C_{max}$, at a certain pitch angle and tip-speed ratio. The pitch angle is easy to control, and it can be reliably maintained at the optimal efficiency point. However, the tip-speed ratio depends on the incoming wind speed $v$, and therefore is constantly changing. Thus Region 2 control is primarily concerned with varying the turbine speed to track the wind speed. When this approach is used, the controller structure for partial load operation follows the sequential optimal calculation and regulation strategies remarked in Section 1.7.

On the other hand, on utility-scale wind turbines, Region 3 control is typically performed via a separate pitch control loop. In Region 3 the primary objective is to limit the turbine power so that safe electrical and mechanical loads are not exceeded. Power limitation is achieved by pitching the blades or by yawing the turbine out of the wind, both of which can reduce the aerodynamic torque below what is theoretically available from an increase in wind speed. In Region 3 the pitch control loop regulates the rotor speed $\omega_r$ (at the turbine "rated speed") so that the turbine operates at its rated power.

In this way the overall strategy of the wind turbine controller is to use two different controllers for the partial load region and the full load region. When the wind speed is below the rated value, the control system should maintain the pitch angle at its optimal value and control the generator torque in order to achieve the optimal tip-speed ratio (switch to Region 2).

At low wind speeds, i.e., in partial load operation, variable-speed control is implemented to track the optimum point on the $C_Q$-surface for maximizing the power output, which corresponds to the $\lambda_{opt}$ value. The speed of the generator is controlled by regulating the demanded torque $T_{g,d}$ on the generator through the generator torque controller. In partial load operation it is chosen to operate the wind turbine at $\beta = 0$ degrees, since the maximum power coefficient is obtained at this pitch angle, namely

$$
T_{g,d} = \frac{1}{2} \rho \pi R^2 \frac{R^3}{n_g^3 \lambda_{opt}^3} C_{max} \omega_g^2(t) - d_S \left( \frac{1}{n_g^2} + 1 \right) \omega_g(t), \tag{2.24}
$$

where $n_g$ is the gear-ratio of the gear-box connecting the rotor shaft with the electric generator/converter, $R$ is the rotor radius, and $\omega_g(t)$ the electric generator/converter speed (Johnson et al., 2006a). The advantage of this approach is that only the measurement of the rotor or generator speed is required.

On the other hand, for high wind speeds, i.e., in full load operation, the desired operation of the wind turbine is to keep the rotor speed and the generated power at constant values. The main idea is to use the pitch system to control the efficiency of the aerodynamics, while applying the rated generator torque. However, in order to improve tracking of the power reference and cancel steady-state errors on the output power, a power controller is also introduced.

With reference to the speed controller, it is implemented as a PI controller that is able to track the speed reference and cancel possible steady-state errors on the generator speed. The speed controller transfer function $D_s(s)$ has the form

$$
D_s(s) = K_{ps} \left( 1 + \frac{1}{T_{is}} \frac{1}{s} \right), \tag{2.25}
$$

where $K_{ps}$ is the PI proportional gain and $T_{is}$ is the reset rate of the integrator.

The power controller is implemented in order to cancel possible steady-state errors in the output power. This suggests using slow integral control for the power controller, as this will eventually cancel steady-state errors on the output power without interfering with the speed controller. However, it may be beneficial to make the power controller faster to improve accuracy in the tracking of the rated power. The power controller is realized as a PI controller, whose transfer function $D_p(s)$ has the standard form

$$
D_p(s) = K_{pp} \left( 1 + \frac{1}{T_{ip}} \frac{1}{s} \right), \tag{2.26}
$$

where $K_{pp}$ is the proportional gain of the PI regulator, whilst $T_{ip}$ is the reset rate of the integrator.

Note finally that speed and power control can be coupled. However, as shown in Odgaard et al. (2013), they can be considered as decoupled, as their dynamics are different. However, more advanced control techniques can exploit multivariate (or decoupling) control, as addressed in Bianchi et al. (2007), Pao and Johnson (2011). It is worth noting that, from the previous considerations, the research issues of wind turbine control may seem very mature. However, the latest generation of giant offshore wind turbines present new dynamics and control issues.

Moreover, new wind turbine solutions, which use further wind turbine state information from the sensing system, have been suggested, also within EU projects, see, e.g., Plumley et al. (2014), Chatzopoulos and Leithead (2010). This improved state information is used to control the wind turbine blades and at the same time reducing the design bearing fatigue and extreme structural loads that are affecting the structure of the wind turbine (Valencia-Palomo et al., 2015; Khan et al., 2016). This control problem will be solved in a multivariate way, by optimizing the conflicting control objectives of power optimization while keeping the different loads below the design requirements. The control goal is to ensure that the controller will guarantee that extreme load requirements are not violated during eventually emergency stops of the wind turbine, as well as during severe wind gusts. The interesting challenge is to be able to use the rotor system to control the turbine, so that in effect the rotor performs like a "high level" sensor. In other words, the goal is to be able to use the rotor itself (along with the enhanced sensor set) to make the control system perform well. A part of this challenge is to ensure that real-time compensation of loading and gust disturbances is put into effect in a suitable time window, taking account of the close spectral content of the disturbance and control. This becomes a very significant challenge for very large rotor wind turbines ($> 10$ MW) as the required control and disturbance bandwidths become close, a problem similar to the structural filtering and control used in high performance combat aircraft (Shi and Patton, 2015).

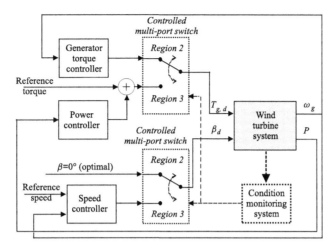

**FIGURE 2.9**   Structure of the switching controller between the working Regions 2 and 3.

## 2.4.2   Wind Turbines Feedback Control

This section provides further information regarding what control strategies are typically used for the torque control and the pitch control blocks in Fig. 2.3 of Section 2.3. As depicted in Fig. 2.3, both control loops typically only use rotor speed feedback. The other sensors and measurements acquired from the wind turbine can be used for fault diagnosis and advanced control purposes, as outlined in Chapters 3 and 4.

As shown in Fig. 1.2, the nominal operating trajectory of the wind turbine is created to satisfy different demands below and above a certain wind speed. Since the classical control approach deals only with SISO transfer functions, and because several references exist, the control task is split into the design of multiple separate compensators, as sketched in Fig. 2.9. The condition monitoring system depicted in Fig. 2.9 determines the actual working point of the wind turbine plant and drives the multiport controlled switches accordingly.

The design of the complete wind turbine controller is thus divided into four main control boxes and design steps, as listed below:

- Controller operating in partial load condition. It refers to the design of the generator torque controller. This controller operates in the partial load (Region 2), and should maximize the energy production while minimizing mechanical stress and actuator usage;
- Controller operating in full load condition. It concerns the speed controller and power controller. These controllers operate in the full load (Region 3), and should track the rated generator speed and limiting the output power;
- Bumpless transfer. It describe the design of the mechanism that eliminates bumps on the control signals, when switching between the controllers in the partial load and full load regions;
- Structural stress damper. It regards the design of structure and drive-train stress damper. The purpose of the module is to dampen drive-train oscillations and reduce structural stress that could affect the wind turbine tower.

The first two items are the main control loops shown in Fig. 2.9, whilst the two remaining tasks concern advanced control issues, which can enhance both the control and system performances, and are sketched in Sections 2.4.4 and 2.4.3, respectively. Note also that the transfer functions outlined throughout this section need to be discretized to allow the implementation of the controllers and filters in real-time conditions. Their implementations will be illustrated in Chapter 6.

In this way, the strategy of the complete controller of Fig. 2.9 is to use two different controllers for the partial load region and the full load region. When the wind speed is below the rated value, the control system should maintain the pitch angle at its optimal value and control the generator torque in order to achieve the optimal tip-speed ratio (switch to Region 2).

Above the rated wind speed the output power is kept constant by pitching the rotor blades, while using a power controller that manipulates the generator torque around a constant value to remove steady-state errors on the output power. This behavior is obtained by setting the two switches in Fig. 2.9 to Region 3. In both regions a drive-train stress damper is exploited to dampen drive-train oscillations actively. Together, the two sets of controllers are able to solve the control task of tracking the ideal power curve in Fig. 2.8. In order to switch smoothly between the two sets of controllers, a bumpless transfer mechanism has to be implemented, as outlined in Section 2.4.4.

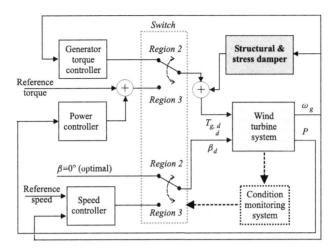

**FIGURE 2.10** Switching controller with structural and stress damper modules.

### 2.4.3 Structural and Drive-Train Stress Damper

Active stress damping solutions are deployed in large HAWTs to mitigate fatigue damage due to drive-train and structural oscillations and vibrations. The idea is to add proper components to the wind turbine control signals to compensate for the oscillations in the drive-train and the tower vibrations. These signals should have frequencies equal to the eigenfrequencies of the drive-train and the wind turbine structure, which can be found by filtering the measurement of the generator speed and the generated power. When the outputs from these filters are added to the generator torque and the pitch command, the phase of the filters must be zero at the resonant frequency to achieve the desired damping effects. These oscillation and vibration dampers are thus implemented to add compensating signals, as sketched in Fig. 2.10.

Second-order filter models for the stress and the structural damping have been proposed and can be applied to dampen the eigenfrequency of both the drive-train and the tower structure (Munteanu and Bratcu, 2008). In general, the filter time constant introduces a zero in the filter that can be used to compensate for time lags in the system. To determine the gain of the filter, the root loci are plotted for the transfer functions from the wind turbine inputs to its outputs. More details on the design of these filters, which are beyond the scope of this monograph, can be found, e.g., in Amit Dixit and Suryanarayanan (2005), Munteanu and Bratcu (2008).

Note that, due to the higher loads at higher wind speeds, it is favorable if the filter gains depend on the point of operation. A simple way of fulfilling this property is to apply different gains in the partial and full load configurations of the wind turbine controller. Therefore Section 2.4.4 outlines the bumpless transfer issue, which must ensure that no bumps exist on the control signals in the switch between two different controllers.

### 2.4.4 Bumpless Transfer

The purpose of this section is to outline how the bumpless transfer mechanism is designed, i.e., how and when to activate the switch illustrated in Fig. 2.10. The considered transition is the one that brings the control system from partial load operation to full load operation, and vice versa. When the control system switches from partial load to full load operation, it is important that this transition is not affecting the control signals, i.e., the generator torque and pitch angle. This procedure is known as a bumpless transfer, and is important because two controllers may not be consistent with the magnitude of the control signal at the time that the transition happens. If a switch between two controllers is performed without bumpless transfer, a bump in the control signal may trigger oscillations between the two controllers, making the system unstable. The transition from partial to full load operation must happen as the wind speed becomes sufficiently large. For stationary wind speeds this usually happens at about 12 m/s. However, it is not convenient to apply the wind speed as the switching condition, since the large inertia of the rotor causes the generator speed and output power to follow significantly later than a rise in the wind speed. Moreover, the wind speed is almost unknown. Therefore, it is more appropriate to exploit the generator speed as switching condition. In particular, the switching from partial to full load condition is achieved when the generator speed $\omega_g(t)$ is greater than the nominal generator speed. On the other hand, the switching from full to partial load condition is applied if both the pitch angle $\beta(t)$ is lower than its optimal value and the generator speed $\omega_g(t)$ is significantly lower than its nominal value. Notice that an hysteresis is usually introduced to ensure a minimum time between each transition.

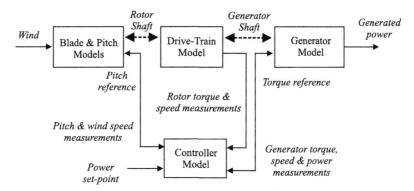

**FIGURE 2.11** The block diagram of the wind turbine benchmark.

Due to the switching condition on $\beta(t)$ and because the output of the speed controller is saturated not to move below 0, the transition already fulfils the bumpless transfer condition for this control signal. On the other hand, for the generator torque signal a bumpless transfer is assured by adjusting the integral state of the controller, such that the generator torque does not change abruptly. The compensation torque is calculated using

$$T_{g,\text{comp}}(k) = T_{g,1}(k) - T_{g,2}(k) + T_{g,\text{comp}}(k-1), \tag{2.27}$$

where $T_{g,1}(k)$ and $T_{g,2}(k)$ are the torque outputs at the sample $k$ when the switching occurs from the controller 1 to the controller 2. $T_{g,\text{comp}}(k)$ represents the compensation torque ensuring a bumpless transfer.

Note finally that the torque compensation is not important when operating above the rated wind speed, because the power controller has integral action. When operating below rated wind speed the compensation torque is discharged to zero, as it otherwise would result in the optimal tip-speed ratio not being followed.

## 2.5 WIND TURBINE BENCHMARK

The single wind turbine benchmark model considered in this monograph is described in detail in Odgaard et al. (2014). It has been implemented in Matlab/Simulink environment and proposes a realistic simulator for a wind turbine system, as well as some common fault scenarios. It presents a specific kind of wind turbine: a three-blade horizontal-axis variable-speed pitch-controlled turbine with a full converter generator. The considered components have been reduced with respect to the representation of Fig. 2.2. The tower supports the nacelle that contains the control systems and the equipment to convert energy from the rotating blades fixed in the hub to electrical energy for the grid. On the nacelle an anemometer provides the measurements of the current wind speed at the hub height.

The internal components of the nacelle consist of a gear-box that connects the rotor main shaft to the generator, adapting its torque and speed values; the generator, that converts the energy into the electrical form; a converter and a transformer that connect the turbine to the grid; finally, the controller adjusts the pitch angle and the generator torque in order to follow the power reference.

The block diagram of Fig. 2.11 recalls again how the main components are connected each other and the input/output variables which indicate the relationship among the blocks.

### 2.5.1 Wind Turbine Benchmark Model

The wind turbine overall model consists of four submodels: the wind model, the blade and pitch model, the drive-train model, and the generator model.

The wind is considered a stochastic process, with the additive contributions of the effects of wind shear and tower shadows. A complete description of the wind model is reported in Dolan and Lehn (2006). Therefore the wind speed $v_w$ is the sum of four components:

$$v_w(t) = v_m(t) + v_s(t) + v_{ws}(t) + v_{ts}(t), \tag{2.28}$$

where $v_m$ is the mean speed, $v_s$ is the stochastic component (a Gaussian white noise), $v_{ws}$ is the wind shear effect that takes into account the speed variation due to the distance from the earth surface, and $v_{ts}$ is the tower shadow effect that includes the loss of speed due to the passing of the blade in front of the tower. Actually, four wind speeds are involved in the benchmark model, one for the wind $v_{hub}$ acting on the hub, and one for each of the three blades $v_{w1}$, $v_{w2}$, and $v_{w3}$. It is worth noting that the wind shear and the tower shadow terms are considered only in the blade wind speeds.

For a specific blade $i$, the wind shear term can be written in analytical form as

$$v_{ws_i}(t) = \frac{2v_m(t)}{3R^2}\left(\frac{R^3\alpha}{3H}\chi + \frac{R^4}{4}\alpha\frac{\alpha - 1}{2H^2}\chi^2\right) + \frac{2v_m(t)}{3R^2}\left(\frac{R^5}{5}\frac{(\alpha^2 - \alpha)(\alpha - 2)}{6H^3}\chi^3\right), \tag{2.29}$$

where $\chi = \cos(\theta_{r_i})$, $\theta_{r_i}$ is the angular position of the $i$th blade, $R$ is the radius, whilst $\alpha$ and $H$ are two aerodynamic parameters. The term relative to the tower effect is represented by the equation

$$v_{ts_i}(t) = \frac{m\bar{\theta}_{r_i}(t)}{3r^2}(\Psi - \nu) \tag{2.30}$$

with

$$\Psi = 2a^2 \frac{R^2 - r_0^2}{(R^2 + r_0^2)\sin^2(\bar{\theta}_{r_i}(t)) + k^2}, \tag{2.31}$$

$$\nu = 2a^2k^2 \frac{(r_0^2 - R^2)r_0^2\sin(\bar{\theta}_{r_i}(t) + k^2)}{R^2\sin^2(\bar{\theta}_{r_i}(t)) + k^2}, \tag{2.32}$$

$$m = 1 + \frac{\alpha(\alpha - 1)r_0^2}{8H^2}, \tag{2.33}$$

$$\bar{\theta}_{r_i}(t) = \theta_{r_i}(t) + \frac{2\pi(i - 1)}{3} - \text{floor}\left(\frac{\theta_{r_i}(t) + \frac{2\pi(i-1)}{3}}{2\pi}\right)2\pi, \tag{2.34}$$

in which $r_0$ is the radius of the blade hub, $k$ and $a$ are two aerodynamic parameters.

The so-called blade and pitch model is fed by the wind speed and it is based on the aerodynamic law of Eq. (2.1). The final value of the aerodynamic torque at the rotor shaft is equally provided by the three blades, hence

$$\tau_r(t) = \sum_{i=1}^{3} \frac{\rho\pi R^3 C_q(\lambda(t), \beta(t))v_w^2(t)}{6}. \tag{2.35}$$

The actual value of the blade pitch angle $\beta$ differs from the reference signal $\beta_r$ provided by the controller by means of a second order transfer function

$$\frac{\beta(s)}{\beta_r(s)} = \frac{\omega_n^2}{s^2 + 2\zeta\omega_n s + \omega_n^2}, \tag{2.36}$$

where $\zeta$ is the damping coefficient, $\omega_n$ is the natural pulsation. It represents a commonly adopted solution in the modeling of hydraulic piston servo systems. In nominal conditions the parameters of the transfer function have the same value for each of the three blades.

The drive-train model describes the power flow through the gear-box from the rotor to the generator. The gear-box involves a torque decrement and a speed increment on the generator shaft. The following two-mass first order differential equations represent the drive-train model included in the wind turbine benchmark model:

$$J_r\dot{\omega}_r = \tau_r(t) - K_{dt}\theta_\Delta(t) - (B_{dt} + B_r)\omega_r(t) + \frac{B_{dt}}{N_g}\omega_g(t), \tag{2.37}$$

$$J_g\dot{\omega}_g = \frac{\eta_{dt}K_{dt}}{N_g}\theta_\Delta(t) + \frac{\eta_{dt}B_{dt}}{N_g}\omega_r(t) - \left(\frac{\eta_{dt}B_{dt}}{N_g^2} + B_g\right)\omega_g(t) - \tau_g(t), \tag{2.38}$$

$$\dot{\theta}_\Delta(t) = \omega_r(t) - \frac{1}{N_g}\omega_g(t), \tag{2.39}$$

where $J_r$ and $J_g$ are the moments of inertia of the shafts, $K_{dt}$ is the torsion stiffness, $B_{dt}$ is the torsion damping factor, $B_g$ is the generator shaft viscous friction, $N_g$ is the gear ratio, $\eta_{dt}$ is the efficiency coefficient, and $\theta_\Delta$ is the torsion angle.

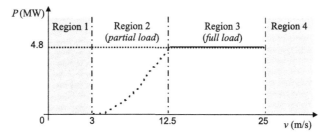

**FIGURE 2.12** Reference power curve with four different working regions.

The dynamics of the generator and the converter is described by means of a first order transfer function

$$\frac{\tau_g(s)}{\tau_{gr}(s)} = \frac{\alpha_g}{s + \alpha_g}. \tag{2.40}$$

The torque reference signal $\tau_{gr}$ comes from the controller, $\alpha_g$ is a generator parameter.

Finally, the generated power is computed as the product of the torque and the speed of the generator, decreased by its efficiency $\eta_g$, and its is described by the following relation:

$$P_g(t) = \eta_g \omega_g(t) \tau_g(t). \tag{2.41}$$

## 2.5.2 Wind Turbine Controller Model

As already remarked in Section 2.4, the controller has the main task of regulating the power generated by the turbine on the basis of the wind speed. Fig. 2.12 depicts the reference power curve that can be divided into four different regions. In the first region ($v_w < 3$ m/s) the wind speed is too low to generate power and the turbine is in idle state. The second region ($3$ m/s $< v_w < 12.5$ m/s) can be defined as *partial load*, or power optimization region, as the controller forces the turbine power to its maximal value for that wind speed range. When the wind speed reaches the value of 12.5 m/s, the maximal power corresponds to the nominal turbine power of 4.8 MW, so in the third region ($12.5$ m/s $< v_w < 25$ m/s) the controller keeps the generated power constant. In the fourth region ($v_w > 25$ m/s) the wind speed is too high and the turbine is stopped to prevent damage. More details on the common industrial wind turbine controllers can be found in Johnson et al. (2006b).

In the partial load region, the optimum power value is obtained by keeping the blade pitch angles constantly equal to zero, $\beta = 0$ degrees, and by controlling the generator torque reference $\tau_{gr}$ to its optimal value in order to get the optimal tip-speed ratio $\lambda_{opt}$ in Eq. (2.42), where

$$\lambda = \frac{\omega_r R}{v_w}. \tag{2.42}$$

This value is found by means of the $C_P$ map, as the tip-speed ratio can act on the maximal power coefficient $C_{P_{max}}$, after fixing the pitch angle to zero. The discrete reference torque signal becomes

$$\tau_{gr}[k] = \frac{1}{2} \rho A R^3 \frac{C_{P_{max}}}{\lambda_{opt}^3} \left( \frac{\omega_g[k]}{N_g} \right)^2, \tag{2.43}$$

where $A$ is the area swept by the blades. In the constant power region the controller is switched to a PI regulator that acts on the pitch angles of the blades forcing the generator speed to its nominal value $\omega_{nom}$ that implies the generation of the nominal power. In this working condition the discrete controller law has the following form:

$$\beta_r[k] - \beta_r[k-1] + K_p e[k] + (K_i T_s - K_p) e[k-1], \tag{2.44}$$

where $K_p$, $K_i$ are the PI coefficients, and $e[k]$ is the error on the generator speed.

Finally, the switching logic of the controller is regulated by means of a bumpless mechanism, as depicted in Fig. 2.9. In particular, as it will be outlined in Chapter 6, the condition for the switch from Region 2 to Region 3 is described by the following rule:

$$\left( P_g[k] > P_r[k] \right) \& \left( \omega_g[k] > \omega_{nom} \right), \tag{2.45}$$

**TABLE 2.1** Signals acquired from the wind turbine benchmark.

| Variable | $\beta_{1,m1}$ | $\beta_{1,m2}$ | $\beta_{2,m1}$ | $\beta_{2,m2}$ | $\beta_{3,m1}$ | $\beta_{3,m2}$ |
|---|---|---|---|---|---|---|
| Pitch # | 1 | 1 | 2 | 2 | 3 | 3 |
| Sensor # | 1 | 2 | 1 | 2 | 1 | 2 |
| Variable | $\omega_{r,m1}$ | $\omega_{r,m2}$ | $\omega_{g,m1}$ | $\omega_{g,m2}$ | | |
| Speed | Rotor | Rotor | Generator | Generator | | |
| Encoder # | 1 | 2 | 1 | 2 | | |
| Variable | $\tau_{g,m}$ | $P_{g,m}$ | $v_{w,m}$ | $\tau_{r,m}$ | | |
| Measure | Generator torque | Generated power | Wind speed | Rotor torque | | |
| Model | Generator | Generator | Anemometer | Estimated | | |

whilst from Region 3 to Region 2 the condition is described by the rule

$$\omega_g[k] < \omega_{nom} - \omega_\Delta, \tag{2.46}$$

with $\omega_\Delta$ being a hysteresis offset (Odgaard et al., 2014).

### 2.5.3 The Measurement Model

The measurements available to the controller come directly from several sensors or, in one case, they are obtained via estimation. In particular, for each of the three blades, a redundant couple of sensors measures the current pitch angle. Then a couple of sensors measure the speed of the rotor and another one the speed of the generator, while a single sensor is available for the wind speed at hub height and another one for the generator torque. The wind torque measurements are estimated exploiting the hub anemometer. Table 2.1 reports a summary of the measured variables. The model of the measurements consists of the sum of the actual value with a white Gaussian noise.

### 2.5.4 Wind Turbine Fault Scenario

In the benchmark model three kinds of actual faults can be simulated, namely sensor, actuator, and system faults. They are modeled as additive or multiplicative faults and involve different degrees of severity, so that they can yield turbine shutdown in case of serious fault, or they can be accommodated by the controller, if the risk for the system safety is low.

Regarding the considered sensor faults, they affect the measurements of the pitch angles and the measurements of the rotor speed in the form of a fixed value or a scaling error. They represent a common fault scenario of wind turbines, but their severity is low and they should be easy to identify and accommodate. In particular, an electrical or mechanical fault in the pitch sensors, if not handled, results in the generation of a wrong pitch reference system by the controller with the consequence of a loss in the generated power. The speed of the rotor is measured by means of two redundant encoders, an offset faulty signal can affect these measurements when the encoder does not detect the updated marker, while a gain factor faulty signal represents the reading of excessive markers in each loop, due to dirt on the rotating part.

The considered actuator faults are modeled either as a fixed value or a changed dynamics of the transfer function. They affect the converter torque actuator, as well as the pitch actuator. In the former case, the fault is located in the electronics of the converter, while in the latter case the fault is on the hydraulic system: it models the pressure drop in the hydraulic supply system (e.g., due to leakage in a hose or a blocked pump) or the excessive air content in the oil that causes the variation of the compressibility factor. The severity of these fault is of medium/high level.

Finally, the considered system fault concerns the drive-train in the form of a slow variation of the friction coefficient in time due to wear and tear (months or year, but for benchmarking reasons in the model it has been accelerated up to some seconds). It results in a combined faulty signal affecting the rotor speed and the generator speed. It can be listed as a high severe fault, as it can produce the breakdown of the drive-train, but in a long time. More details on such a fault and on its diagnosis at gear-box level are reported in Hameed et al. (2009).

Table 2.2 summarizes the considered faults, with a brief description of their typology and topology.

**TABLE 2.2** Turbine benchmark model fault scenarios.

| Fault # | 1 | 2 | 3 |
|---|---|---|---|
| Typology | Fixed value | Scaling error | Fixed value |
| Sensor # | Blade 1 | Blade 2 | Blade 1 |
| Fault # | 4 | 5 | 6 |
| Typology | Fixed value | Scaling error | Dynamics |
| Sensor # | Pitch 1 | Generator 2 | Actuator 2 |
| Fault # | 7 | 8 | 9 |
| Typology | Dynamics | Fixed value | Dynamics |
| Sensor # | Pitch 3 | Converter | Drive-train |

**TABLE 2.3** Wind turbine benchmark model parameters.

| $R$ | $\rho$ | $\zeta$ | $\omega_n$ | $B_{dt}$ | $B_r$ |
|---|---|---|---|---|---|
| 57.5 m | 1.225 kg/m$^{-3}$ | 0.6 | 11.11 rad s$^{-1}$ | 775.49 N m s rad$^{-1}$ | 7.11 N m s rad$^{-1}$ |
| $B_g$ | $N_g$ | $K_{dt}$ | $\eta_{dt}$ | $J_g$ | $\eta_g$ |
| 45.6 N m s rad$^{-1}$ | 95 | $2.7 \times 10^9$ N m rad$^{-1}$ | 0.97 | 390 kg/m$^2$ | 0.98 |
| $\alpha_g$ | $K_i$ | $K_p$ | $K_d$ | $\omega_{nom}$ | $P_{ref}$ |
| 50 rad s$^{-1}$ | 1 | 4 | 0 | 162 rad s$^{-1}$ | 4.8 MW |

### 2.5.5  Model Parameters

The parameters adopted in the benchmark model described in Section 2.5.1 are summarized in Table 2.3. It is worth noting that these parameter values will be modified in order to assess the robustness and reliability features of the considered FDI and FTC schemes addressed in Chapters 3 and 4.

### 2.5.6  Wind Turbine Benchmark Overall Model

With these assumptions the complete model of the wind turbine system considered in this monograph can be represented by means of a nonlinear continuous-time function $\mathbf{f}_{wt}$ that describes the evolution of the turbine state vector $\mathbf{x}_{wt}$ excited by the input vector $\mathbf{u}$:

$$\begin{cases} \dot{\mathbf{x}}_{wt}(t) = \mathbf{f}_{wt}\left(\mathbf{x}_{wt}, \mathbf{u}(t)\right), \\ \mathbf{y}(t) = \mathbf{x}_{wt}(t), \end{cases} \tag{2.47}$$

where the state of the system $\dot{\mathbf{x}}_{wt}$ is considered equal to the monitored system output $\mathbf{y}$, i.e., the rotor speed, generator speed, and generated power:

$$\mathbf{x}_{wt}(t) = \mathbf{y}(t) = \left[\omega_{g,m1}, \omega_{g,m2}, \omega_{r,m1}, \omega_{r,m2}, P_{g,m}\right]. \tag{2.48}$$

The input vector is described as follows:

$$\mathbf{u}(t) = \left[\beta_{1,m1}, \beta_{1,m2}, \beta_{2,m1}, \beta_{2,m2}, \beta_{3,m1}, \beta_{3,m2}, \tau_{g,m}\right] \tag{2.49}$$

and contains the measurements of the pitch angles from the three sensor couples, as well as the measured torque. These vectors are sampled for obtaining a number of $N$ input–output data $\mathbf{u}[k]$ and $\mathbf{y}[k]$ with $k = 1, \ldots, N$, and sample time $T = 0.01$ s in order to show the achieved results presented in Chapter 5.

## 2.6  WIND FARM BENCHMARK

The wind farm benchmark model considered in this monograph has been proposed in Odgaard and Stoustrup (2013a) by the same authors who developed the wind turbine benchmark model. It consists of nine wind turbines arranged in a square grid of three rows and three columns. The distance between two adjacent turbines is 7 times the rotor diameter $R$. Two measuring masts (anemometers) are placed in front of the first line of turbines, at a distance of 10 times $R$, providing the

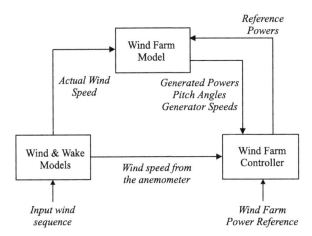

**FIGURE 2.13** Block diagram of the wind farm benchmark model.

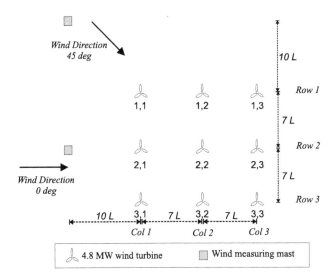

**FIGURE 2.14** Layout of the wind farm benchmark with nine wind turbines.

measurements of the undisturbed wind speed. The considered turbines are 4.8 MW three-blade HAWTs, represented by a simpler model with respect to the previously described turbine. Each of them is provided with a controller, but also a wind farm controller is included in the benchmark model. Three common fault scenarios can be simulated. The complete wind farm model consists of three main submodels: the wind and wake model, the plant model, and the controller model, interacting as sketched in Fig. 2.13. The layout of the wind farm with nine turbines on the square grid and the masts along the wind directions are sketched in Fig. 2.14.

The distance between the wind turbines in both directions is 7 times the rotor diameter $L$. Two measuring masts are located in front of the wind turbines, one in each of the wind directions considered in this benchmark model, e.g., 0 and 45 degrees. The wind speed is measured by these measuring masts which are located at a distance of 10 times $L$ in front of the wind farm. The wind turbines of the farm are defined by their row and column indices in the coordinate system, as illustrated in Fig. 2.14. The farm uses generic 4.8 MW wind turbines, which are three-blade horizontal-axis pitch-controlled variable-speed wind turbines. Each of the wind turbines is described by simplified models including control logics, variable parameters and three states (Odgaard and Stoustrup, 2013a).

## 2.6.1 Wind and Wake Model

The wind and wake model provides the wind speed for each of the nine turbines, contained in the vector $\mathbf{v}_w$, as well as for a measuring mast $v_{w,m}$. They are determined starting from a certain wind sequence (two different wind sequences are included in the simulator) and their elaboration takes into account the delay and the interaction among the turbines depending on wind direction. In particular, the wake effect is described as reported in Jensen (1983) by means of a static deficit coefficient of 0.9. Finally, the turbulence is modeled by an additive Gaussian white noise.

### 2.6.2 Wind Farm Benchmark Overall Model

The plant model represents the nine wind turbines with the same submodel for each of them. It receives as input the $\mathbf{v}_w$ vector and the $\mathbf{P}_r$ vector containing the nine reference signals from the controller. The outputs are the vectors $\mathbf{P}_g$, $\boldsymbol{\beta}$, $\boldsymbol{\omega}_g$ that contain the generated powers, pitch angles, and generator speeds, respectively, for each of the nine turbines. Inside the turbine submodel, the current wind speed is elaborated by means of a lookup table in order to compute the available power $P_w(t)$. Then the generated power is computed according to

$$P_g(t) = P_c(t) + \gamma_p \sin(2\pi\sigma_p t), \tag{2.50}$$

where the first term $P_c(t)$ is equal to the current lower value between the filtered available power $\hat{P}_w$ and the filtered reference power $\hat{P}_r$:

$$P_c(t) = \begin{cases} \hat{P}_w(t), & \text{if } \hat{P}_w(t) < \hat{P}_r(t), \\ \hat{P}_r(t), & \text{if } \hat{P}_w(t) > \hat{P}_r(t). \end{cases} \tag{2.51}$$

The second term of Eq. (2.50) represents the oscillations caused by the drive-train, whose amplitude is $\gamma_p$ and frequency $\sigma_p$.

The filtered signals $\hat{P}_w(t)$ and $\hat{P}_r(t)$ differ from the input variables by means of a first order transfer function in the following form:

$$\hat{P}_w(s) = \frac{\tau_w(v_w)}{s + \alpha_w(v_w)} P_w(s), \tag{2.52}$$

$$\hat{P}_r(s) = \frac{\tau_p}{s + \tau_p} P_r(s), \tag{2.53}$$

where the parameter $\tau_p$ is a fixed value, while the parameters $\tau_w$ and $\alpha_w$ depend on the wind speed and are computed by means of a lookup table.

Regarding the pitch angle output, the model is similar to that of the turbine benchmark model of Eq. (2.36), but the transfer function between the reference pitch signal and the actual pitch angle has been reduced to a first-order transfer function in the form

$$\beta(s) = \frac{\tau_\beta}{s + \tau_\beta} P_r(s). \tag{2.54}$$

Then the generator speed of each turbine is modeled as

$$\omega_g(t) = f_\omega(P_c(t)) \left(1 + \frac{\gamma_\omega}{\omega_{g,max}} \sin(2\pi\sigma_p t)\right), \tag{2.55}$$

where $f_\omega$ is computed by means of a lookup table and the oscillation term, due to the drive-train, has an amplitude equal to the ratio between the parameter $\gamma_\omega$ and the maximal generator speed $\omega_{g,max}$.

Finally, the wind farm controller forces each turbine to follow a reference power signal $P_r[k]$ that is one-ninth of the wind farm power reference. Moreover, in order to avoid fast variation of the control signal, the wind farm power reference is low-pass filtered to obtain $\hat{P}_{wf,r}$. The controller is modeled as discrete-time system and uses a sample frequency of 10 Hz (Odgaard and Stoustrup, 2013a):

$$P_r[k] = \frac{1}{9} \hat{P}_{wf,r}[k]. \tag{2.56}$$

### 2.6.3 Wind Farm Fault Scenario

Three common fault scenarios are considered (Odgaard and Stoustrup, 2013a). They affect the output variables of the plant system. These faults are difficult to detect considering the single wind turbine, but they can be identified at wind farm level.

**TABLE 2.4** Wind farm benchmark fault scenarios.

| Fault # | 1 | 2 | 3 |
|---|---|---|---|
| Type | Blade surface debris | Blade misalignment | Drive-train wear & tear |
| Topology | Aerodynamic system | Pitch system | Generator system |

**TABLE 2.5** Wind farm benchmark model parameters.

| $R$ | $L$ | $\tau_p$ | $\gamma_p$ |
|---|---|---|---|
| 57.5 m | 115 m | 1.2 rad s$^{-1}$ | 1000 W |
| $\sigma_p$ | $\tau_\beta$ | $\omega_{g,max}$ | $\gamma_\omega$ |
| 10 Hz | 1.6 rad s$^{-1}$ | 158 rad s$^{-1}$ | 0.4 |

The first fault considered represents debris build-up on the blade surface. Its effect is a change of the aerodynamic of the affected turbine and the consequent decreasing of the generated power that is modeled by a scaling factor of 0.97 applied to the generated power signal. An analysis of this kind of fault is reported in Johnson et al. (2006b).

The second fault is a misalignment of one blade caused by an imperfect installation. The effect is an offset between the actual and the measured pitch angle of the affected turbine. This fault can excite structural modes and creates undesired vibrations that can damage the system severely. The faulty signal involves an offset of 0.3 degrees on the pitch angle.

The third fault represents the wear and tear in the drive-train. It has been demonstrated (Odgaard and Stoustrup, 2012c) that such a fault is difficult to detect at wind turbine level, and the current trend is to analyze the frequency spectra of different vibration measurements. In this benchmark model the fault affects the generated power, increasing the amplitude of its oscillation by 26% of the nominal value, and the generator speed, increasing the amplitude of its oscillation by 130%.

Table 2.4 summarizes the considered faults, with a brief description of their typology and topology.

### 2.6.4 Model Parameters

Finally, also in this case Table 2.5 reports the parameters used in the wind farm benchmark model.

### 2.7 FAULT ANALYSIS

The purpose of this section is to illustrate the procedure for identifying possible faults on the wind turbine and determine their effect on the system behavior. A number of the analyzed faults are then selected for further discussion in Chapter 5, and redundancies in the system are identified to determine the detectable faults, as addressed in Chapter 3. Finally, the remedial actions that must be conducted to stop the propagation of the faults are established, as described in Chapter 4. The method utilized to structure the fault analysis is inspired by Stamatis (2003a), and the steps in the analysis and design procedures are outlined in Fig. 2.15.

Note that the diagnosis design and the supervisor design can be performed only when an active fault-tolerant control system is considered, as highlighted in Chapter 4.

Fig. 2.15 shows the steps in the procedure and their interconnections, but to give an insight into the purpose of every step in the analysis, the list of the following items is analyzed in detail:

**Model Partitioning.** The wind turbine model is divided into submodels suitable for analysis and identification of the possible component faults in each subsystem;

**Fault Propagation Analysis.** The fault propagation analysis propagates the component faults through the system and determines their end-effects at system level;

**Fault Assessment.** The fault assessment assesses the faults identified in the fault propagation analysis, by determining their occurrence and their impact on the performance of the wind turbine control system. The fault assessment furthermore determines the end-effects which must be handled in this project. Finally, the component faults are traced back from the end-effects to the associated component faults;

**Structural Analysis.** The structural analysis determines analytical redundancy relations in the system and determines the detectable faults;

**Fault Specification.** The fault specification specifies the dynamics of the faults identified in the fault assessment;

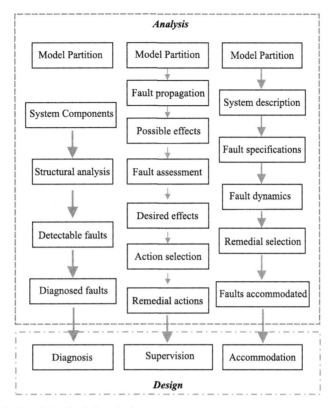

**FIGURE 2.15** Overview of the methods exploited for fault analysis purposes.

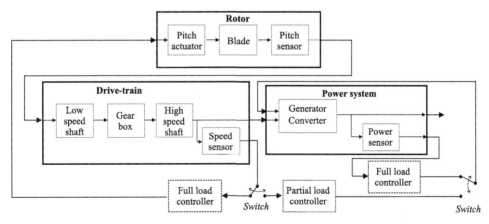

**FIGURE 2.16** Overview of the three submodels of the wind turbine benchmark used for fault analysis purposes.

**Remedial Action Selection.** The remedial action selection determines the action that must be taken to stop the propagation of the faults. This could be performed by switching between redundant sensors or reconfiguring the controllers. Additionally, the requirements to the fault diagnosis and fault-tolerant control system are determined.

The most important task used for performing the fault analysis relies on the model partitioning. It consists of dividing the system model described in Section 2.3 into appropriate submodels, which can be analyzed separately. This division makes it possible to identify all possible component faults in each submodel.

As already highlighted in Section 2.3, the wind turbine model can be divided into submodels based on separate functionalities: rotor, drive-train, and power system. Fig. 2.16 shows the components included in each of the submodels and their interconnections. Note that internal controllers are located in each pitch actuator and converter unit, which can have a significant impact on the effects of the component faults. One direct consequence is that an offset in the output value of an actuator is compensated for by the internal controller, so that no fault is seen at the output.

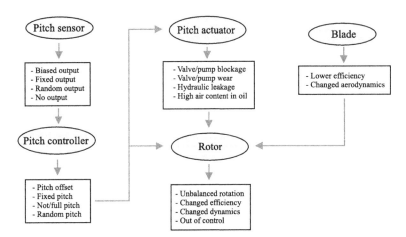

**FIGURE 2.17** FMEA scheme illustrating the propagation of the fault effects in the rotor system.

The analysis is performed in closed-loop, where the wind turbine is controlled according to two different strategies as explained in Section 2.3. This is elaborated in the reference controller design described in Section 2.4.2. To distinguish between the different control strategies, the configurations of the controller in the two regions are marked with different colors.

In order to provide an overview of the closed-loop system, the configurations of the controller in partial load and full load operations are included in Fig. 2.16.

### 2.7.1 Failure Mode and Effect Analysis

The purpose of this section is to describe how component faults propagate through the system, by describing the effects on surrounding components of the system. By considering the scheme sketched in Fig. 2.16, this is accomplished by making a Failure Mode and Effect Analysis (FMEA) for the three main components of the wind turbine system; rotor, drive-train, and power system, identified in the previous section. FMEA is a commonly accepted technique for making a fault propagation analysis, as described, e.g., in Stamatis (2003a), Blanke et al. (2006).

In FMEA it is assumed that no fault handling exists in the wind turbine control system, even though this is not true. However, since each wind turbine manufacturer has its own fault handling system, it is difficult to assume something general. Furthermore, it is assumed that the nominal control system is fault-free. In the following, the main component of the wind turbine system are considered, and the fault propagation analysis is performed according to the scheme of Fig. 2.16.

**Rotor.** The FMEA scheme for the rotor is sketched in Fig. 2.17. Even though the wind turbine benchmark uses three pitch actuators, three pitch sensors, and three blades, only one block is represented in the diagram, since they are identical and cause the same effects. Therefore, according to Fig. 2.17, a pitch actuator controls the blade that is connected to a pitch sensor. This configuration has some consequences for the effects of the pitch sensor faults, since, for example, a fixed output from this sensor turns the blade to one of its extreme positions depending on the sensor output compared to the pitch angle reference signal. The detection of the internal fault resulting in no output from a pitch sensor is assumed to already exist, since no measurement is sent to the controller in this situation. The control system is fed by the last valid measurement from the sensor when this fault occurs. Note that Fig. 2.17 summarizes all possible faults and failures affecting the different subsystems comprising the pitch sensor, actuator, controller, and blade systems. In this way the FMEA diagram of Fig. 2.17 for the rotor system describes the possible faults affecting the rotor submodels and their propagation effects on the rotor itself.

**Drive-Train.** The drive-train consists of a low-speed shaft, high-speed shaft, and gear-box. The speed controller controls the speed of the high-speed shaft in the full load operation of the wind turbine. The FMEA scheme for the drive-train system is sketched in Fig. 2.18. The high- and low-speed shafts can be both affected by bearing damages. A damaged bearing produces an uneven rotation of the drive-train, since it is usually assumed that it is modeled as a friction effect, which depends on the angle of the shaft itself. According to Section 2.3, the speed sensor is an encoder that can be affected by a proportional error. Therefore Fig. 2.18 highlights the FMEA diagram for the drive-train system, which describes the faults affecting the drive-train submodels and their propagation effects on the drive-train itself.

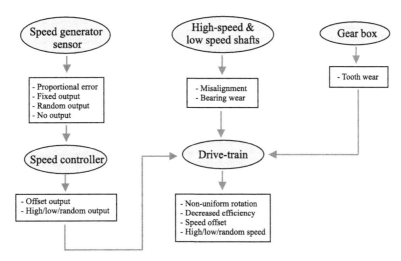

**FIGURE 2.18** FMEA diagram showing the propagation of the fault effects in the drive-train system.

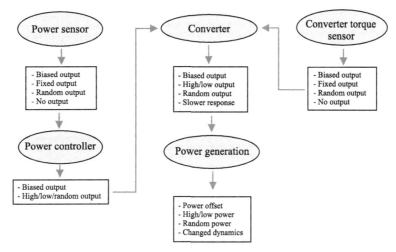

**FIGURE 2.19** FMEA diagram showing the propagation of the fault effects in the power generation system.

**Power System.** The power generation system comprises two types of sensors, namely converter torque and power sensors. Beside these sensors, the power system consists also of a generator and converter. Note, however, that no faults in the generator are considered here. The FMEA scheme for the power system is shown in Fig. 2.19. Beside the power controller, which guarantees that the wind turbine produces the desired power in the full load region, the converter can also have an internal controller. Even though multiple converters and converter torque sensors can be considered, only one device is represented for each of these systems, since they are identical and cause the same effects. Therefore Fig. 2.19 illustrates the FMEA scheme for the power system and describes the faults affecting the power submodels and their propagation effects on the power system itself.

Note finally that this section has described how component faults propagate to end-effects in the submodels of the wind turbine operating in closed loop. On the other hand, Section 2.7.2 analyzes the wind turbine faults by considering the severity of their end-effects and possible occurrences of the component faults.

## 2.7.2 Fault Specifications and Requirements

The purpose of this section is to evaluate the severity of the end-effects and the occurrence of the faults as described in Section 2.5, and to determine which of these should be considered for the task of fault diagnosis and fault accommodation. Furthermore, causal relations are established to trace back the faulty components from the end-effects.

In particular, in this section, definitions of occurrence and severity are provided first. The occurrence rates can be determined from statistics about the distribution of failures described in the literature. In order to find the severities of the

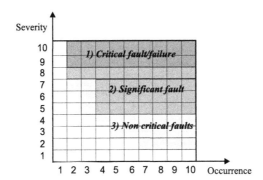

**FIGURE 2.20** Graph used to represent fault priority.

**TABLE 2.6** Failure probability with respect to ranking and rate.

| Ranking | Frequency | Fault Rate |
|---------|-----------|------------|
| 10 | Very high | $\geq 20\%$ |
| 9 | Inevitable fault | 15–20% |
| 8 | High | 10–15% |
| 7 | Frequent fault | 5–10% |
| 6 | Moderate | 4–5% |
| 5 | Occasional fault | 3–4% |
| 4 | Low | 2–3% |
| 3 | Rare fault | 1–2% |
| 2 | Remote | 0.1–1% |
| 1 | Unlikely fault | < 0.1% |

end-effects, simulations with fault injections are conducted to determine the severity indices. These simulations are carried out on the wind turbine model controlled by the reference controller described in Chapter 2.4. Moreover, the simulations and the evidence of these fault occurrence are illustrated in Chapter 5 in order to maintain the structure of the monograph. The section is finalized by a discussion on the severity and occurrence analysis and a selection of the fault to be handled.

Faults are classified based on their likelihood of occurrence and the severity of their end-effect on scales from 1 to 10. These can be combined in the Severity Occurrence Index (SOI), which is obtained through multiplication of the severity and occurrence values (Stamatis, 2003a). The idea is that faults with the highest SOI should have the highest attention. However, in order to highlight the most severe failures of a system, a classification of the faults based on their gravity is illustrated in Fig. 2.20, and it is often used instead, which is based on a different approach proposed in Stamatis (2003a).

Therefore, according to the representation of Fig. 2.20, faults with priority 1 (dark grey) and 2 (medium grey) require actions, but for faults with priority 3 (light grey) actions are not required but only preferred.

The fault classification is performed according to the indices, described in the following:

**Occurrence.** It is the frequency of the fault and is quantified on a scale from 1 (unlikely) to 10 (inevitable). The occurrence ranking is defined in Table 2.6. Note that the occurrence scale has a relative meaning. In fact, statistics about the distribution of failures experienced by wind turbines should be considered. However, the failure rates found in the related literature often regard the subsystems rather than individual components of the system, which have to be estimated.

**Severity.** It refers to the potential harm due to the faults affecting the system. The severity ranking is represented in Table 2.7. It is worth noting that the severity ranking found in the literature has originated from the American automotive companies, when FMEA began to be incorporated into their product development process from the early 1980s. Therefore, industry engineers have adopted and adapted this tool over the years, and the description of Table 2.7 should be considered with respect to wind turbines.

In order to finalize the severity and the occurrence analysis, all the end-effects are compared, and the faults are identified, according to the strategies described in Chapter 3. This analysis and comparison are based on the FMEA analysis performed

**TABLE 2.7** Fault severity with respect to wind turbine operations.

| Ranking | Fault Effect | Wind Turbine Conditions |
|---------|--------------|-------------------------|
| 10 | Extremely high severity | Inoperative |
| 9 | Very high severity | Very degraded operations |
| 8 | Primary functions loss | Very limited operations |
| 7 | High severity | Very reduced performance |
| 6 | Moderate severity | Reduced performance |
| 5 | Low severity | Limited performance |
| 4 | Very low severity | Graceful degradation |
| 3 | Minor severity | Minor degradation |
| 2 | Very minor severity | Limited degradation |
| 1 | None | Nominal condition |

in Section 2.7.1. This task is accomplished by tracing back the faults, which cause the most severe end-effects. Among these faults, the most frequent ones should be handled and managed by the accommodation strategy.

From the severity indices computed in Chapter 5 it follows that several faults have severe effects on the wind turbine, and should therefore be handled. However, in order to reduce the set of possible faults to be handled and considered, this monograph is focused on the faults affecting the pitch sensors, pitch actuators, and generator speed sensors. The rationale of this choice is that faults occurring on the pitch system, e.g., affecting the rotor balance, are generally difficult to detect based on the typical outputs, such as generator speed and output power. However, these faults result in increased fatigue loads on the wind turbine structure. Moreover, Chapter 5 will show that changed dynamics of the pitch system, caused by low pressure or high air content, may result in an unstable closed-loop system. On the other hand, with reference to the measurement of the generator speed, the wind turbine control systems rely on the control regions as shown in Section 2.4. Therefore, it is essential to diagnose and accommodate any possible malfunctions regarding this particular measurement.

The faults to be diagnosed and accommodated that are considered in this monograph are described in Sections 2.5 and 2.6. However, since a random output from a sensor is unlikely to occur, this fault is not considered. Moreover, with reference to fixed or no outputs from a pitch sensor, also these faults are neglected, as it is assumed that the wind turbine cannot be controlled satisfactory if this measurement is not available.

## 2.8 SUMMARY

This chapter revised the most important modeling and control issues for wind turbines from a systems and control engineering point of view. A walk around the wind turbine control loops discussed the goals of the most common solutions and overviews the typical actuation and sensing available on commercial turbines. The chapter intended to provide also an updated and broader perspective by covering not only the modeling and control of individual wind turbines, but also outlining a number of areas for further research, and anticipating new issues that can open up new paradigms for advanced control approaches. In summary, wind energy is a fast growing industry, and this growth has led to a large demand for better modeling and control of wind turbines. Uncertainty, disturbance, and other deviations from normal working conditions of the wind turbines make the control challenging, thus motivating the need for advanced modeling and a number of so-called sustainable control approaches that should be explored to reduce the cost of wind energy. By enabling this clean renewable energy source to provide and reliably meet the world's electricity needs, the tremendous challenge of solving the world's energy requirements in the future will be enhanced. The wind resource available worldwide is large, and much of the world's future electrical energy needs can be provided by wind energy alone if the technological barriers are overcome. The application of sustainable controls for wind energy systems is still in its infancy, and there are many fundamental and applied issues that can be addressed by the systems and control community to significantly improve the efficiency, operation, and lifetimes of wind turbines. Finally, this chapter analyzed the occurrence of the faults with reference to the severity of their end-effects, in order to perform a motivated selection of the faults that will be diagnosed and accommodated.

Chapter 3

# Fault Diagnosis for Wind Turbine Systems

## 3.1 INTRODUCTION

The model-based approach to fault diagnosis in technical processes has been receiving more and more attention over the last four decades, in the contexts of both research and real plant application.

Stemming from this activity, a great variety of methods are found in current literature, based on the use of mathematical models of the technical process under diagnosis and exploiting advanced control theory.

Model-based fault diagnosis methods usually use residuals which indicate changes between the process and the model. One general assumption is that the residuals are changed significantly so that a detection is possible. This means that the residual size after the appearance of a fault is large and long enough to be detectable.

This chapter provides an overview on different fault diagnosis strategies, with particular attention to the FDI techniques related to the energy conversion system applications described in this book.

All the methods considered require that the technical process can be described by a mathematical model. As there is almost never an exact agreement between the model used to represent the process and the plant, the model–reality discrepancy is of primary interest.

Hence the most important issue in model-based fault detection is concerned with the accuracy of the model describing the behavior of the monitored system. This issue has become a central research theme over recent years, as modeling uncertainty arises from the impossibility of obtaining complete knowledge and understanding of the monitored process.

The main focus of this chapter is the mathematical description aspects of the process whose faults are to be detected and isolated. The chapter also studies the general structure of the controlled system, its possible fault locations and modes. Residual generation is then identified as an essential problem in model-based FDI, since, if it is not performed correctly, some fault information could be lost. The general framework for the residual generation is also recalled.

Residual generators based on different methods, such as input–output, state and output observers, parity relations and parameter estimations, are just special cases in this general framework. In the following, some commonly used residual generation and evaluation techniques are discussed and their mathematical formulation presented.

Finally, the chapter presents and summarizes special features and problems regarding the different methods.

### 3.1.1 Plant and Fault Models

According to the definitions given in Chapter 1, model-based FDI can be defined as the *detection*, *isolation*, and *identification* of faults on a system by means of methods which extract features from measured signals and use *a priori* information on the process available in term of a mathematical models. Faults are thus detected by setting fixed or variable thresholds on residual signals generated from the difference between actual measurements and their estimates obtained by using the process model.

A number of residuals can be designed with each having sensitivity to individual faults occurring in different locations of the system. The analysis of each residual, once the threshold is exceeded, then leads to fault isolation.

Fig. 3.1 shows the general and logic block diagram of model-based FDI system. It comprises two main stages of residual generation and residual evaluation. This structure was first suggested by Chow and Willsky (1980) and now is widely accepted by the fault diagnosis community.

The two main blocks shown in Fig. 3.1 perform the following tasks. The residual generation module generates residual signals using the available inputs and outputs from the process under diagnosis. This residual (or fault symptom) should

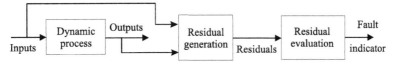

**FIGURE 3.1** Structure of model-based fault diagnosis module.

*Fault Diagnosis and Sustainable Control of Wind Turbines.* DOI: 10.1016/B978-0-12-812984-5.00003-1

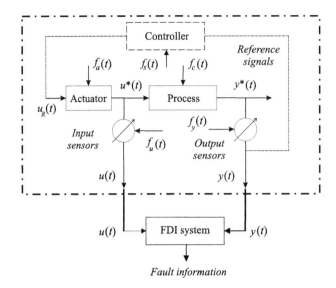

**FIGURE 3.2** The monitored process and the fault diagnosis scheme.

indicate that a fault has occurred. It should normally be zero or close to zero under no fault condition, whilst distinguishably different from zero when a fault occurs. This means that the residual is characteristically independent of process inputs and outputs, in ideal conditions. Referring to Fig. 3.1, this block is called *residual generation*.

Once the residual signals have been generated, the residual evaluation module examines residuals for the likelihood of faults and a decision rule is then applied to determine if any faults have occurred. The *residual evaluation* block, shown in Fig. 3.1, may perform a simple threshold test (geometrical methods) on the instantaneous values or moving averages of the residuals. On the other hand, it may consist of statistical methods, e.g., generalized likelihood ratio testing or sequential probability ratio testing (Isermann, 1997a; Willsky, 1976; Basseville, 1988; Patton et al., 2000).

Many works in the field of quantitative model-based FDI have focused on the residual generation problem, since the decision-making problem can be considered relatively straightforward if residuals are well-designed. In the following, a number of different strategies oriented to solve the quantitative residual generation problem have been recalled, with reference to Multiple-Input Single-Output (MISO) and Multiple-Input Multiple-Output (MIMO) dynamic processes.

The first step in FDI model-based approach consists of providing a mathematical description of the system under investigation that describes the possible fault cases, as well.

The detailed scheme for FDI techniques here presented is depicted by Fig. 3.2. The main components are the *Plant* under diagnosis, the *Actuators*, and *Sensors*, which can be further subdivided as *input* and *output* sensors, and finally, the *Controller*. In the following, the process behavior will be monitored by analyzing its input $u(t)$ and output $y(t)$ measurements and the signals from the controller $u_R(t)$, which are supposed completely available for FDI purposes. Moreover, as shown in Fig. 3.2, the behavior of any controller that drives the system is inherently taken into consideration.

It is worth noting that, when the signals $u_R(t)$ from the controller or measurements of plant inputs $u(t)$ are not available, the controller plays an important role in the design of the FDI scheme, as a robust controller may desensitize faults effects and make diagnosis difficult.

Once the actual process inputs and outputs $u^*(t)$ and $y^*(t)$ (usually not available) are measured by the input and output sensors, FDI theory can be treated as an observation problem of $u(t)$ and $y(t)$. Concerning the occurrence of malfunctions, *location of faults,* and their modeling, the system under diagnosis can be separated into different parts which can be affected by faults as illustrated in Fig. 3.2. With respect to former works (see, e.g., in Patton et al., 1989a; Gertler, 1998; Patton et al., 2000), it is necessary to distinguish between input and output sensors.

Fig. 3.2 also shows that the input and output signals $u^*(t)$ and $y^*(t)$ are acquired in order to obtain the measurements $u(t)$ and $y(t)$ from the sensors. Fig. 3.2 also shows the situation where the controller can be affected by faults, since the monitored process consists of a closed-loop plant. However, because of technological reasons (e.g., the control action is performed by a digital computer), when the actuator is considered as a part or a component of the whole controller device, the former can be treated as subsystem where faults are likelier to occur whilst the latter remains free from faults. Under these assumptions, as shown in Fig. 3.2, when the monitored process is considered in view of fault location, since input and output measurements are supposed completely available for FDI purposes, the controller behavior in the design of a fault diagnosis scheme can be neglected, as can the interconnection between control system and the process.

Under the hypothesis of the superposition principle, the monitored process dynamics are described by the following Linear Time-Invariant (LTI) dynamic system in the state-space form

$$\begin{cases} \dot{x}(t) = A\,x(t) + B\,u^*(t), \\ y^*(t) = C\,x(t), \end{cases} \tag{3.1}$$

where $x(t) \in \Re^n$ is the system state vector, $u^*(t) \in \Re^r$ is the input signal vector driven by actuators, and $y^*(t) \in \Re^m$ is the real system output vector, not directly available. $A$, $B$, and $C$ are system matrices with appropriate dimensions obtained by modeling or identification procedure.

With reference to Fig. 3.2, a component fault vector $f_c(t)$ affects process dynamics as follows:

$$\dot{x}(t) = A\,x(t) + B\,u^*(t) + f_c(t). \tag{3.2}$$

In some cases, component faults come from a change in the system parameters, e.g., a change in entries of the $A$ matrix. For example, a change in the $i$th row and the $j$th column of the $A$ matrix leads to a fault vector $f_c(t)$ described as

$$f_c(t) = I_i\,\Delta a_{ij}\,x_j(t), \tag{3.3}$$

where $x_j(t)$ in the $j$th element of the vector $x(t)$ and $I_i$ is an $n$-dimensional vector with all zeroes except a "1" as the $i$th element.

As stated previously, since the actual process output $y^*(t)$ is not directly available, a sensor is used to acquire a measure of the system outputs. Moreover, generally speaking, a sensor can be also used to measure the system inputs $u^*(t)$ (e.g., for uncontrolled system). Therefore, by neglecting sensor dynamics, faults on input and output sensors are modeled with additive signals, respectively, as

$$\begin{cases} u(t) = u^*(t) + f_u(t), \\ y(t) = y^*(t) + f_y(t), \end{cases} \tag{3.4}$$

where the vectors $f_u(t) = [f_{u_1}(t), \ldots, f_{u_r}(t)]^T$ and $f_y(t) = [f_{y_1}(t), \ldots, f_{y_m}(t)]^T$ are chosen to describe a fault situation.

Usually, as shown in Chapter 5, fault modes can be described by step and ramp signals in order to model abrupt and incipient (hard to detect) faults, representing bias and drift, respectively.

Moreover, as remarked in Chapter 2, for technical reasons, sensor output signals are generally affected by measurement noise. Fault-free sensor signals $u(t)$ and $y(t)$, with additive noise can be modeled as

$$\begin{cases} u(t) = u^*(t) + \tilde{u}(t), \\ y(t) = y^*(t) + \tilde{y}(t), \end{cases} \tag{3.5}$$

in which the sequences $\tilde{u}(t)$ and $\tilde{y}(t)$ are usually described as white, zero-mean, uncorrelated Gaussian processes. In this case, taking into account the effects of faults and noise, Eq. (3.4) can be rewritten in the form

$$\begin{cases} u(t) = u^*(t) + \tilde{u}(t) + f_u(t), \\ y(t) = y^*(t) + \tilde{y}(t) + f_y(t). \end{cases} \tag{3.6}$$

Measurement descriptions as in Eq. (3.5) are also known as Error-In-Variable (EIV) models (Kalman, 1982b, 1990).

With reference to a controlled system, according to the scheme in Fig. 3.2, the signals $u^*(t)$ are the actuator response to the command signals $u_R(t)$. A purely algebraic actuator (i.e., with gain equal to 1) can be described by

$$u^*(t) = u_R(t) + f_a(t), \tag{3.7}$$

where, similarly to input–output sensor fault situation, $f_a(t) \in \Re^r$ is the actuator fault vector.

In general, as shown in Fig. 3.2, if the actuation signals $u^*(t)$ are assumed to be measurable, by neglecting input and output sensor noises, the process model with fault can be described by the following system equation:

$$\begin{cases} \dot{x}(t) = A\,x(t) + f_c(t) + B\,u^*(t), \\ y(t) = C\,x(t) + f_y(t), \\ u(t) = u^*(t) + f_u(t). \end{cases} \tag{3.8}$$

On the other hand, Fig. 3.2 represents the situation where the $u_R$ signals are measured only by the input sensors. Finally, considering the general case, a system affected by all possible faults can be described by the following state-space model:

$$\begin{cases} \dot{x}(t) = A\,x(t) + B\,u^*(t) + L_1\,f(t), \\ y(t) = C\,x(t) + L_2\,f(t), \\ u(t) = u^*(t) + L_3\,f(t), \end{cases} \tag{3.9}$$

where entries of the vector $f(t) = [f_a^T,\, f_u^T,\, f_c^T,\, f_y^T]^T \in \Re^k$ correspond to specific faults. In practice, it is reasonable to assume that the fault signals are described by *unknown* time functions. The matrices $L_1$, $L_2$, and $L_3$ are known as faulty entry matrices, which describe how the faults enter the system. The vectors $u(t)$ and $y(t)$ are the available and measurable inputs and outputs, respectively. Both vectors are assumed known for FDI purposes.

It is worth noting that the distribution of the fault in the diagram of Fig. 3.2 can be described as input–output transfer matrix representation in the following form:

$$y(s) = G_{yu^*}(s)u^*(s) + G_{yf}(s)\,f(s), \tag{3.10}$$

with $s$ being the derivative operator, whilst the transfer matrices $G_{yu^*}(s)$ and $G_{yf}(s)$ are defined as

$$\begin{cases} G_{yu^*}(s) = C\,(s\,I - A)^{-1}\,B, \\ G_{yf}(s) = C\,(s\,I - A)^{-1}L_1 + L_2. \end{cases} \tag{3.11}$$

Both general models for FDI, described by Eqs. (3.9) and (3.10) in the time and frequency domain, respectively, have been widely accepted in the fault diagnosis literature (Patton et al., 1989a, 2000; Chen and Patton, 2012; Gertler, 1998). Under these assumptions, the general model-based FDI problem treated here can be performed on the basis of knowing only the measured sequences $u(t)$ and $y(t)$.

Frequency domain descriptions are typically applied when the effects of faults, as well as the disturbances, have frequency characteristics which differ from each other, and thus information in the frequency spectra serve as criteria to distinguish the faults (Ding and Frank, 1990; Massoumnia et al., 1989). On the other hand, since state-space descriptions provide general and mathematically rigorous tools for system modeling and robust residual generation, for both the deterministic (noise free measurements) and stochastic case (measurements affected by noises), the system matrices $A$, $B$, and $C$ in Eq. (3.9) in proper forms can be obtained by multivariate modeling or identification procedures (Ljung, 1999).

Although most systems to be monitored are actually nonlinear, linear system modeling and identification methods are mainly recalled here. There is certainly an increasing interest in the use of nonlinear methods (nonlinear observers, extended Kalman filters, fuzzy-logic methods, and neural networks). However, as the feature of system supervision is to monitor the operation and performance of the system with respect to an expected point of operation, linear system methods can be very valid. Deviations from expected behavior can be used to monitor system performance changes, as well as component malfunctions.

### 3.1.2 Residual Generation General Scheme

This section recalls the general structure of the residual generator for fault diagnosis. The basic methods will be described briefly whilst their presentation and application to wind turbines will be shown in Chapter 5.

The residual generator module already introduced in Fig. 3.1 can be interpreted as illustrated in Fig. 3.3 (Basseville, 1988).

In the above structure the auxiliary redundant signal $z(t)$ is generated by the function $W_z(u(\cdot),\, y(\cdot))$ and, together with the measurement $y(t)$, the symptom signal $r(t)$ is computed by means of $W_y(z(\cdot),\, y(\cdot))$.

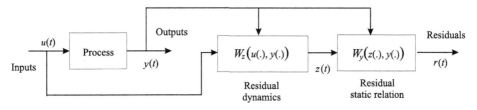

**FIGURE 3.3** General structure of the residual generator for FDI.

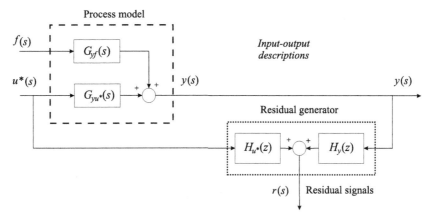

**FIGURE 3.4** Residual generator in the form of input–output transfer matrices.

In the fault-free case the relations of Eq. (3.12) are fulfilled:

$$\begin{cases} z(t) = W_z\big(u(\cdot), y(\cdot)\big), \\ r(t) = W_y\big(z(\cdot), y(\cdot)\big) = 0. \end{cases} \tag{3.12}$$

When a fault occurs in the plant, the residual $r(t)$ will be different from zero.

The simplest residual generator that will be also considered in Chapter 5, and it is obtained when the system $W_z$ is a plant identical model $z(t) = W_z\big(u(\cdot)\big)$, or it is an input–output description for the actual plant simply obtained from the system simulator described in Chapter 2.

In the former case the measurement $y(t)$ is not required in $W_z$ because it is a process simulator. The signal $z(t)$ represents the simulated output, and the residual is directly computed as $r(t) = z(t) - y(t)$. An extension to the model-based residual generation is to replace $W_z\big(u(\cdot)\big)$ by $W_z\,(u(\cdot), y(\cdot))$, i.e., an *output estimator* fed by both system input and output. In such a case function $W_z$ generates an estimate of the function of the output,

$$W_z\,(u(\cdot),\,y(\cdot)) = M\,y(t), \tag{3.13}$$

whilst function $W_y$ can be defined as

$$W_z\,(z(\cdot),\,y(\cdot)) = W\,(z(t) - M\,y(t)), \tag{3.14}$$

with $W$ being a weighting matrix.

Concluding, no matter which type of method is used, the residual generation process is nothing but a mapping whose inputs consist of process inputs and outputs.

On the other hand, by considering an input–output approach, Fig. 3.4 represents a general structure for all residual generators using the input–output transfer matrix description; see, e.g., Patton and Chen (1991).

With reference to Eqs. (3.10) and (3.11), the input–output residual generator function is expressed mathematically by the generalized representation

$$r(s) = \begin{bmatrix} H_{u^*}(s) & H_y(s) \end{bmatrix} \begin{bmatrix} u^*(s) \\ y(s) \end{bmatrix} = H_{u^*}(s)\,u^*(s) + H_y(s)\,y(s), \tag{3.15}$$

where $H_{u*}(s)$ and $H_y(s)$ are continuous-time transfer matrices that can be designed using stable linear systems. The functions $u^*(s)$, $y(s)$, $r(s)$, and $f(s)$ are the Laplace transforms of the corresponding continuous-time signals.

According to the definition, the residual function $r(t)$ has to be designed to become almost zero for the fault-free case and significantly different from zero in case of failures. This means that

$$r(t) = 0 \text{ if and only if } f(t) = 0. \tag{3.16}$$

In order to satisfy Eq. (3.16), the design of the transfer matrices $H_{u*}(s)$ and $H_y(s)$ must satisfy the constraint conditions defined by the relation

$$H_{u*}(s) + H_y(s) G_{yu*} = 0. \tag{3.17}$$

It is worth noting that different residual generators can be obtained by using different parametrizations of $H_{u*}(s)$ and $H_y(s)$ as shown, e.g., in Chen and Patton (2012).

After generating the residual, the simplest and most widely used way to fault detection is achieved by directly comparing residual signal $r(t)$ or a residual function $J(r(t))$ with a fixed threshold $\varepsilon$ or a threshold function $\varepsilon(t)$ defined by the relations

$$\begin{cases} J(r(t)) \leq \varepsilon(t) & \text{in fault-free conditions,} \\ J(r(t)) > \varepsilon(t) & \text{in the faulty cases,} \end{cases} \tag{3.18}$$

where the fault-free and the faulty conditions depend on $f(t)$, that is, the general fault vector defined in Eq. (3.9). If the residual exceeds the threshold, a fault may have occurred.

It will be shown in Chapter 5 that this test works especially well with fixed thresholds $\varepsilon$, if the process operates approximately in a steady state and it reacts after relatively large feature, i.e., after either a large sudden or a long-lasting gradually increasing fault.

On the other hand, adaptive thresholds $\varepsilon(t)$ can be exploited which depend on plant operating conditions, for example, when $\varepsilon(t)$ is expressed as a function of plant inputs (Chen and Patton, 2012).

### 3.1.3 Residual Evaluation for Change Detection

When the residual generation stage has been performed, the second step requires the examination of symptoms in order to determine if any faults have occurred.

As shown by Eq. (3.18), a decision process may consist of a simple threshold test on the instantaneous values of moving averages of residuals. On the other hand, due to the presence of noise, disturbances, and other unknown signals acting upon the monitored system, the decision making process can exploit statistical methods. In this case the measured or estimated quantities, such as signals, parameters, state variables, or residuals, are usually represented by stochastic variables as

$$r(t) = \{r_i(t)\}_i^q, \tag{3.19}$$

with mean value and variance defined as (Willsky, 1976)

$$\bar{r}_i = E\{r_i(t)\}, \qquad \bar{\sigma}_i^2 = E\{[r_i(t) - \bar{r}_i]^2\}, \tag{3.20}$$

which represent the nominal values for the fault-free process.

Analytic symptoms are then obtained as changes

$$\Delta r_i = E\{r_i(t) - \bar{r}_i\}, \qquad \Delta \sigma_i = E\{\sigma_i(t) - \bar{\sigma}_i\}, \tag{3.21}$$

with reference to the normal values. Usually, the time instant $t > t_f$ represents the unknown instant of the fault occurrence.

In order to separate nominal from faulty behavior, usually a fixed threshold $\Delta r_{tol}$, defined as

$$\Delta r_{tol} = \varepsilon \bar{\sigma}_r, \qquad \varepsilon \geq 2, \tag{3.22}$$

has to be selected. By a proper choice of $\varepsilon$, a good compromise has to be made between the detection of small faults (missed fault rates) and false alarm rates.

Another class of methods can be exploited for detecting residual changes due to faults. Therefore, techniques of change detection, e.g., a likelihood ratio test, Bayes decision, or a run sum test, are commonly used (Isermann, 1984;

Basseville and Benveniste, 1986; Basseville and Nikiforov, 1993). Moreover, fuzzy or adaptive thresholds may improve the binary decision (Chen and Patton, 2012; Patton et al., 2000).

Finally, when several variables change, classification methods can be used. In a multidimensional space, the symptom vector, defined as

$$\Delta r = \begin{bmatrix} \Delta r_1 & \Delta r_2 & \ldots & \Delta r_q \end{bmatrix}, \tag{3.23}$$

belongs to a $q$-dimensional space, and its direction depends on the fault occurrence.

In this case the process of residual evaluation consists of determining the direction, as well as the distance, of $\Delta r$ from the origin. Geometrical distance methods (Carpenter and Grossberg, 1987; Tou and Gonzalez, 1974) or artificial neural networks (Himmelblau et al., 1991; Meneganti et al., 1998) can be hence applied.

The generation and evaluation of analytic symptoms concludes the task of fault-detection within the framework of model-based fault diagnosis represented in Fig. 3.1.

## 3.2   RESIDUAL GENERATION MODEL-BASED APPROACHES

The generation of symptoms is the main issue in model-based fault diagnosis. A variety of methods are available in the literature for residual generation, and this chapter briefly presents some of the most common methods.

Most of the residual generation techniques are based on both continuous and discrete system models, however, in this monograph, attention is mainly focused on continuous-time dynamic models.

The model-based residual generation schemes relying on parity space (relation) methods (Gertler and Singer, 1990; Patton and Chen, 1991; Gertler and Monajemy, 1993; Delmaire et al., 1999) and observer-based approaches (Beard, 1971; Frank, 1993; Frank and Ding, 1997; Patton and Chen, 1997; Willsky, 1976; Basseville, 1988) will be considered and summarized (Isermann and Ballé, 1997; Patton et al., 2000) for their application to the fault diagnosis of wind turbine systems.

### 3.2.1   Parity Space Methods

The basic idea of the parity relations approach is to provide a proper check of the parity (consistency) of the measurements acquired from the monitored system. In the early development of fault diagnosis, the parity vector (relation) approach was applied to static or parallel redundancy schemes (Potter and Suman, 1977), which may be obtained directly from measurements (hardware redundancy) or from analytical relations (analytical redundancy). A survey of these methods can be found in Ray and Luck (1991). In the case of hardware redundancy, two methods can be exploited to obtain redundant relations. The first requires the use of several sensors having identical or similar functions to measure the same variable. The second approach consists of dissimilar sensors to measure different variables but with their outputs being relative to each other.

Even if these techniques have been successfully applied for fault diagnosis (Potter and Suman, 1977; Daly et al., 1979), the attention of this section is focused on analytical forms of redundancy.

A straightforward model-based method of fault detection is to take a model $G_M(s) = \frac{\hat{A}(s)}{\hat{B}(s)}$ and to run it in parallel to the process described by

$$G_P(s) = \frac{A(s)}{B(s)}, \tag{3.24}$$

thereby forming an error vector $r(s)$ in the following form:

$$r(s) = \left( \frac{A(s)}{B(s)} - \frac{\hat{A}(s)}{\hat{B}(s)} \right) u(s). \tag{3.25}$$

The methodology described here is represented in Fig. 3.5.

However, as for observers, the model parameters and structure of the monitored process have to be known *a priori*. With reference to Fig. 3.3, if the following relations hold:

$$G_M(s) = G_P(s), \quad \text{i.e.,} \quad \frac{\hat{A}(s)}{\hat{B}(s)} = \frac{A(s)}{B(s)}, \tag{3.26}$$

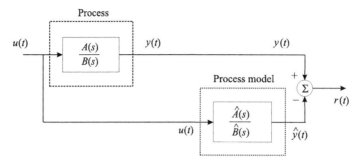

**FIGURE 3.5** Parity equation method.

for additive input $f_u(s)$ and output $f_y(s)$ faults, the $r(s)$ error has the form

$$r(s) = \frac{A(s)}{B(s)} f_u(s) + f_y(s). \tag{3.27}$$

With respect to Fig. 3.5, another possibility is to generate a polynomial error in the form

$$r(s) = \hat{A}(s)\, y(s) - \hat{B}(s)\, u(s) = B(s)\, f_u(s) + A(s)\, f_y(s). \tag{3.28}$$

In both cases, different time responses are obtained for an additive input or output fault.

Eqs. (3.27) and (3.28) that generate the residual signals are called *parity equations* (Gertler, 1991) under the assumptions of fault occurrence and of exact agreement between process and model. However, within the parity equations, the model parameters are assumed to be known and constant, whereas the parameter estimations can vary the parameters of $\hat{A}(s)$ and $\hat{B}(s)$ in order to minimize the residuals. Moreover, for the generation of specific characteristics of the parity vector $r(s)$ and for obtaining fault detection and isolation properties, the residuals can be filtered according to matrix $G_f(s)$ to compute the vector $r_f(s)$ (Patton et al., 2000) in the form

$$r_f(s) = G_f(s)\, r(z). \tag{3.29}$$

Eqs. (3.29), (3.27) and (3.28) can be therefore used to implement and design the residual generation system, in order to meet fault detection and isolation specifications, too (Gertler, 1998). However, for SISO processes only one residual can be generated, and it is therefore not easy to distinguish between different faults.

On the other hand, more freedom in the design of parity equations can be obtained when for SISO processes intermediate signals can be measured (see Fig. 3.3), or for MIMO systems. As an extension of the parity equation method, the parity relation concept presented here can be generalized (Chow and Willsky, 1984; Lou et al., 1986; Patton and Chen, 1994) and then extended to state-space descriptions, as shown in Gertler (1998) for discrete-time models. In the discrete-time case, the redundancy relations are now specified mathematically as follows.

Given the discrete-time system under diagnosis in the form

$$\begin{cases} x(t+1) = Ax(t) + Bu(t), \\ y(t) = Cx(t), \end{cases} \tag{3.30}$$

by substituting the second relation of Eq. (3.30) into the first one and delaying several times, the following system is obtained:

$$\begin{bmatrix} y(t) \\ y(t+1) \\ y(t+2) \\ \vdots \end{bmatrix} = \begin{bmatrix} C \\ CA \\ CA^2 \\ \vdots \end{bmatrix} x(t) + \begin{bmatrix} 0 & 0 & 0 & \ldots \\ CB & 0 & 0 & \ldots \\ CAB & CB & 0 & \ldots \\ \vdots & \vdots & \vdots & \ddots \end{bmatrix} \begin{bmatrix} u(t) \\ u(t+1) \\ u(t+2) \\ \vdots \end{bmatrix}, \tag{3.31}$$

$$Y_f(t) = Tx(t) + QU_f(t). \tag{3.32}$$

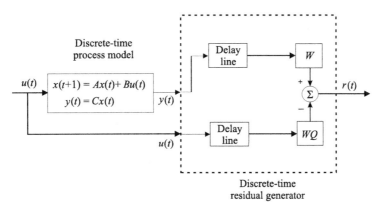

**FIGURE 3.6** Parity equation methods for discrete-time MIMO processes.

In order to remove the nonmeasurable states $x(t)$ and to obtain a parity vector useful for FDI, Eq. (3.31) is multiplied by $W$, such that the following relation holds:

$$WT = 0. \tag{3.33}$$

This leads to residuals in the form

$$r(t) = WY_f - WQU_f(t), \tag{3.34}$$

as shown in Fig. 3.6. The filtered input and output vectors $U_f$ and $Y_f$ are obtained by delaying the corresponding signals. The design of the matrix $W$ gives some freedom to generate a structured set of residuals.

One possibility is to select the elements of $W$ such that one measured variable has no impact on a specific residual. Then this residual remains small in the case of an additive fault on this variable, and the other residuals increase (Patton and Chen, 1994; Chen and Patton, 2012).

Finally, because of the previous results, it is clear therefore that some correspondence exists between parity relation and observer-based methods. This aspect was firstly pointed out by Massoumnia (Massoumnia, 1986) and later demonstrated by Frank and Wunnenberg (Wünnenberg, 1990; Patton et al., 1989a). The problem was reexamined in detail in Patton and Chen (1994), and the equivalence under different conditions and in different meanings was discussed. It was shown that the discrete-time parity relation approach is equivalent to the use of a dead-beat observer. This implies that the discrete-time parity relation scheme provides less design flexibility when compared with methods which are based on observers without any restriction.

A further comparison between observer-based and parity space techniques was proposed (Delmaire et al., 1999). Both methods were first explored for SISO systems and therefore extended the comparison to MIMO systems. The comparison was performed using linear discrete-time models. In particular, considering MIMO systems described by estimated input–output discrete-time forms (e.g., ARX or Auto Regressive Moving Average eXogenous (ARMAX) models) of Eqs. (3.27) and (3.28) leads to a representation in which parameter redundancy cannot be avoided. To overcome this drawback, Delmaire et al. proposed using observers designed from identified canonical state-space forms (Delmaire et al., 1999). Moreover, in the case of parameters redundancy, multiple identification of some parameters may occur, leading to inconsistent estimates, which might produce inconsistent FDI decisions (Delmaire et al., 1999).

### 3.2.2 Observer-Based Methods

The basic idea behind the observer-based techniques is to estimate the outputs of the system from the measurements by using either Luenberger observers in a deterministic setting or Kalman filters in a noisy environment. The output estimation error (or its weighted value) is therefore used as residual.

It is worth noting that when an observer is exploited for FDI purposes, estimation of the outputs is necessary, whilst estimation of the state vector is usually not needed (Chen and Patton, 2012). Moreover, the advantage of using the observer is the flexibility in the selection of its gains, which leads to a rich variety of FDI schemes (Frank, 1994; Frank and Ding, 1997; Chen et al., 1996; Liu and Patton, 1998).

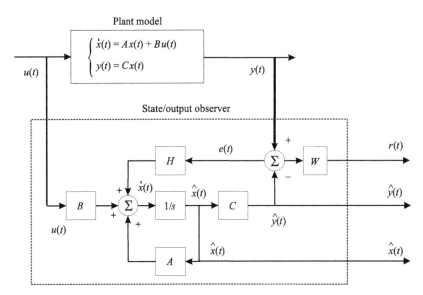

**FIGURE 3.7** Dynamic process and output observer for residual generation.

In order to obtain the structure of a (generalized) observer, a linear time-invariant dynamic model for the plant under consideration in a state-space form is considered:

$$\begin{cases} \dot{x}(t) = A\,x(t) + B\,u(t), \\ y(t) = C\,x(t), \end{cases} \tag{3.35}$$

with $u(t) \in \Re^r$, $x(t) \in \Re^n$, and $y(t) \in \Re^m$.

Assuming that all matrices $A$, $B$, and $C$ are perfectly known, an observer is used to reconstruct the system variables based on the measured inputs and outputs $u(t)$ and $y(t)$:

$$\begin{cases} \dot{\hat{x}}(t) = A\,\hat{x}(t) + B\,u(t) + H\,e(t), \\ e(t) = y(t) - C\,\hat{x}(t). \end{cases} \tag{3.36}$$

The observer scheme described by Eq. (3.36) is reported in Fig. 3.7.

For the state estimation error $e_x(t)$, it follows from Eq. (3.36) that

$$\begin{cases} e_x(t) = x(t) - \hat{x}(t), \\ \dot{e}_x(t) = (A - H\,C)\,e_x(t). \end{cases} \tag{3.37}$$

The state error $e_x(t)$ (and the error $e(t)$) vanishes asymptotically, that is,

$$\lim_{t \to \infty} e_x(t) = 0, \tag{3.38}$$

if the observer is stable, which can be achieved by proper design of the observer feedback $H$.

If the process is influenced by disturbance and faults, by considering the relations of Eqs. (3.9), the overall system is described by the following model:

$$\begin{cases} \dot{x}(t) = A\,x(t) + B\,u(t) + Q\,v(t) + L_1\,f(t), \\ y(t) = C\,x(t) + R\,w(t) + L_2\,f(t), \end{cases} \tag{3.39}$$

where $v(t)$ represents the nonmeasurable disturbance vector at the input, $w(t)$ the nonmeasurable disturbance vector at the output, and $f(t)$ describes fault signals at the input and output acting through $L_1$ and $L_2$, respectively. These terms can represent actuator, process, input and output sensor additive faults acting on the considered wind turbine system.

For the state estimation error, the following relations hold if the disturbances $v(t) = 0$ and $w(t) = 0$:

$$\dot{e}_x(t) = (A - HC)e_x(t) + L_1 f(t) - H L_2 f(t), \tag{3.40}$$

and the output error $e(t)$ has the form

$$e(t) = C e_x(t) + L_2 f(t). \tag{3.41}$$

The vector $f(t)$ represents *additive faults* because they influence $e(t)$ and $x(t)$ by being terms in a sum. When sudden and permanent faults $f(t)$ occur, the state estimation error will deviate from zero. Moreover, $e_x(t)$ and $e(t)$ show dynamic behaviors, which are different for $L_1 f(t)$ and $L_2 f(t)$. Both $e_x(t)$ and $e(t)$ can be taken as residuals.

In particular, the residual $e(t)$ is the basis for different fault detection methods based on output estimation. For the generation of residual with special properties, the design of the observer feedback matrix $H$ can be of interest (Chen and Patton, 2012; Liu and Patton, 1998).

Limiting conditions are the stability and the sensitivity against disturbances $v(t)$ and $w(t)$. If the signals are affected by noise, the Kalman filter must be used instead of classical observers (Jazwinski, 1970).

The robustness and reliability features of the designed residual generators represent the key point when the proposed solutions have to be applied to real processes. If faults appear as changes $\Delta A$ or $\Delta B$ of the parameters, the process behavior becomes

$$\begin{cases} \dot{x}(t) = (A + \Delta A)x(t) + (B + \Delta B)u(t), \\ y(t) = Cx(t), \end{cases} \tag{3.42}$$

while the state $e_x(t)$ and the output estimation $e(t)$ errors are described by the following relations:

$$\begin{cases} \dot{e}_x(t) = (A - HC)e_x(t) + \Delta Ax(t) + \Delta Bu(t), \\ e(t) = Ce_x(t). \end{cases} \tag{3.43}$$

The changes $\Delta A$ and $\Delta B$ can represent both the so-called model–reality mismatch and *multiplicative faults* (Isermann, 1997a; Patton et al., 2000). In the latter case, the changes in the residuals depend on the parameter changes, as well as input and state variable changes. Hence the influence of parameter changes on the residuals is not as straightforward as in the case of the additive faults $f(t)$.

The following output observer-based fault detection scheme and its configuration are briefly recalled in the following (Isermann, 1997a; Chen and Patton, 2012; Patton et al., 2000), since they will enhance the design of the FDI schemes applied to wind turbines.

The main solution for residual generation relies on the use of output observers for the reconstruction of the output signals, if the estimate of the state variable $\hat{x}(t)$ is not of primary interest. In this context it is worth mentioning the paper (Chen et al., 1996) concerning the design of output observers for robust FDI using eigenstructure assignment method.

Therefore, through a linear transformation

$$z(t) = T x(t), \tag{3.44}$$

the state-space representation of the observer has the form

$$\dot{\hat{z}}(t) = F \hat{z}(t) + J u(t) + G y(t), \tag{3.45}$$

and the residual is determined by the relation

$$r(t) = W_z \hat{z}(t) + W_y y(t). \tag{3.46}$$

The state estimation error has the form

$$e_x(t) = \hat{z}(t) - z(t) = \hat{z}(t) - T x(t), \tag{3.47}$$

and the residuals $r(t)$ are then designed to be independent of the process states $x(t)$, the known input $u(t)$, and the unknown inputs $v(t)$ and $w(t)$. In this way the residuals are dependent only on fault signals $f(t)$ (Patton and Chen, 2009; Chen et al., 1996; Gertler, 1998; Patton et al., 2000).

Finally, a generalization of the techniques above leads to residual generation schemes based on *dedicated observers for MIMO processes* that exploit the following properties of the output observers:

1. *Observer excited by one output.* One observer is driven by one sensor output. The other outputs $\hat{y}(t)$ are reconstructed and compared with measured outputs $y(t)$. This allows detecting single output sensor faults (Clark, 1978).
2. *Bank of observers, excited by all outputs.* Several observers are designed for a definite fault signal and detected by hypothesis test (Willsky, 1976).
3. *Bank of observers, excited by single outputs.* Several observers for single sensor outputs are used. The estimated outputs $\hat{y}(t)$ are compared with the measured outputs $y(t)$. This allows detecting multiple sensor faults (DOS, Dedicated Observer Scheme) (Clark, 1978).
4. *Bank of observers, excited by all outputs except for one.* Similarly as before, but here each observer is excited by all outputs except for one sensor output, which is supervised (GOS, Generalized Observer Scheme) (Wünnenberg and Frank, 1987; Frank, 1993).
5. *Fault detection filters.* The feedback $H$ of the state observer in Eq. (3.36) is chosen so that particular fault signals $L_1 f(t)$ change in a definite direction and fault signals $L_2 f(t)$ in a definite plane (Speyer, 1999).

### 3.2.3 Filtering Methods

With reference to Eq. (3.5), when the signal-to-noise ratios $\|u^*(t)\|_2^2/\|\tilde{u}(t)\|_2^2$ and $\|y^*(t)\|_2^2/\|\tilde{y}(t)\|_2^2$ are low, a bank of Kalman Filters (KFs) must be employed to improve the performance of the FDI system. Even in this situation the mathematical formulation of a classical KF is similar to the one described in Section 3.2.2 (Chen and Patton, 2012).

The essential difference concerns the feedback matrix, which becomes time-dependent and is computed by solving a Riccati equation. The solution of this equation requires knowing the covariance matrices of the input and the output noises, which can be identified by means of, e.g., the dynamic Frisch scheme (Diversi and Guidorzi, 1998).

With reference to the time-invariant, discrete-time, linear dynamic system described by Eq. (3.1), and by following a design based on bank structures, as remarked in Section 3.2.2, the $i$th KF for the $i$th output has the structure (Jazwinski, 1970):

$$x_F^i(t+1|t) = A\,\hat{x}^i(t|t) + B\,u(t), \tag{3.48}$$

$$\hat{y}^i(t+1|t) = C_i\,\hat{x}^i(t+1|t), \tag{3.49}$$

$$P(t+1|t) = A\,P(t|t)\,A^T + Q, \tag{3.50}$$

$$K_i(t+1) = P(t+1|t)\,C_i^T\,\left[C_i\,P(t+1|t)\,C_i^T + R\right]^{-1}, \tag{3.51}$$

$$\hat{x}^i(t+1|t+1) = \hat{x}^i(t+1|t) + K_i(t+1)\left[y_i(t+1) - \hat{y}^i(t+1|t)\right], \tag{3.52}$$

$$P(t+1|t+1) = [I - K_i(t+1)\,C_i]\,P(t+1|t)\,[I - K_i(t+1)C_i]^T + K_i(t+1)\,R\,K_i^T(t+1). \tag{3.53}$$

It is worth noting that the variables $\hat{x}^i(t+1|t)$ and $\hat{y}^i(t+1|t)$ are the one step prediction of the state and of the output of the process, respectively. $\hat{x}^i(t|t)$ is the state estimate given by the filter, $C_i$ the $i$th row of the output distribution matrix $C$, $P(t+1|t)$ is the covariance matrix of the one step prediction error $x(t+1) - \hat{x}^i(t+1|t)$, whilst $P(t|t)$ is the covariance matrix of the filtered state error $x(t) - \hat{x}^i(t|t)$. $Q$ is the covariance matrix of the input vector noise $\tilde{u}(t)$ and $R$ is the variance of the $i$th component of the output noise $\tilde{y}(t)$. $K_i(t+1)$ is the time-variant gain of the filter and $y_i(t)$ is the $i$th component of the measured output $y(t)$.

It can be proved that the innovation $e_i(t+1) = y_i(t+1) - \hat{y}^i(t+1|t) = y_i(t+1) - C_i\,\hat{x}^i(t+1|t)$ is a zero-mean white process when all the assumptions regarding the system of Eq. (3.1) and the statistical characteristics of the noises described by Eq. (3.4) are completely fulfilled. A Riccati equation is obtained by substituting Eq. (3.50) into Eq. (3.53). The solution of this equation converges to a steady-state solution when the pair $(A, C_i)$ is completely observable and the pair $(A, D)$ is completely reachable, where $D$ is a matrix such that $Q = D\,D^T$.

In the presence of a fault on the $i$th output ($f_{y_i}(t) \neq 0$), the stochastic properties (mean value, variance, whiteness, etc.) of the innovation process $e_i(t)$ change abruptly so that the fault detection can be based on these variations (Basseville, 1988).

Finally, note how multiple faults in outputs can be isolated since a fault on the $i$th output affects only the innovation of the KF driven by the $i$th output and all the innovation of the filters with unknown input.

In a similar way to the approach relying in KF, an Extended Kalman filter (EKF) can be proposed to solve the FDI/FDD problem of fault diagnosis in wind turbine systems (Kalman, 1990). Usually, the methodology is based on joint parameter and state estimation techniques and consists in providing an (optimal) estimate of the fault, which could be used for FTC, as described in Chapter 4.

Also in this case, the following nonlinear state-space model in the discrete-time framework is considered in the form of

$$\begin{cases} x(t+1) = h_i\left(u(t),\, x(t),\, f_i(t)\right) + v(t), \\ y(t) = g\left(u(t),\, x(t)\right) + w(t), \end{cases} \tag{3.54}$$

where $h_i(\cdot)$ and $g(\cdot)$ are nonlinear functions, whilst $f_i(t)$ refers to the $i$th fault function (unknown input) to be diagnosed. The index $i$ is used to outline the fact that the estimate of the $i$th filter is used to estimate the fault $f_i(t)$. The stochastic inputs $v$ and $w$ denote the process and measurement noises, respectively, which are assumed to be uncorrelated white noise processes with covariance matrices

$$Q(t) = E\left\{v(t)\,v(t)^T\right\} \text{ and } R(t) = E\left\{w(t)\,w(t)^T\right\}. \tag{3.55}$$

The initial estimates of state and covariance matrix are denoted by

$$\overline{x}_0 = E\{x_0\} \text{ and } P_0 = E\left\{(x_0 - \overline{x}_0)\,(x_0 - \overline{x}_0)^T\right\}. \tag{3.56}$$

Following the method proposed in Norgaard et al. (2000), the problem of recursively estimating the augmented state vector $x$ can be formulated as a nonlinear filtering problem that minimizes the conditional mean-square-error, i.e.,

$$\hat{x}(t) = \operatorname{argmin} E\left\{\tilde{x}(t)^T\,\tilde{x}(k)|Y^{t-1}\right\}, \tag{3.57}$$

where $\tilde{x}(t) = x(t) - \hat{x}(t)$ represents the state estimation error, whilst $Y^{t-1} = \left\{y_0,\, y_1,\, \ldots,\, y^{k-1}\right\}$ is a matrix containing the past measurements. The state estimate $\hat{x}(k)$ is equivalent to the conditional mean of the Gaussian probability density function $p\left(x(t)/Y^{(t-1)}\right) \sim \mathcal{N}\left(\hat{x}(t),\, P(t)\right)$ such that

$$\hat{x}(t) = E\left\{x(t)|Y^{(t-1)}\right\} \tag{3.58}$$

where

$$P(t) = E\left\{\left(x(t) - \hat{x}(t)\right)\left(x(t) - \hat{x}(t)\right)^T |Y^{(k-1)}\right\} \tag{3.59}$$

refers to the state covariance matrix that is used to quantify the uncertainty of the estimate. The estimation algorithm can then be formulated as the following nonlinear observer-based scheme:

$$\begin{cases} \hat{x}(t+1) = f_i\left(u(t),\, \hat{x}(t),\, f_i(t)\right) + K(t)\,e(k), \\ \hat{y}(t) = g\left(u(t),\, \hat{x}(t)\right), \end{cases} \tag{3.60}$$

where $K(t)$ is a nonstationary gain to be computed and $e(t) = y(t) - \hat{y}(t|t-1)$ is the innovation sequence associated to the covariance matrix $P_{ee}$,

$$P_{ee} = E\left\{\left(y(t) - \hat{y}(t)\right)\left(y(t) - \hat{y}(t)\right)^T |Y^{k-1}\right\}. \tag{3.61}$$

Based on the previous estimate of the state $\hat{x}(t|t)$ with covariance $\hat{P}(t|t)$, the filter computes at a subsequent time-step an optimal estimate of the state $\hat{x}(t+1|k)$ and its covariance matrix $\hat{P}(t+1|k)$ whenever observations become available. This leads to the following update equations:

$$\begin{cases} \hat{x}(t+1) = \hat{x}(t) + K(t)\,e(t), \\ P(t+1) = P(t) - K(t)\,P_{ee}(t)\,K^T(t). \end{cases} \tag{3.62}$$

The expression of $K(t)$ is given in the form

$$K(t) = P_{xy}(t)\,P_{ee}^{-1}(t), \tag{3.63}$$

where $P_{xy}$ denotes the predicted cross-correlation matrix defined in the following form:

$$P_{xy} = E\left\{\left(x(t) - \hat{x}(t)\right)\left(y(k) - \hat{y}(k)\right)^T |Y^{t-1}\right\}. \tag{3.64}$$

Since the above statistical expectations are generally difficult to obtain, some kind of approximation must be used, like for the EKF case, which exploits a first-order Taylor linearization. However, even if the EKF estimator seems to be adapted, some well-known drawbacks exist in practice, i.e., the parameter estimates can converge slower than the state estimates and, in general, only local convergence can be expected. Based on the work reported in Norgaard et al. (2000), this motivated using an approximation of the nonlinear function $h_i(\cdot)$ by means of a multidimensional extension of Stirling's interpolation formula (Zolghadri, 1996).

Although this method presents some optimality proofs, the key feature remains the *a priori* choice of the covariance matrices $Q$ and $R$. The matrix $Q$ controls the flexibility of the model, whereas the measurement covariance matrix $R$ controls the flexibility of the measurement equations. In the most practical cases, the optimization of $Q$ and $R$ is done by iteratively testing different values and evaluating the results over a test period.

In practice this tuning problem is often tackled as an *ad hoc* process involving a very large number of manual trials. In view of this difficulty it has been chosen to automatically tune these matrices by means of an optimization method (Simani et al., 2003a). The performance index to be minimized corresponds to the root-mean-square of the state estimate errors subjected to positivity constraints of $Q$ and $R$ matrices, that is,

$$J(t) = \left(\frac{1}{N}\sum_{t_0}^{t_f}(\tilde{x}^T \Pi \tilde{x})\right)^{\frac{1}{2}} \quad \text{s.t.} \quad \begin{cases} Q > 0, R > 0, \\ R = diag(r_i), \\ Q = diag(q_i). \end{cases} \tag{3.65}$$

For convenience, the additional constraints $Q = diag(q_i)$ and $R = diag(r_i)$ are included in the optimization algorithm. $\Pi$ is a weighting matrix introduced to manage separately each component of the vector $\tilde{x}$. $t_0$ and $t_f$ are respectively the initial and final discrete time of the tuning interval, and $N$ denotes the number of data points in the tuning interval. Because of the multiparameter, the nonlinear and the discrete nature of this optimization problem, Particle Swarm Optimization (PSO) algorithms can be exploited to derive numerical solutions.

### 3.2.4 Nonlinear Geometric Approach Method to FDI

This section addresses the NonLinear Geometric Approach (NLGA) to FDI that was proposed in Bonfè et al. (2006). The classical NLGA technique is summarized below. Moreover, a procedure to obtain suitable NLGA filters for the estimation of the fault affecting the wind turbine systems that can be used for the AFTC task is recalled in Chapter 4.

The NLGA approach to the nonlinear FDI problem was originally suggested in De Persis and Isidori (2000), and formally developed in De Persis and Isidori (2004b). It consists of finding, by means of a coordinate change in both the state and output spaces, an observable subsystem which, if possible, is affected by the fault and not affected by disturbance. In this way necessary and sufficient conditions for the FDI problem to be solvable are given. Finally, a residual generator can be designed on the basis of the model of the observable subsystem. In this work the complete NLGA strategy with its further extensions and developments is applied to the nonlinear model of the wind turbine.

In more detail the NLGA approach considered here requires a nonlinear system model in the form:

$$\begin{cases} \dot{x} = n(x) + g(x)c + \ell(x)f + p(x)d, \\ y = h(x), \end{cases} \tag{3.66}$$

in which $x \in \mathcal{X}$ (an open subset of $\Re^n$) is the state vector, $c(t) \in \Re^{\ell_c}$ is the control input vector, $f(t) \in \Re$ is the fault, $d(t) \in \Re^{\ell_d}$ the disturbance vector (embedding also the faults which have to be decoupled), and $y \in \Re^m$ the output vector. $n(x)$, $\ell(x)$, the columns of $g(x)$, and $p(x)$ are smooth vector fields; and $h(x)$ is a smooth map.

Therefore, if $P$ represents the distribution spanned by the column of $p(x)$, the NLGA method can be described by means of the following steps (De Persis and Isidori, 2004b):

1. Determine the minimal conditioned invariant distribution containing $P$ (denoted by $\Sigma_*^P$);
2. By using $(\Sigma_*^P)^\perp$, i.e., the maximal conditioned invariant codistribution contained in $P^\perp$, determine the largest observability codistribution contained in $P^\perp$, denoted by $\Omega^*$;

**3.** If $\ell(x) \notin \Omega^*$, continue to the next step, otherwise the fault is not detectable.

**4.** If the condition of the previous step is satisfied, find a surjection $\Psi_1$ and a function $\Phi_1$ fulfilling $\Omega^* \cap \text{span}\{dh\} = \text{span}\{d(\Psi_1 \circ h)\}$ and $\Omega^* = \text{span}\{d(\Phi_1)\}$, respectively. The functions $\Psi(y)$ and $\Phi(x)$, defined as

$$\Psi(y) = \begin{pmatrix} \bar{y}_1 \\ \bar{y}_2 \end{pmatrix} = \begin{pmatrix} \Psi_1(y) \\ H_2 y \end{pmatrix} \quad \text{and} \quad \Phi(x) = \begin{pmatrix} \bar{x}_1 \\ \bar{x}_2 \\ \bar{x}_3 \end{pmatrix} = \begin{pmatrix} \Phi_1(x) \\ H_2 h(x) \\ \Phi_3(x) \end{pmatrix} \quad (3.67)$$

are (local) diffeomorphisms, where $H_2$ is a selection matrix (i.e., a matrix in which any row has all 0 entries but one, which is equal to 1), $\Phi_1(x)$ represents the measured part of the state which is affected by $f$ and not affected by $d$, and $\Phi_3(x)$ represents the unmeasured part of the state which is affected by $f$ and by $d$.

It is worth noting that $\Sigma_*^P$ can be computed by means of the following recursive algorithm:

$$\begin{cases} S_0 = \bar{P}, \\ S_{k+1} = \bar{S} + \sum_{i=0}^{m} [g_i, \bar{S}_k \cap \ker\{dh\}], \end{cases} \quad (3.68)$$

where $m$ is the number of inputs, $\bar{S}$ represents the involutive closure of $S$, $[g, \Delta]$ is the distribution spanned by all vector fields $[g, \tau]$, with $\tau \in \Delta$, and $[g, \tau]$ is the Lie bracket of $g$ and $\tau$. It can be shown that if there exists a $k \geq 0$ such that $S_{k+1} = S_k$, the algorithm (3.68) stops and $\Sigma_*^P = S_k$ (De Persis and Isidori, 2004b).

Once $\Sigma_*^P$ has been determined, $\Omega^*$ can be obtained by exploiting the following algorithm:

$$\begin{cases} Q_0 = (\Sigma_*^P)^\perp \cap \text{span}\{dh\}, \\ Q_{k+1} = (\Sigma_*^P)^\perp \cap \sum_{i=0}^{m} [L_{g_i} Q_k + \text{span}\{dh\}], \end{cases} \quad (3.69)$$

where $L_g \Gamma$ denotes the codistribution spanned by all covector fields $L_g \omega$, with $\omega \in \Gamma$, and $L_g \omega$ is the derivative of $\omega$ along $g$.

If there exists an integer $k^*$ such that $Q_{k^*} = Q_{k^*+1}$, $Q_{k^*}$ is indicated as o.c.a. $((\Sigma_*^P)^\perp)$, where o.c.a. stands for observability codistribution algorithm. It can be shown that $Q_{k^*} = $ o.c.a. $((\Sigma_*^P)^\perp)$ represents the maximal observability codistribution contained in $P^\perp$, i.e., $\Omega^*$ (De Persis and Isidori, 2004b). Therefore, with reference to the model of Eq. (3.66), when $\ell(x) \notin (\Omega^*)^\perp$, the disturbance $d$ can be decoupled and the fault $f$ is detectable.

In the new (local) coordinates defined previously the system of Eq. (3.66) is described by the relations in the following form:

$$\begin{cases} \dot{\bar{x}}_1 = n_1(\bar{x}_1, \bar{x}_2) + g_1(\bar{x}_1, \bar{x}_2) c + \ell_1(\bar{x}_1, \bar{x}_2, \bar{x}_3) f, \\ \dot{\bar{x}}_2 = n_2(\bar{x}_1, \bar{x}_2, \bar{x}_3) + g_2(\bar{x}_1, \bar{x}_2, \bar{x}_3) c + \ell_2(\bar{x}_1, \bar{x}_2, \bar{x}_3) f + p_2(\bar{x}_1, \bar{x}_2, \bar{x}_3) d, \\ \dot{\bar{x}}_3 = n_3(\bar{x}_1, \bar{x}_2, \bar{x}_3) + g_3(\bar{x}_1, \bar{x}_2, \bar{x}_3) c + \ell_3(\bar{x}_1, \bar{x}_2, \bar{x}_3) f + p_3(\bar{x}_1, \bar{x}_2, \bar{x}_3) d, \\ \bar{y}_1 = h(\bar{x}_1), \\ \bar{y}_2 = \bar{x}_2, \end{cases} \quad (3.70)$$

with $\ell_1(\bar{x}_1, \bar{x}_2, \bar{x}_3)$ not identically zero.

Denoting $\bar{x}_2$ by $\bar{y}_2$ and considering it as an independent input, the so-called $\bar{x}_1$-subsystem written in the following form:

$$\begin{cases} \dot{\bar{x}}_1 = n_1(\bar{x}_1, \bar{y}_2) + g_1(\bar{x}_1, \bar{y}_2) c + \ell_1(\bar{x}_1, \bar{y}_2, \bar{x}_3) f, \\ \bar{y}_1 = h(\bar{x}_1) \end{cases} \quad (3.71)$$

is affected by the single fault $f$ and decoupled from the disturbance vector $d$. This subsystem is exploited for the design of the residual generator for the FDI of the fault $f$.

## 3.3 RESIDUAL GENERATION DATA-DRIVEN APPROACHES

The problem of identifying an unknown system given samples of its behavior is well-known (Söderström and Stoica, 1987; Ljung, 1999) to be ill-posed in the sense of Hadamard (Hadamard, 1964), as its solution is neither unique nor depends continuously on the given data.

When *a priori* knowledge on the characteristics of the unknown system is available, the identification procedure can be enhanced. This knowledge may act as a set of constraints shaping the space of possible models so that identification problem in this new space becomes more tractable. As an example, the regularity of the unknown system can be translated into smoothness constraints of some kind, transforming the identification problem into a minimization problem (Tikhonov and Arsenin, 1977; Morozov, 1984). This point of view can be successfully applied to estimate algebraic and dynamic affine systems from noisy samples, by assuming certain good properties of the noise and of the sampling process (Söderström and Stoica, 1987; Ljung, 1999). The data-driven methods described in this section start from the results based on the algebraic case with the purpose of showing the possibility of extending estimation methods to dynamic systems, determining the whole family of models compatible with noisy sequences.

As often happens to new disciplines, Systems Theory borrowed some tools and viewpoints from existing and well-established fields. Thus the identification of static and dynamical systems, i.e., the determination of models from noisy data, has relied heavily on techniques developed by statisticians, who have traditionally considered it mandatory to associate a unique model to every available set of data, whether or not contaminated by noise. Kalman (1982b, 1982a, 1984) reconsidered this problem pointing out how the association of a single model to uncertain data if often based on the introduction of additional information, unrelated to the data, i.e., of prejudices. While the introduction of such prejudices can be convenient in some practical cases, it is, of course, very important to evaluate the family of solutions that can be found without introducing prejudices or, at least, by introducing only mild ones.

The data-driven approach described in this chapter has started from the algebraic results with the purpose of investigating the possibility of extending estimation schemes to dynamical systems, determining the whole family of models compatible with noisy sequences. The results obtained differ from the expectations of the authors in that, as it is proved in the following sections, a single model is, in general, compatible with the data. This result is not in contrast with Kalman's; in fact, in the dynamic case, the additional information necessary to obtain a single model is carried by the correlations established among the samples by the dynamic nature of the generating process.

This section also addresses the problem of the identification of both linear and nonlinear dynamic systems. In the case of nonlinear dynamic systems the identification will be performed by exploiting parametric nonlinear models, such as affine, neural, and fuzzy models.

### 3.3.1 Recursive Identification Approaches

In most practical cases the process parameters are not known at all, or they are not known exactly enough. On the other hand, the process can change over time, due to varying working conditions or due to wearing or ageing situations. Then, in these situations, the process model can be determined with data-driven parameter estimation methods, by measuring input and output signals, $u(t)$ and $y(t)$, if the basic structure of the model is known (Isermann, 1997a; Patton et al., 2000). This strategy can be considered also as a data-driven adaptive approach to FDI, which can be extended to FTC, as shown in Chapter 4.

This approach is based also on the assumption that the faults are reflected in the physical system parameters and the basic idea is that the parameters of the actual process are estimated on–line using well-known parameter estimations methods. The results are thus compared with the parameters of the reference model; obtained initially under fault-free assumptions. Any discrepancy can indicate that a fault may have occurred. In the following two different techniques can be compared. They exploit different models for describing the input–output behavior of the monitored system.

The first approach relies on the so-called Equation Error (EE) method, i.e., the SISO process is described by a discrete-time model of order $n$ that is written in the vector form

$$y(t) = \Psi^T \Theta, \tag{3.72}$$

where

$$\Theta^T = [a_1, \ldots, a_n, b_1, \ldots, b_n] \tag{3.73}$$

is the parameter vector and

$$\Psi^T = \left[ y(t-1), \ldots, y(t-n), u(t-1), \ldots, u(t-n) \right] \tag{3.74}$$

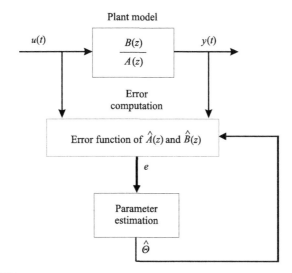

**FIGURE 3.8** System identification for FDI.

corresponds to the discrete-time data vector. This scheme assumes that the faults and the disturbance can affect the process parameters, which can be thus used for the change detection task.

According to the representation sketched in Fig. 3.8, the equation error $e(t)$ of Eq. (3.75) is introduced as

$$e(t) = y(t) - \Psi^T \Theta, \tag{3.75}$$

or, if

$$\frac{y(t)}{u(t)} = \frac{B(z)}{A(z)}, \tag{3.76}$$

with $z$ being the complex variable and $\frac{B(z)}{A(z)}$ the transfer function of the process, the equation error via Z-transformation becomes

$$e(t) = \hat{B}(z)u(t) - \hat{A}(z)y(t), \tag{3.77}$$

in which $\hat{A}(z)$ and $\hat{B}(z)$ correspond to the estimates of the parameters of the polynomials in $A(z)$ and $B(z)$.

The parameter identification approach can be based on the well-known Least-Squares (LS) estimate

$$\hat{\Theta} = [\Psi^T \Psi]^{-1} \Psi^T y \tag{3.78}$$

that is achieved by minimizing the sum of squares and computed according to the relation

$$\begin{cases} J(\Theta) = \sum_t e^2(t) = e^T e, \\ \frac{d\,J(\Theta)}{d\,\Theta} = 0. \end{cases} \tag{3.79}$$

As described in, e.g., Patton et al. (2000), Isermann (1992), the Least-Squares estimate can be also expressed in recursive form (RLS) with respect to the estimates at the instant $t$, with $t = 0, 1, 2, \ldots,$

$$\hat{\Theta}(t+1) = \ddot{\Theta}(t) + \gamma(t)\left[y(t+1) - \Psi^T(t+1)\hat{\Theta}(t+1)\right], \tag{3.80}$$

where

$$\begin{cases} \gamma(t) = \frac{1}{\Psi^T(t+1)P(t)\Psi(t+1)+1} P(t)\Psi(t+1), \\ P(t+1) = \left[I - \gamma(t)\Psi^T(t+1)\right] P(t). \end{cases} \tag{3.81}$$

For improved estimates, filtering methods can be exploited. In particular, as remarked in Section 3.2.3, when the measurements acquired from the process under diagnosis are affected by noise or uncertainty, filtering techniques can be used for parameter estimation (Jazwinski, 1970).

When it is assumed that faults directly affect the process output, an output–error (OE) method can be proposed. Under this assumption, instead of the EE computed in Eq. (3.75), the OE is described in the form

$$e(t) = y(t) - \hat{y}(\Theta, t), \tag{3.82}$$

where

$$\hat{y}(\Theta, z) = \frac{\hat{B}(z)}{\hat{A}(z)} u(z) \tag{3.83}$$

represents the model output that can also be used, as depicted in Fig. 3.8. Unfortunately, in this case, direct calculation of the parameter estimate $\Theta$ is not possible, because $e(t)$ is nonlinear in the parameters, as highlighted in Eq. (3.83).

Therefore the loss function of Eq. (3.82), as well as Eq. (3.75), has to be minimized by numerical optimization methods. The computational effort is then much larger, and online real-time application is in general impossible. However, relatively precise parameter estimates may be obtained.

If a fault within the process changes one or several parameters by $\Delta \Theta$, the output signal changes for small deviations according to

$$\Delta y(t) = \Psi^T(t) \Delta \Theta(t) + \Delta \Psi^T(t) \Theta(t) + \Delta \Psi^T(t) \Delta \Theta(t), \tag{3.84}$$

and the parameter estimator indicates a change $\Delta \Theta$.

Generally, the process parameters $\Theta$ depend on physical process coefficients $p$ (like stiffness, damping factor, resistance, ...), as represented by

$$\Theta = f(p) \tag{3.85}$$

via nonlinear algebraic relations. If the inverse of the relationship,

$$p = f^{-1}(\Theta), \tag{3.86}$$

exists (Patton et al., 2000), the changes $\Delta p$ of the process coefficients can be calculated. These changes in the coefficients are in many cases directly related to faults. Thus, although the knowledge of $\Delta p$ facilitates the fault diagnosis problem, it is not necessary for fault detection only. Parameter estimation can also be applied to nonlinear static process models (Isermann, 2005).

### 3.3.2 Artificial Intelligence Methods

This section recalls data-driven approaches that are based on fuzzy systems and neural networks, and are used to implement the fault diagnosis block. In this section a brief introduction on the general structure of a fault diagnosis system relying on fuzzy systems and neural networks is proposed. In particular, their architectures of open-loop NARX systems are reported, since they represent, in combination with proper training algorithms, the exploited solutions for the implementation of the fuzzy systems and neural network fault estimators.

With reference to fuzzy system modeling, the design of the fault diagnosis module is achieved by means of Takagi–Sugeno (TS) prototypes. Indeed, the unknown relationships between noisy measurements and faults are provided by fuzzy models, which consist of a number of rules connecting the inputs with the output of the system under investigation, on the basis of a knowledge of its dynamics in form of IF $\Longrightarrow$ THEN relations, processed by fuzzy reasoning (Babuška, 2012). In fact, the approximation of nonlinear Multiple-Input Single-Output (MISO) systems (but also extension to MIMO systems can be considered) can be achieved by the Takagi–Sugeno (TS) fuzzy reasoning, as reported, e.g., in Fantuzzi and Rovatti (1996), Rovatti (1996). According to TS modeling approach proposed in Takagi and Sugeno (1985b), the consequents become crisp functions of the input, while the antecedents remain fuzzy propositions, therefore the fuzzy rule takes the form of

$$\begin{aligned} R_i : IF \quad &\text{(fuzzy combination of inputs)} \quad THEN \\ &\text{output} = f_i(\text{inputs}), \end{aligned} \tag{3.87}$$

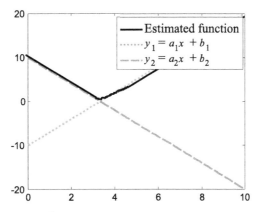

**FIGURE 3.9** TS Fuzzy Inference System output example.

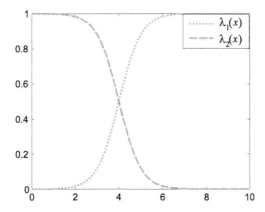

**FIGURE 3.10** TS Fuzzy Inference System membership functions.

where $i$ indicates the number of rules. The antecedent does not differ from the Mamdani rules, with a combined membership function $\lambda_i(x)$ that takes into account the logical connectives expressed by linguistic propositions. The rule consequent function $f_i$ have a defined structure: it is the instance of parametrized function in the affine linear form

$$y_i = a_i^T x + b_i, \tag{3.88}$$

where $a_i$ is a parameter vector and $b$ is a scalar offset, while $y_i$ is the $i$th rule output. The number of rules is supposed to be equal to the number of clusters $n_C$ used for partitioning the data into regions where local linear relations can be assumed (Babuška, 2012). Furthermore, the antecedent of each rule defines the degree of fulfilment for the corresponding consequent model, so that the rule global model can be seen as a fuzzy composition of linear local models.

Thus the TS inference takes the form of the simple algebraic expression

$$\hat{y} = \frac{\sum_{i=1}^{n_C} \lambda_i(x)\, y_i}{\sum_{i=1}^{n_C} \lambda_i(x)}. \tag{3.89}$$

The estimated output $\hat{y}$ is the weighted average of linear functions of the measured input, where the weights are the combined degree of fulfilment of the system input. An example of TS models as nonlinear function approximator is depicted in Figs. 3.9 and 3.10 for a Single-Input Single-Output (SISO) system, with $n_C = 2$.

With reference again to Figs. 3.9 and 3.10, the output of a TS Fuzzy Inference System (FIS) is characterized by two rules with consequent parameters $a_1 = -3$, $b_1 = 10$, $a_2 = 3$, and $b_2 = -10$. The antecedent membership functions $\lambda_1$ and $\lambda_2$ are sigmoidally shaped, as shown in Fig. 3.10, with functions and parameters

$$\begin{cases} \lambda_1(x) = \frac{1}{e^{-2(x-4)}}, \\ \lambda_2(x) = 1 - \frac{1}{e^{-2(x-4)}}. \end{cases} \tag{3.90}$$

It is worth noting that the nonlinear system under investigation can have either a static or dynamic behavior. In the latter case the considered model input vector $x$ can contain current, as well as previous, samples of the system input or output.

Indeed, in order to introduce the time dependence into the model of Eq. (3.87) the consequents are considered as discrete-time linear AutoRegressive models with eXogenous input (ARX) of order $o$, in which the regressor vector takes the form of

$$x(k) = \left[y(k-1), \ldots, y(k-o), u(k), \ldots, u(k-o)\right]^T, \tag{3.91}$$

where $u$ and $y$ are the actual system input and output vectors, and $k$ is the time step. The affine parameters of Eq. (3.88) can be grouped into

$$a_i = [\alpha_1^{(i)}, \ldots, \alpha_o^{(i)}, \delta_1^{(i)}, \ldots, \delta_o^{(i)}]^T, \tag{3.92}$$

where the $a^{(i)}$ coefficients are associated to the output samples, and the $\delta^{(i)}$ are associated to the input.

An effective approach to the design of an FIS as approximator of a complex nonlinear system begins with the partitioning of the available data into subsets characterized by simpler (linear) behavior. A cluster can be defined as a group of data that are more similar to each other rather than to the members of another cluster. The similarity among data can be expressed in terms of their distance from a particular item, exploited as the cluster prototype. Fuzzy clustering provides an effective tool to obtain a partitioning of data in which the transitions among subsets are smooth, rather than abrupt.

Indeed, fuzzy clustering allows an item to belong to several clusters simultaneously, with different degrees of fulfilment, whereas the classic crisp clustering relies on mutually exclusive subsets. Different clustering methods have been proposed in the literature, see, e.g., the review (Jain et al., 1999) or the more recent works (Jun et al., 2011 and Graaff and Engelbrecht, 2012).

Typically, the available data consist of noisy measurements acquired from the system. They are grouped into the data matrix $Z$, whose columns are the vectors $z$ containing the measurements of a single observation of the system under analysis:

$$Z = \begin{bmatrix} z_{11} & \cdots & z_{1N} \\ \vdots & \ddots & \vdots \\ z_{n1} & \cdots & z_{nN} \end{bmatrix}, \tag{3.93}$$

where $n$ is the data dimension and $N$ is the number of available observations.

Most fuzzy clustering algorithms are based on the optimization of the $c$-means goal function $J(Z, U, V)$ performed as follows:

- Define $Z$ as the data matrix above;
- Define $U = [\mu_{ik}]$, the so-called fuzzy partition matrix that contains the values of the membership function for the couple ($i$th measurement, $k$th cluster);
- Define $V = [v_1, \ldots, v_{n_C}]$ containing the cluster prototypes that have to be determined and represent the centers from which the distance of each measurement can be calculated.

The widespread $c$-means goal function adopted in this work was formulated in Bezdek (2013) in the following form:

$$J(Z, U, V) = \sum_{i=1}^{n_C} \sum_{k=1}^{N} (\mu_{ik})^m D_{ikA}^2, \tag{3.94}$$

with $m > 1$ being the weighting exponent and

$$D_{ikA}^2 = \|z_k - v_i\|_A^2 = (z_k - v_i)^T A (z_k - v_i), \tag{3.95}$$

that is, a squared inner product distance norm, with $i = 1, \ldots, n_C$ and $k = 1, \ldots, N$. The matrix $A$ determines the cluster shape.

The minimization algorithm exploits a series of Picard iterations consisting in the updating of the cluster prototypes and of the partition matrix until the stopping criterion is met (Babuška, 2012).

An important point concerns the determination of the optimal number of clusters $n_C$, as the clustering algorithm operates on the assumption of a certain number of clusters, regardless of whether they are really present in the data or not. Once the partition matrix has been estimated, the antecedent degrees of fulfilment are easily derived by interpolation or curve fitting methods.

Then the design of the FIS assumes the form of an identification problem addressed to estimate the consequent parameters $a_i$ and $b_i$ of Eq. (3.88) in a noisy environment. The identification scheme adopted in this work was proposed in Simani et al. (1999c) and successfully exploited in the approximation of nonlinear functions through the piecewise affine models (Fantuzzi et al., 2002). This approach is based on the minimization of the prediction errors of the individual TS local models understood as $n_C$ independent problems. Their solutions rely on the so-called Frisch scheme (Beghelli et al., 1990) that is usually exploited in connection with the identification of Errors-In-Variables models (Fantuzzi et al., 2002).

Considering a discrete-time MISO system, the noise is supposed to affect the input $u$, as well as the output $y$, measurements in the form of the additive signals $\tilde{u}$, $\tilde{y}$ on the noise-free unmeasurable quantities $u^*$, $y^*$:

$$\begin{cases} u(k) = u^*(k) + \tilde{u}(k), \\ y(k) = y^*(k) + \tilde{y}(k). \end{cases} \tag{3.96}$$

Thus, considering the $i$th TS consequent as in Eq. (3.89) and the associated dynamic local ARX model of order $o$ with the regressors grouped into the vector $x$ as in Eq. (3.91), the acquisition of $N_i$ noisy measurements of input and output samples permits the construction of the $i$th data matrix $X^{(i)}$ defined as

$$X^{(i)} = \begin{bmatrix} y(k) & x^T(k) & 1 \\ y(k+1) & x^T(k+1) & 1 \\ \vdots & \vdots & \vdots \\ y(k+N_i-1) & x^T(k+N_i-1) & 1 \end{bmatrix}. \tag{3.97}$$

The $i$th covariance matrix $\Sigma^{(i)}$ from the acquired data can be computed as

$$\Sigma^{(i)} = X^{(i)^T} X^{(i)} \geq 0, \tag{3.98}$$

which is a positive-definite matrix consisting of the sum of two terms:

$$\Sigma^{(i)} = \Sigma^{(i)^*} + \bar{\tilde{\Sigma}}^{(i)}, \tag{3.99}$$

where $\Sigma^{(i)^*}$ refers to the noise free signals, while $\bar{\tilde{\Sigma}}^{(i)}$ is the noise covariance matrix, which depends on the unknown noise variances $\bar{\tilde{\sigma}}_u$, $\bar{\tilde{\sigma}}_y$ through the expression

$$\bar{\tilde{\Sigma}}^{(i)} = diag\left[\bar{\tilde{\sigma}}_y I, \bar{\tilde{\sigma}}_u I, 0\right]. \tag{3.100}$$

The solution of the identification problem above requires estimating $\bar{\tilde{\sigma}}_u$ and $\bar{\tilde{\sigma}}_u$, which can be performed by solving the expression in the form

$$\Sigma^{(i)^*} = \Sigma^{(i)} - \tilde{\Sigma}^{(i)} \tag{3.101}$$

with

$$\tilde{\Sigma}^{(i)} = diag\left[\tilde{\sigma}_y I, \tilde{\sigma}_u I, 0\right] \tag{3.102}$$

with respect to the variables $\tilde{\sigma}_u$ and $\tilde{\sigma}_y$.

In case all the assumptions regarding the Frisch scheme (Simani et al., 1999c) are satisfied, there exists one common point belonging to all the surfaces $\Gamma^{(i)} = 0$ determined as the root locus of Eq. (3.101) that represents the actual noise variance values $(\bar{\tilde{\sigma}}_u, \bar{\tilde{\sigma}}_y)$. However, in real cases, the Frisch assumptions are commonly violated, so that a unique solution cannot be obtained. In these situations the identification aims at finding the nearest point to all the surfaces.

After the computation of the variances, the covariance noise matrix can be built as in Eq. (3.100), and the linear parameters in each cluster (therefore in each TS consequent) can be finally determined as a solution of the following expression:

$$\left(\Sigma^{(i)} - \bar{\tilde{\Sigma}}^{(i)}\right) a_i = 0. \tag{3.103}$$

Alongside the fuzzy models a different data-driven approach, based on neural networks, has been proposed in order to implement the fault diagnosis block. In this section, after a brief introduction on the general structure, the properties and

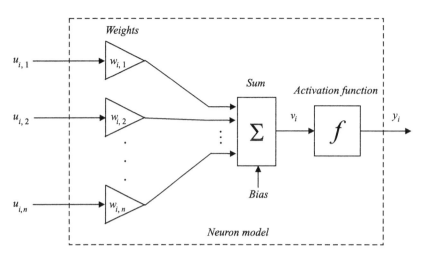

**FIGURE 3.11** The $i$th neuron model.

the functioning of a neural network, as well as the architecture, are recalled. They will be exploited for implementing the neural network fault estimators.

In this work a set of neural network estimators is designed and trained in order to reproduce the behavior of the systems under investigation, thus accomplishing the modeling and identification task. The structure of the $i$th single neuron (Haykin, 2009) is also called *perceptron*. It features an MISO system where the output $y_i$ is computed as a function $f$ of the weighted sum $v_i$ of all the $n_i$ neuron inputs $u_{i,1}, \ldots, u_{i,n_i}$, with the associated weights $w_{i,1}, \ldots, w_{i,n_i}$. The function $f$, referred to as *activation function*, represents the engine of the neuron, as shown in Fig. 3.11.

A structural categorization of neural networks concerns the way in which their elements are connected to each other (Liu, 2012). In a *feed-forward network*, also called *multilayer perceptron*, neurons are grouped into unidirectional layers. The first of them, namely the *input* layer, is fed directly by the network inputs, then each successive *hidden* layer takes the inputs from the neurons of the previous layer and transmits the output to the neurons of the next layer, up to the last *output* layer, in which the final network outputs are produced. Therefore, neurons are connected from one layer to the next, but not within the same layer. The only constraint is the number of neurons in the output layer, which has to be equal to the number of actual network outputs. On the other hand, *recurrent networks* (Medsker and Jain, 1999) are multilayer networks in which the output of some neurons is fed back to neurons belonging to previous layers, thus the information flows in forward and backward directions, allowing a dynamic memory inside the network.

A noteworthy intermediate solution is provided by the multilayer perceptron with a tapped delay line, which is a feed-forward network whose inputs come from a delay line. This kind of network represents a suitable tool to model, or predict, the evolution of a dynamic system. In particular, the open-loop NARX network belongs to this category as its inputs are delayed samples of the system inputs and outputs. Indeed, if properly trained, an NARX network can estimate the current (or the next) system output on the basis of the acquired past measurements of system inputs and outputs.

Generally speaking, considering an MIMO system, the elaborations of the open-loop NARX network follow the law

$$\hat{y}(k) = f_{net}\left(u(k), \ldots, u(k - d_u), y(k - 1), \ldots, y(k - d_y)\right), \tag{3.104}$$

where $\hat{y}$ is the estimate of the system output, $u$ and $y$ are the measured system inputs and outputs, $k$ is the time step, $d_u$ and $d_y$ are the delays of inputs and outputs, respectively. $f_{net}$ is the function realized by the network that depends on the layer architecture, number of neurons, their weights, and their activation functions. The functioning of an open-loop NARX network used as estimator is depicted in Fig. 3.12.

It is worth noting that when only input measurements are available, an NARX network can become a recurrent network by closing the loop feeding back the network outputs to the inputs, as shown in Fig. 3.13.

The parameters on which the designer can act concern the overall architecture (number of neurons, connections between layers), while the value of the weights inside each neuron are derived from the network training.

A neural network is a learning system requiring an initial training procedure that adjusts the weights to improve the network performance. When the network task is the estimation of a nonlinear function, the training is performed by presenting to the network a set of examples of proper behavior, consisting of the inputs and the desired outputs (targets) for the relative inputs. Training can be implemented in two different ways:

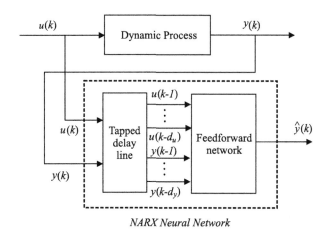

**FIGURE 3.12**  Open-loop NARX network.

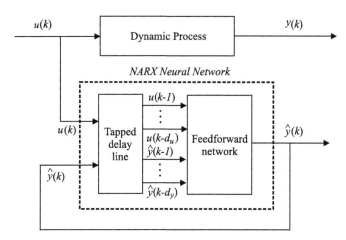

**FIGURE 3.13**  Closed-loop NARX network.

- **Incremental mode**, where each input–target couple generates an updating of the network weights;
- **Batch mode**, where all inputs and targets are applied to the network before the weights are updated.
  Although this kind of training requires more memory storage capability, with respect to the incremental mode, it is characterized by a faster convergence and produces smaller errors, thus it will be considered in the following.

The training objective is the minimization of a performance function $E$, which depends on the weight vector $w$.

Generally speaking, considering a number $P$ of available example patterns consisting in the input–target pairs $(u_p, t_p)$, with $p = 1, \ldots, P$, with $\hat{y}_p$ representing the output generated by the network fed by $u_p$, the $p$th error vector can be expressed as

$$e_p = \left[t_p - \hat{y}_p\right] = \left[e_{p,1}, \ldots, e_{p,M}\right]^T, \tag{3.105}$$

with $p = 1, \ldots, P$ and $M$ being the number of outputs. Furthermore, the global error vector $\bar{e}$ collects each $e_p$ as

$$\bar{e} = \left[e_{1,1}, \ldots, e_{1,M}, \ldots, e_{P,1}, \ldots, e_{P,M}\right]^T. \tag{3.106}$$

Consequently, the performance function becomes

$$E(w) = \frac{1}{P} \sum_{p=1}^{P} \left(t_i - \hat{y}_i\right)^2 = \frac{1}{P} \sum_{p=1}^{P} \sum_{m=1}^{M} e_{p,m}^2, \tag{3.107}$$

where the dependence of $E$ on the $N$ parameters grouped in the vector $w = [w_1, \ldots, w_N]^T$ is implicit in the generated output $\hat{y} = \hat{y}(w)$.

**TABLE 3.1** Updating rules and algorithms.

| Optimization Method | Adaptation Law |
|---|---|
| Gradient Descent | $w_{k+1} = w_k - \alpha\, g_k$ |
| Newton–Raphson | $w_{k+1} = w_k - H_k^{-1}\, g_k$ |
| Gauss–Newton | $w_{k+1} = w_k - \left( J_k^T J_k \right)^{-1} J_k\, \bar{e}_k$ |
| Levenberg–Marquardt | $w_{k+1} = w_k - \left( J_k^T J_k + \mu I \right)^{-1} J_k\, \bar{e}_k$ |

Any standard numerical optimization algorithm can be used to update the parameters in order to minimize $E$. Among these, the most common are iterative and making use of characteristic matrices, such as the gradient $g$ (or the Hessian $H$) of the performance function, or the Jacobian $J$ of the estimation error, defined as:

$$g = \frac{\partial E(w)}{\partial w} = \left[ \frac{\partial E}{\partial w_1}, \ldots, \frac{\partial E}{\partial w_N} \right]^T, \tag{3.108}$$

$$H = \begin{bmatrix} \frac{\partial^2 E}{\partial w_1^2} & \cdots & \frac{\partial^2 E}{\partial w_1 w_N} \\ \vdots & \ddots & \vdots \\ \frac{\partial^2 E}{\partial w_N w_1} & \cdots & \frac{\partial^2 E}{\partial w_N^2} \end{bmatrix}, \tag{3.109}$$

$$J = \begin{bmatrix} \frac{\partial_{1,1}^e}{\partial w_1} & \cdots & \frac{\partial_{1,1}^e}{\partial w_N} \\ \frac{\partial_{1,2}^e}{\partial w_1} & \cdots & \frac{\partial_{1,2}^e}{\partial w_N} \\ \vdots & \ddots & \vdots \\ \frac{\partial_{P,M}^e}{\partial w_1} & \cdots & \frac{\partial_{P,M}^e}{\partial w_N} \end{bmatrix}. \tag{3.110}$$

The successive iterations of these algorithms consist of the updating of the parameters and the calculation of the new value of the performance function, until a stopping criterion is met. The updating rules of the most common optimization algorithms (i.e., the gradient descent, Newton, Gauss–Newton, and Levenberg–Marquardt algorithm) are reported in Table 3.1.

Table 3.1 summarizes the parameters of the updating rules of the most common optimization algorithms, aimed at the minimization of the performance function $E$; $k$ is the iteration index, $\alpha$ is the learning rate, and $\mu$ the combination coefficient.

It can be demonstrated that the gradient descent algorithm, for a sufficiently small learning rate $\alpha$ value, is asymptotically convergent: around the solution $g$, the gradient is close to zero and the weights do not meaningfully change. Otherwise, Newton and Gauss–Newton algorithms provide a faster convergence, but they both involve the computation of the inverse of a matrix which may not be invertible, causing instability in the procedure. Moreover, the Hessian matrix entails a burdensome computational effort, as it contains the second order derivative terms.

The *Levenberg–Marquardt* algorithm, originally proposed in Marquardt (1963), introduces an approximation of the Hessian matrix as $H \approx J^T J + \mu I$, where the first term of the sum is the Jacobian approximation (also exploited in Gauss–Newton algorithm) and the second term, driven by the combination coefficient $\mu > 0$, ensures the invertibility of the resulting matrix. Therefore, the Levenberg–Marquardt algorithm provides both fast and stable convergence and it represents a suitable tool to train a neural network. Indeed, as shown in Chapter 5, the neural network fault estimator blocks have been trained exploiting this method.

The training of a neural network based on the Levenberg–Marquardt algorithm, as explained in Hagan and Menhaj (1994), uses a technique called *back-propagation* training in order to compute the Jacobian matrix for the updating rule. Its name refers to the backward processing that starts from the output layer of the network towards the first layer, after a previous forward computation of neuron outputs.

### 3.3.3 Fault Diagnosis Technique Integration

Several FDI techniques have been developed, and their application shows different properties with respect of the diagnosis of different faults in a process. In order to achieve a reliable FDI technique, a good solution consists in a proper integration of several methods which take advantage of the different procedures (Isermann, 1994; Isermann and Ballé, 1997). Furthermore, a comprehensive approach to fault diagnosis should exploit a knowledge-based treatment of all available analytical and heuristic information. This successful approach can be performed by an integrated method to knowledge-based fault diagnosis.

Regarding fuzzy logic for residual generation, as stated in Section 3.1.2, model-based FDI consists of two stages, residual generation and decision making. The first block is exploited to generate residuals by means of the available inputs and outputs from the monitored system.

For the first step, classical fault diagnosis model-based methods can exploit state-space of input–output dynamic models of the process under investigation. Within this framework faults are supposed to appear as changes on the system state or output caused by malfunctions of the components, as well as of the sensors. Such fault indices are often monitored using estimation techniques. The main problem with these techniques is that the precision of the process model affects the accuracy of the detection and isolation system, as well as the diagnostic sensibility. On the other hand, wind turbine systems are nonlinear, as highlighted in Chapter 2, and may not be modeled by using a single model for all operating conditions.

Since a mathematical model is a description of system behavior, accurate modeling for a complex nonlinear system is very difficult to achieve in practice. Sometimes for some nonlinear systems it can be impossible to describe them by analytical equations. Moreover, sometimes the system structure or parameters are not precisely known and, if diagnosis has to be based primarily on heuristic information, no qualitative model can be set up. Because of these assumptions, fuzzy system theory seems to be a natural tool to handle complicated and uncertain conditions (Babuška, 1998). Therefore, instead of exploiting complicated nonlinear models obtained by modeling techniques, it is also possible to describe the wind turbine process behavior by a collection of local affine fuzzy and nonfuzzy models (Leontaritis and Billings, 1985b, 1985a; Takagi and Sugeno, 1985a), whose parameters are obtained by identification procedures.

The second stage of model-based FDI consists of a logic decision process that transforms residual signal information (quantitative knowledge) into qualitative statements (faulty or normal working conditions). Therefore the problem of decision-making can be treated in a novel way by means of fuzzy logic.

As noise contamination and uncertainty affect the residuals even in fault-free conditions, they fluctuate and become unequal to zero. This common situation, which may hide the fault effects, can be handled by means of the fuzzy logic framework. The interesting feature of fuzzy logic is that it represents a powerful tool for describing vague and imprecise fact and is therefore suited for applications where complete information about fault and system is not available to the designer.

Even if much effort has been spent on trying to decrease the uncertainty associated with quantitative residual generation, it is impossible to fully eliminate the effect of uncertainty. On the basis of this limitation, the residual evaluation problem consists of making the correct decision with respect to uncertain information. Fuzzy logic can be a suitable tool for this task. For instance, a lot of processes can be managed by humans heuristically since an analytical description is impossible to use. Fuzzy logic can express expert knowledge in the form of a rule-based knowledge format. The introduction of fuzzy logic can thus improve decision making in order to provide reliable FDI methods which are applicable for real wind turbine systems.

It should finally be pointed out how the fuzzy approach in FDI can solve the problem at two levels: first, fuzzy descriptions are used to generate symptoms and then, the fault detection and isolation is achieved using again fuzzy logic (Dexter and Benouarets, 1997; Isermann, 1998).

On the other hand, with reference to neural networks in fault diagnosis, quantitative model-based fault diagnosis generates symptoms on the basis of the analytical knowledge of the process under investigation. In most cases, however, this does not provide enough information to perform an efficient FDI, i.e., to indicate the location and the mode of the fault.

A typical integrated fault diagnosis system uses both analytical and heuristic knowledge of the monitored system. This knowledge can be processed in terms of residual generation (analytical knowledge) and feature extraction (heuristic knowledge). The processed knowledge is then provided to an inference mechanism which can comprise residual evaluation, symptom observation, and *pattern recognition*. In particular, when the process model is only known to a certain extent of precision, a pattern recognition method can provide a convenient approach to solve the fault identification problem, i.e., to determine the size of the fault (Himmelblau, 1978; Pau, 1981).

NNs have been used successfully in pattern recognition, as well as system identification, and they have been proposed as a possible technique for fault diagnosis, too. NNs can handle nonlinear behavior and partially known process because they learn the diagnostic requirements by means of the information of the training data. NNs are noise tolerant and their ability to generalize knowledge and adapt during use are extremely interesting properties (Hoskins and Himmelblau, 1988; Dietz et al., 1989; Venkatasubramanian and Chan, 1989; McDuff and Simpson, 1990; Chen et al., 1990).

Some example processes were considered, in which FDI was performed by an NN using input and output measurements. In these works the NN is trained to identify the fault from measurement patterns; however, the classification of individual measurement pattern is not always unique in dynamic situations, therefore the straightforward use of NN in fault diagnosis of wind turbine plants is not practical, and other approaches should be investigated. An NN could be exploited in order to find a dynamic model of the monitored system or connections from faults to residuals. In the latter case, the NN is used as a pattern classifier or nonlinear function approximator. In fact, artificial neural networks are capable of approximating a large class of functions for fault diagnosis of an industrial plant.

Under these considerations, in this chapter, the identification of fuzzy and nonfuzzy models for the system under diagnosis, as well as the application of NN as a function approximator, will be shown. Quantitative and qualitative approaches have a lot of complementary characteristics which can be suitably combined together to exploit their advantages and to increase the robustness of quantitative techniques. The suggested combination can also minimize the disadvantages of the two procedures; in particular, it is important that partial knowledge deriving from qualitative reasoning is reduced by quantitative methods. Hence the main aim of further research on model-based fault diagnosis consists in finding the way to properly combine these two approaches together to provide highly reliable diagnostic information.

After these remarks it is worth highlighting the use of neuro-fuzzy approaches to FDI. Identification of multivariate processes can be interpreted as a problem of approximation to an input–output mapping. The mathematical model used in traditional methods is sensitive to modeling errors, parameter variation, noise, and disturbance (Chen and Patton, 2012; Patton et al., 2000). Process modeling has limitations, especially when the system is complex and uncertain, and the data are ambiguous and not information rich.

As previously stated, NN are known to approximate any nonlinear even dynamic function, given suitable weighting factors and architecture. Moreover, online training makes it possible to change the FDI system easily in cases where changes are made in the physical process or the control system. NN can generalize when presented with inputs not appearing in the training data and make intelligent decisions in cases of noisy or corrupted data. They are also readily applicable to multivariate systems and have a highly parallel structure, which is expected to achieve a higher degree of fault tolerance. An NN can operate simultaneously on qualitative and quantitative data. NNs can be very useful when no mathematical model of the system is available, i.e., analytical models cannot be applied.

It is clear that almost all the physical processes are dynamic in nature. Combining dynamic elements such as filters and delays yields a powerful modeling technique. But the NN operates as a "black box" with no qualitative/quantitative information available of the model it represents. Usually, engineers and operators want to visualize how the system is working and what rules govern its operation. There is also ambiguity about the performance of the NN in case of unexpected situation (Korbicz et al., 1999).

Fuzzy logic systems, on the other hand, have the ability to mimic the sensing, generalizing, processing, operating, and learning abilities of a human operator. They offer a linguistic model of the system dynamics which can be easily understood by certain rules. They also have inherent abilities to deal with imprecise or noisy data. Fuzzy logic can be used with neural networks (Chiang et al., 2001). A fuzzy neuron has the same basic structure as the artificial neuron, except that some or all of its components and parameters may be described through fuzzy logic. A fuzzy neural network is built on fuzzy neurons or on standard neurons but dealing with fuzzy data. A fuzzy neural network is a connectionist model for the implementation and inference of fuzzy rules. There are many different ways to fuzzify an artificial neuron, which results in a variety of fuzzy neurons and fuzzy networks (Chiang et al., 2001; Nelles, 2001).

Different neuro-fuzzy structures can be therefore designed to combine the advantages of both neural networks and fuzzy logic (Patton et al., 1999; Calado et al., 2001). These structures have been successfully applied to a wide range of applications from industrial processes to financial systems, because of the ease of rule base design, linguistic modeling, application to complex and uncertain systems, inherent nonlinear nature, learning abilities, parallel processing and fault-tolerance abilities (Wu and Harris, 1996; Ayoubi, 1995). However, successful implementation depends heavily on prior knowledge of the system and the training data. There are three common methods of combining neural networks with the fuzzy logic:

1. Fuzzification of the inputs or outputs of the neural networks;
2. Fuzzification of the interconnections of conventional neural networks;
3. Using neural networks in fuzzy models where neurons provide the necessary membership functions and rule base.

All of the neuro-fuzzy (NF) modeling structures combine, in a single framework, both numerical and symbolic knowledge about the process. Automatic linguistic rule extraction is a useful aspect of NF, especially when little or no prior knowledge about the process is available (Brown and Harris, 1994; Jang and Sur, 1995). For example, an NF model of a nonlinear dynamical system can be identified from the empirical data. This modeling approach can give us some insight about the nonlinearity and dynamical properties of the system.

The most common NF systems are based on two types of fuzzy models (Takagi and Sugeno, 1985a; Sugeno and Kang, 1988) and (Mamdani, 1976; Mamdani and Assilian, 1995) combined with NN learning algorithms. TS models use local linear models in the consequents, which are easier to interpret and can be used for control and fault diagnosis (Füssel et al., 1997; Isermann and Ballé, 1997). Mamdani models use fuzzy sets or rules as consequents and therefore give a more qualitative description. The B-spline neural network (with triangular basis functions) is the simplest of all Mamdani NF structures, but the large consequent rule set means that the method is not easy to use due to low transparency.

Many neuro-fuzzy structures have been successfully applied to a wide range of applications from industrial processes to financial systems, because of the ease of rule base design, linguistic modeling, application to complex and uncertain systems, inherent nonlinear nature, learning abilities, parallel processing and fault-tolerance abilities. However, successful implementation depends heavily on prior knowledge of the system and the empirical data (Ayoubi, 1995).

NF networks by their intrinsic nature can handle a limited number of inputs and can usually be identified in a not very transparent way from the empirical data. Transparency corresponds here to a more meaningful description of the process, i.e., fewer rules with appropriate membership functions. In Adaptive Neuro-Fuzzy Inference Systems (ANFIS) (Jang, 1993; Jang and Sur, 1995) a fixed structure with grid partition is used. Antecedent and consequent parameters are identified by a combination of least-squares estimates and gradient based methods, the so-called called *hybrid learning rule*. This method is fast and easy to implement for low dimensional input spaces. It is more prone to losing the transparency and the local model accuracy because of the use of the error back-propagation, which is a global and not locally nonlinear optimization procedure. One possible method to overcome this problem can be to find the antecedents and rules separately, e.g., clustering and constrain the antecedents, and then apply optimization.

Hierarchical NF networks can be used to overcome the dimensionality problem by decomposing the system into a series of MISO and/or SISO systems called *hierarchical systems* (Tachibana and Furuhashi, 1994). The local rules use subsets of input spaces and are activated by higher level rules. The criteria on which to build an NF model are based on the requirements for fault diagnosis and the system characteristics. The function of the NF model in the FDI scheme is also important, i.e., preprocessing data, identification (residual generation) or classification (decision making/fault isolation). For example, an NF model with high approximation capability and disturbance rejection is needed for identification so that the residuals are more accurate. Whereas, in the classification stage, an NF network with more transparency is required.

In the remainder of this section, the problem of structure identification for NF models is briefly addressed. For complexity reduction and transparency, structure identification methods can be applied to find appropriate input partition, rules, and membership functions (MFs). Methods like Evolutionary Algorithms (EA), Classification and Regression Trees CART (Jang, 1994), clustering and unsupervised NN (e.g., like the Kohonen feature maps) can be used. Once the structure is determined, i.e., the rules and input membership functions, the consequent parameters can be identified by optimization techniques like Least-Squares Estimation. The Product Space Clustering approach can be used (Babuška, 1998) for structure identification of TS and Mamdani fuzzy models. For an MISO nonlinear dynamic system with $p$ inputs, the Product Space $X \times Y \subset \Re^{p+1}$ is divided in subspaces in which linear models can approximate the nonlinear system. The LOcally LInear MOdel Tree (LOLIMOT) algorithm developed in Nelles and Isermann (1996) can be used to identify a fuzzy model with dynamic linear models as consequents. When using such structure identification techniques, a major issue is the sensitivity to uneven distribution of data. For example, in most clustering algorithms, more clusters are created in regions with more data. A possible solution to this is problem may be to initialize the algorithm with a large number of clusters.

Fig. 3.14 describes an FDI scheme including different residual generation and evaluation strategies. Several models are constructed to identify the faulty and the fault-free behavior of the system under diagnosis:

$$r_i(t) = f(u(t), \ldots, u(t-n), y(t), \ldots, y(t-n)) \tag{3.111}$$

with $i = 1, \ldots, m$. Each residual $r_i(t)$ in Eq. (3.111) is ideally sensitive to one particular fault in the system. These residual functions can be obtained by both data-driven and model-based approaches, as described in the previous sections. In practice, however, as a consequence of noise and disturbances, residuals can be sensitive to more than one fault. To take into account the sensitivity of residuals to various faults and noise, in the scheme of Fig. 3.14, an NF classifier is exploited. In this case an NF network is used which processes the residuals to indicate the fault. This NF model is constructed with following set of rules:

$$\textbf{If } r_1 \text{ is small } \ldots r_j \text{ is large } r_m \text{ is small } \textbf{then } \text{fault}_r \text{ is large.} \tag{3.112}$$

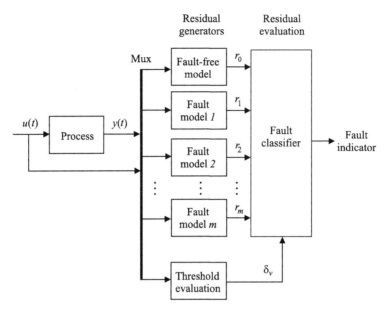

**FIGURE 3.14** FDI scheme based on different residual generation and evaluation approaches.

As alternative to geometric methods for residual evaluation, Eq. (3.113) employs a fuzzy threshold $\delta_v$ in order to take into account the imprecision of the residual generator at different regions in the input space:

$$\delta_v(u) = \frac{\sum_{i=1}^{C} \delta_i \, \eta_i(u)}{\sum_{i=1}^{C} \eta_i(u)}, \tag{3.113}$$

where $C$ is the total number of regions (or clusters) with different sensitivity to faults and a multidimensional fuzzy set $\eta_i$ defines the fuzzy boundary of $i$th such region. This approach depends heavily on the availability of the faulty and fault-free data and it is more difficult to isolate faults that appear in the dynamics.

Residuals $r_i$ can also be generated by nonlinear dynamic prototypes of the plant that approximate the nonlinear process under diagnosis. These models can be obtained by means of model-based or data-driven techniques, as described in the previous sections. In the case of fuzzy descriptions they can be derived via, e.g., product space clustering (Babuška, 1998) or tree-like (LOLIMOT) algorithms (Nelles and Isermann, 1996). Each local model is a linear approximation of the process in a subspace and the selection of the local model is fuzzy. The output of such a model can be described by

$$y(t) = \frac{\sum_{i=1}^{C} \mu_i(x(t)) \, y_i(t)}{\sum_{i=1}^{C} \mu_i(u_s)}, \tag{3.114}$$

where $x(t)$ is a suitable combination of the input and output signals, whilst $y_i(t)$ is the $i$th output of the local linear (or affine) model given by

$$y_i(t) = \sum_{k=1}^{n} b_{i,k} \, u(t-k) + \sum_{k=1}^{n} a_{i,k} \, y(t-k) + c_i, \tag{3.115}$$

with $a_{i,k}$, $b_{i,k}$, and $c_i$ being the parameters of the $i$th model, and $x(t)$ subspace defines the operating point, $\mu_i$ is the degree to which the $i$th local model is valid at this operating point.

From $a_{i,k}$, $b_{i,k}$ and $c_i$, physical parameters like time constants, static gains, offsets, etc. (Füssel et al., 1997), can be extracted for each operating point and can be compared with the parameters estimated online, as recalled in Section 3.3.1. This approach heavily depends on the accuracy of the nonlinear dynamic model described above. Also the output error should be minimum when operated in parallel to the system. Moreover, this method requires that there is sufficient excitation at each operating point for online estimation of parameters. This residual generation scheme for FDI is sketched in Fig. 3.15.

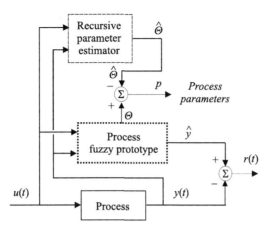

**FIGURE 3.15** Residual generation scheme with parameter estimation for FDI.

## 3.4 ROBUST RESIDUAL GENERATION ISSUES

Although the analytical redundancy method for residual generation has been recognized as an effective technique for detecting and isolating faults, the critical problem of unavoidable modeling uncertainty has not been fully solved. The main problem regarding the reliability of FDI schemes is the modeling uncertainty which is due, for example, to process noise, parameter variations, and nonlinearities.

On the other hand, all model-based methods use a model of the monitored system to produce the symptom generator. If the system is not complex and can be described accurately by the mathematical model, FDI is directly performed by using a simple geometrical analysis of residuals. In real industrial systems, however, modeling uncertainty is unavoidable. The design of an effective and reliable FDI scheme for residual generation should take into account of modeling uncertainty with respect to the sensitivity of the faults. Therefore, the task of the design of an FDI system is thus to generate residuals which are *robust* and *reliable* (Frank, 1994; Frank and Ding, 1997; Patton and Chen, 1994). Several papers addressed this problem. For example, optimal robust parity relations were proposed by Chow and Willsky (1984), Chung and Speyer (1998), Speyer (1999), Lou et al. (1986) and the threshold selector concept was introduced by Emami-Naeini et al. (1988). Robust FDI using the disturbance decoupling technique was also used by Patton and Chen (1994), Chen et al. (1996). In Patton and Chen (1994), Chen et al. (1996) this approach represents an interesting contrast to the method proposed in Chow and Willsky (1984) which seems to minimize the modeling uncertainty over several points of operation. In Patton and Chen (1994), Chen et al. (1996) this problem was solved by estimating the optimum unknown input distribution matrix over a range of operating points and exploiting the eigenstructure assignment approach (Patton and Chen, 1994; Chen and Patton, 2012).

The model-based FDI technique requires a high accuracy mathematical description of the monitored system. The better the model represents the dynamic behavior of the system, the better will be the FDI precision. If an FDI method can be developed which is insensitive to modeling uncertainty, a very accurate model is not necessarily needed. All uncertainties can be are summarized as disturbances acting on the system. Although the disturbance vector is unknown, its distribution matrix can be obtained by an identification procedure. Under this assumption, the "disturbance decoupling" principle can be exploited to design a robust FDI scheme.

In order to summarize the approach to the robustness and reliability problems, the state-space model of the monitored system should be considered (Patton and Chen, 1993):

$$\begin{cases} \dot{x}(t) = (A + \Delta A)\,x(t) + (B + \Delta B)\,u(t) + E_1\varepsilon(t) + R_1 f(t), \\ y(t) = Cx(t) + E_2\varepsilon(t) + R_2 f(t), \end{cases} \tag{3.116}$$

where $\varepsilon(t)$ is the disturbance vector, and $E_1$ and $E_2$ are the known or unknown input distribution matrices. The matrices $\Delta A$ and $\Delta B$ are the parameter errors or variations which represent modeling errors.

The transfer matrix description between the output $y(t)$ and input $u(t)$ of the system of Eq. (3.116) is then represented in the following form:

$$y(s) = (G_u(s) + \Delta G_u(s))\,u(s) + G_\varepsilon(s)\varepsilon(s) + G_f(s)\,f(s), \tag{3.117}$$

where $\Delta G_u(s)$ is used to describe modeling errors, whilst both $\Delta G_u(s)$ and $G_\varepsilon(s)$ represent modeling uncertainty.

With reference to the residual generator of Eq. (3.4) and described by the relation of Eq. (3.15), the $s$-domain residual vector has to be rewritten in the following form:

$$r(s) = H_y(s) G_f(s) f(s) + H_y(s) G_\varepsilon(s) \varepsilon(s) + H_y(s) \Delta G_u(s) u(s). \tag{3.118}$$

With respect to the relation of Eq. (3.15), the terms $H_y(s)G_\varepsilon(s)$ and $H_y(s)\Delta G_u(s)$ cannot be deleted. Both faults and modeling uncertainty (disturbance and modeling error) affect the residual, and hence discrimination between these two effects is difficult.

The principle of disturbance decoupling for robust and reliable residual generation requires that the residual generator satisfies the following condition:

$$H_y(s)G_\varepsilon(s) = 0 \tag{3.119}$$

in order to achieve total decoupling between residual $r(s)$ and disturbance $\varepsilon(s)$. This property can be achieved by using the unknown input observer (Chen et al., 1996; Frank et al., 2000), optimal (robust) parity relations (Chow and Willsky, 1984; Frank et al., 2000) or alternatively the eigenstructure assignment approach (Liu and Patton, 1998; Patton and Chen, 2000; Duan et al., 2002).

Hence, for disturbance decoupling approaches in FDI, the aim is to completely eliminate the disturbance effect from the residual. However, the complete elimination of disturbance effects may not be possible due to the lack of degrees of freedom. Moreover, it may be problematic in some cases because the fault effect may also be eliminated. Hence an appropriate criterion for robust residual design should take into account the effects of both modeling error and faults. There is a trade-off between sensitivity to faults and robustness to modeling uncertainty and hence robust residual generation can be considered as a *multiobjective optimization problem* (Chen and Patton, 2012). It consists of the maximization of fault effects and the minimization of uncertainty effects. Therefore the approach to the design of optimal residuals can require the satisfaction of a set of objectives. These objectives are essential for achieving robust diagnosis of incipient faults. If such joint optimization problems, which can be also expressed in the frequency domain, were reformulated for satisfying a set of inequalities on the performance indices, Genetic Algorithms (GA) (Davis, 1991) and Linear Matrix Inequalities (LMI) (Boyd et al., 1994) can be successfully exploited to search the optimal solution (Chen and Patton, 2012, 2001).

Disturbance decoupling can also be achieved using frequency domain design techniques. As an example, the robust fault detection problem can be managed by using the standard $H_\infty$ filtering formulation (Hou and Patton, 1996; Frank and Ding, 1997). With this method, the minimization of the disturbance effect on the residual is formulated as a standard $H_\infty$ filtering problem (Chen and Patton, 2000; Frank et al., 2000). On the other hand, the so-called $H_\infty/H_-$ approach can be also exploited (Chen and Patton, 2000).

Among the many ways for eliminating or minimizing disturbance and modeling error effects on the residual and hence for achieving robustness in FDI (Patton et al., 2000) $H_\infty$ optimization is a robust design method with the original motivation firmly rooted in the consideration of various uncertainties, especially the modeling errors. It is reasonable to seek an application of this technique in the robust design of FDI systems for wind turbine systems. Therefore, the $H_\infty$ optimization method can be successfully exploited for robust residual generation of FDI.

The early work of using $H_\infty$ optimization techniques for robust FDI was based on the use of factorization approach (Ding and Frank, 1990; Ding et al., 2000). The factorization-based $H_\infty$ optimization technique is useful in solving FDI problems. However, the more elegant and advanced $H_\infty$ optimization methods are based on the use of the Algebraic Riccati Equation (ARE) (Zhou et al., 1996a). Mangoubi et al. (1992) first solved the robust FDI estimation problem using the ARE approach via the use of $H_\infty$ and $\mu$ robust estimator synthesis methods developed by Appleby et al. (1991). A direct formulation of the FDI problem as a robust $H_\infty$ filter design problem with the solution of an ARE was given in Edelmayer et al. (1997). To deal with modeling errors as well as disturbances in robust FDI design, in Niemann and Stoustrup (1996) it was introduced modeling error blocks into the standard $H_\infty$ observer design. The weighting factors are then introduced in the problem formulation for finding an optimal FDI solution. This is further extended to nonlinear systems where the nonlinearity is treated in the same way as a modeling error block (Stoustrup and Niemann, 1998; Stoustrup et al., 1997).

The majority of studies discussed so far involve the use of a slightly modified $H_\infty$ filter for the residual generation, i.e., the design objective is to minimize the effect of disturbances and modeling errors on the estimation error and subsequently on the residual. However, robust residual generation is different from the robust estimation because it does not only require the disturbance attenuation. The residual has to remain sensitive to faults whilst the effect of disturbance is minimized. In Sauter et al. (1997) this problem was studied, where the fault sensitivity was enhanced by applying an optimal postfilter to the "primary residual." The problem of enhancing fault sensitivity while increasing robustness against disturbances and modeling errors was studied extensively in Sadrnia et al. (1997). The essential idea is to reach an acceptable compromise between disturbance robustness and fault sensitivity. In the beginning, an observer with very small disturbance sensitivity bound is designed via an ARE. Then the fault sensitivity is checked. If the fault sensitivity is too small, the disturbance robustness requirement should be relaxed, i.e., to design another optimal observer with an increased disturbance sensitivity bound. This procedure is likely to be repeated several times. The final goal is to find a design which provides the maximum ratio between fault sensitivity and disturbance sensitivity.

In Chen and Patton (2012, 2000) a robust residual generation problem within the standard $H_\infty$ filtering framework was formulated, i.e., the problem of generating the residual whose sensitivity to disturbances is minimized. To facilitate reliable FDI, the residual sensitivity to faults has to be maintained (or maximized) in addition to the minimization of the disturbance sensitivity. This problem was solved via the minimization of the difference between the residual and the fault against the disturbance and the fault, i.e., the objective was to replicate the fault using the residual. In this way, the residual sensitivity to the fault was indirectly maximized. The residual sensitivity to the modeling error can be minimized if the modeling error is approximately represented by the disturbance vector with the estimated distribution matrix (Chen and Patton, 2012). However, the modeling error can be handled directly using standard $H_\infty$. In Chen and Patton (2012) the way of including the modeling error in the robust residual design within the standard $H_\infty$ framework was shown.

Generally speaking, the robust FDI approach can be approached in different ways. It is therefore important to mention the design principle of residual generators under a certain performance index (Basseville, 1997; Frank et al., 2000). This is indeed a reasonable extension of the unknown input residual generator design, in which, instead of full decoupling, a compromise between the robustness and sensitivity is made. It is worth focusing the attention to this scheme, due to its important role in theoretical studies and its relationship to the residual evaluation and integrated design of FDI systems. Since the goal of residual generation is to enhance the robustness of the residual to the model uncertainty without loss of its sensitivity to the faults, the minimization of performance index in the form of (Frank et al., 2000)

$$J = \frac{\|\frac{\partial r}{\partial d}\|}{\|\frac{\partial r}{\partial f}\|} \text{ or } J = \|\frac{\partial r}{\partial d}\| \text{ with } \|\frac{\partial r}{\partial f}\| > \alpha \tag{3.120}$$

is widely recognized as a suitable design objective. Associated to the norm used, the type of the residual generator and the mathematical tool adopted, a number of optimization approaches have been developed (Frank et al., 2000). In Ding et al. (2000) a unified solution for a number of optimization problems was originally derived and thus provided a satisfactory solution to the above-defined optimization problem ten years after it was first proposed. In Frank et al. (2000) a brief review of the state-of-art of the solutions can be found, whilst Hou and Patton (1996, 1997), Frank et al. (2000) firstly introduced the $H_\infty/H_-$ method. According to the norm selected, by minimizing the performance index of Eq. (3.120) over a specified range, an approximate decoupling design can be achieved (Ding and Frank, 1990; Patton and Hou, 1997; Frank and Ding, 1997; Ding et al., 1999). Moreover, the approximated design for optimal disturbance decoupling can also be solved in the time domain (Wünnenberg, 1990; Chen et al., 1993). On the other hand, with reference to the modeling errors in Eq. (3.118) represented by the term $\Delta G_u(s)$, the robust problem is more difficult to solve.

Two main techniques have been described in Patton and Chen (1994), Chen et al. (1996). In the first case the uncertainty is taken into account at the residual design stage (Chen et al., 1996); this is known as *active robustness* in fault diagnosis (Patton and Chen, 1994). The active way of achieving a robust solution is to approximate uncertainties, i.e., representing approximate modeling errors as disturbances (Chen and Patton, 2012)

$$\Delta G_u(s) u(s) \approx G_d(s) d(s), \tag{3.121}$$

where $d(s)$ is an unknown vector and $G_d(s)$ is an estimated transfer function. When this approximate structure is exploited to design disturbance decoupling residual generators, robust FDI can be achieved. An active approach applied to the robust AFTC will be addressed in Chapter 4.

The second approach called *passive robustness* makes use of a residual evaluator with adaptive threshold. As a simple example, it is assumed that the residual generation uncertainty of Eq. (3.118) is only represented by modeling errors. The fault-free residual $r(s)$ is represented in the form

$$r(s) = H_y(s) \Delta G_u(s) u(s). \tag{3.122}$$

Under the assumption that the modeling errors are bounded by a value $\delta$ such that the following relation holds:

$$\| \Delta G_u(w) \| \le \delta, \tag{3.123}$$

an adaptive threshold $\varepsilon(t)$ can be generated by a linear system

$$\varepsilon(t) = \delta H_y(s) \, u(s). \tag{3.124}$$

In such case, the threshold $\varepsilon(t)$ is no longer fixed but depends on the input $u(t)$, thus being adaptive to the system operating point. A fault is then detected if the following relation holds:

$$\| \, r(t) \, \| > \| \, \varepsilon(t) \, \| \, . \tag{3.125}$$

A robust FDI technique with the threshold adaptor or selector was originally presented in Clark (1989), Emami-Naeini et al. (1988). This method represents a passive approach since no effort is made to design a robust residual.

Even if disturbance decoupling methods for robust FDI has been studied extensively, their effectiveness regarding real problems has not been fully demonstrated. The main difficulty arises as most of the disturbance only account for a small percentage of the uncertainty in the real system. The presented disturbance decoupling methods cannot be directly applied to the systems with other uncertainties such as modeling errors.

The estimation and approximate representation of modeling errors, as well as other uncertain factors, as the disturbance term provides a practical way to tackle the robustness issue for technical processes, such as the wind turbine systems, as shown in Chapters 4 and 5.

Some concluding remarks can be finally be drawn here. This chapter provided some theoretical study results for the detection and diagnosis of faults in the actuators, systems, and sensors of wind turbine systems through the use of different FDD schemes.

Residual generators were designed from the linear and nonlinear input–output descriptions of the system under diagnosis, and the disturbance decoupling was obtained. Procedures for optimizing the residual generator fault sensitivity and dynamic response were also suggested.

An important aspect of the strategies based on linear residual generators is the simplicity of the technique used to generate these residuals when compared with different schemes. The algorithmic simplicity is a very important aspect when considering the need for verification and validation of demonstrable schemes that will be applied to ream processes. The more complex the computations required to implement the scheme, the higher the cost and complexity in terms of verification and validation.

On the other hand, nonlinear methodologies rely on a design scheme based on the structural decoupling of the disturbance obtained by means of a coordinate transformation in the state space and in the output space. To apply the nonlinear theory, simplified models of the system under investigation can be required. The mixed $\mathcal{H}_- / \mathcal{H}_\infty$ optimization of the trade-off between fault sensitivity, disturbances and modeling errors is now well understood in the theoretical work and is a promising area for application study to wind turbine systems.

The nonlinear fault diagnosis strategies were based also on adaptive filter schemes. In addition to a proper detection and isolation, these methods provided also a fault size estimation. This feature is not usual for a fault detection and isolation method and can be fundamental for the online automatic control system reconfiguration in order to recover a faulty operating condition. Compared with similar methods proposed in the literature, the nonlinear adaptive fault diagnosis technique described here has the advantage of being applicable to more general classes of nonlinear systems and less sensitive to measurement noise, since it does not use input–output signal derivatives.

Suitable filtering algorithms for stochastic systems were also recalled. The knowledge regarding the noise process acting on the system under diagnosis can be exploited by the fault diagnosis method design, hence the proposed scheme provides a possible solution to nonlinear system diagnosis with non-Gaussian noise and disturbance.

The main advantage of nonlinear-based FDD techniques with disturbance decoupling features is represented by the fact that they take into account directly the model nonlinearity and the system reality–model mismatch.

The fault diagnosis techniques that have been outlined in this chapter will be applied to high-fidelity wind turbine simulators, which is able to take into account disturbances and measurement errors acting on the system under investigation. Moreover, the robustness characteristics and the achievable performances of the fault diagnosis approaches described have been carefully considered and investigated.

The effectiveness of the proposed diagnosis schemes was shown by simulations and a comparison with widely used data-driven and model-based FDI and FDD scheme with disturbance decoupling. The reliability and the robustness properties of the designed residual generators to model uncertainty, disturbances and measurements noise will be analyzed via extensive simulations, including the use of Monte Carlo simulation experiments to tune the FDI and FDD parameters.

Finally, the need to bridge the design gap between FDD and recovery mechanisms, i.e., the sustainable schemes is obvious and addressed in Chapter 4, where it is shown how fault diagnosis and fault tolerant control strategies can be properly combined.

## 3.5   SUMMARY

This chapter presented a tutorial treatment on the basis principles of model-based and data-driven FDI with application to technical processes.

The FDI problem has been formalized in a uniform framework by presenting a mathematical description and definition. Within this framework, the residual generation has been identified as a central issue in both model-based and data-driven FDI. By choosing the proper design approach, the FDI task can be performed.

The residual generator was summarized in different residual generation structures. The ways of designing residuals for fault isolation were also discussed. The most commonly used residual generation strategies were recalled by presenting related problems and discussing the applicability of model-based and data-driven FDI methods to dynamic processes such as wind turbine systems.

It is worth noting that the success of fault diagnosis depends on the quality of the residuals. Successful diagnosis requires residual signals which are robust with respect to modeling uncertainty. The robust FDI problem was also discussed in this chapter, and the implementation of robust residual generators will be shown in the following chapters of the book.

Finally, data-driven FDI methods such as fuzzy logic, qualitative modeling, and neural networks were also considered, and the concept of integrated knowledge-based fault diagnosis, utilizing both analytical and heuristic information, was illustrated.

# Chapter 4

# Fault Tolerant Control for Wind Turbine Systems

## 4.1 INTRODUCTION

As remarked in Chapter 3, model-based and data-driven FDI strategies have been studied for over 40 years; however, they still represent an open research domain when considered to be applied to wind turbine systems, and many problems are waiting to be solved. The material presented in this monograph has inevitably had to end before all the interesting topics for future FDI research could be fully explored. In the following sections the authors describe some important topics that should help the reader understand how to move from fault diagnosis to fault tolerance.

Moreover, this chapter presents the fault tolerant control algorithms applied to wind turbine systems. In general, they are based on the signal correction principle, which means that the control system is not modified since the inputs and outputs of the baseline controller are compensated according to the estimated faults. The fault tolerant control algorithms recalled in this chapter rely on the fault diagnosis design addressed in Chapter 3. Passive and active fault tolerant control systems are also discussed and compared, in order to highlight the achievable performances and the complexity of their design procedures. Controller reconfiguration mechanisms are also considered, which are able to guarantee the system stability and satisfactory performance.

As described in Chapter 3, there are different approaches for eliminating or minimizing disturbances and modeling error effects on residuals and hence for achieving robustness in fault diagnosis. However, these techniques were developed for ideal systems or with a special uncertainty structure, and then efforts have been made to include nonideal or more general uncertainty.

In contrast, frequency domain design methods are designed to possess robustness properties. In particular, $H_\infty$ optimization has been developed from the very beginning with the understanding that no design goal of a system can be perfectly achieved without being compromised by an optimization in the presence of uncertainty, hence this technique is very suitable for tackling uncertainty issues.

It is worth noting that Patton et al. (1986) first discussed the possibility of using frequency domain information to design fault diagnosis algorithms. The design of a residual generator in the frequency domain was firstly based on a frequency domain optimal observer and then by using the factorization of the transfer function matrix of the monitored system. These methods were developed and later extended in Ding and Frank (1990). Some important modifications in robust FDI design were made in Gertler (1998) by using the factorization-based $H_\infty$ optimization technique. More elegant and advanced $H_\infty$ optimization methods are based mainly on the use of the Algebraic Riccati Equations (ARE). In particular, the robust fault estimation problem was solved by using Riccati equation approach through the use of $H_\infty$ and $\mu$ robust estimator synthesis methods (Chen and Patton, 2012). These approaches were further extended to time-variant and nonlinear systems.

The majority of studies considered in the related literature involved the use of a slightly modified $H_\infty$ filter for residual generation. That is to say, the design objective was to minimize the effect of disturbances and modeling errors on the estimation error and subsequently on the residual. The residual had to remain sensitive to faults whilst the effect of disturbances had to be minimized. Hence, the essential idea was to reach an acceptable compromise between disturbance robustness and fault sensitivity. The final goal was to find an observer design which provided the maximum ratio between fault sensitivity and disturbance sensitivity in the form

$$J - \sup_{Q(s)} \frac{\| Q(s)\, G_f(s) \|_\infty}{\| Q(s)\, G_d(s) \|_\infty} \tag{4.1}$$

over a frequency range, with $Q(s)\, G_f(s)$ being the transfer matrix between the residual and fault, whilst $Q(s)G_d(s)$ being the transfer matrix between the residual and disturbance.

Solutions for this optimization problem were given and revised, in order to obtain a robust FDI technique (Chen and Patton, 2012). Unfortunately, it was shown that $\| Q(s)\, G_f(s) \|_\infty$ may be smaller than $\| Q(s)\, G_d(s) \|_\infty$ in a certain frequency range even if their ratio in Eq. (4.1) had been maximized.

Fault Diagnosis and Sustainable Control of Wind Turbines. DOI: 10.1016/B978-0-12-812984-5.00004-3

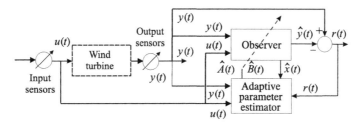

**FIGURE 4.1** Residual generator with adaptive observer.

It should be pointed out that the transfer function matrix $G_d(s)$ can only be defined for disturbances, hence the technique presented can only deal with robustness against disturbances. The robust problem with respect to modeling errors has still not been solved. The only solution suggested is to calculate the residual bound and set an adaptive threshold.

Little progress was made in solving the robust fault diagnosis problem against modeling errors when $\mu$ synthesis with $H_\infty$ optimization is incorporated. Robust FDI design based on $H_\infty$ optimization and $\mu$ synthesis is still in its early development, and some research is still needed. This could be a direction for future research which has great potential.

In connection with frequency domain, fault diagnosis techniques can exploit different data-driven approaches compared to the one presented in Chapter 3. For example, an identification method based on the frequency domain approach for Errors-In-Variable models and its application to the dynamic Frisch scheme estimation technique were presented in Beghelli et al. (1994a, 1997). Such a procedure can provide an accurate estimation of the transfer matrices $Q(s)$, $G_f(s)$, and $G_d(s)$ from input–output measurements affected by white, mutually uncorrelated, and correlated noises.

This general method, using the frequency domain approach, facilitates a unique determination of both the characteristics of the noise affecting the data ($G_d(s)$) and the transfer matrices ($Q(s)$ and $G_f(s)$) of the process under investigation. A comparison between time-domain and frequency-domain approaches can be found in Beghelli et al. (1994b).

As already highlighted in Chapter 3, the system dynamics and parameters may vary or may be perturbed during the system operation. A fault diagnosis system designed for a system model corresponding to nominal system operation may not perform well when applied to the system with perturbed conditions.

To overcome this problem, instead of using complex nonlinear models, as shown in Chapter 3, a residual generator scheme using adaptive observers was proposed. The idea is to estimate and compensate system parameter variations. Fig. 4.1 illustrates the basic principle of this approach. It can be applied to linear systems with parametric variations if stability and convergence conditions are satisfied.

As described in Chapter 3, adaptive residual generation schemes for both linear and nonlinear uncertain dynamic systems using adaptive observers were proposed in the literature (Patton et al., 1989a). Unfortunately, the disadvantage of this approach is the complexity.

In Chen and Patton (2012) an alternative way was presented to generate adaptive symptoms using a method to estimate the bias term in the residuals due to modeling errors, then to compensate it adaptively. This technique decreases the effects of uncertainties on residuals. The approach to estimate such a bias term in residuals rather than computing modeling errors themselves avoids complicated estimation algorithms.

The simultaneous estimation algorithms of the state $\hat{x}(t)$ and parameters $\hat{A}(t)$ and $\hat{B}(t)$, presented in Söderström and Stoica (1987), Ljung (1999), can be also used to generate adaptive residuals. With reference to Fig. 4.1, observer parameters are linked by the relations

$$\begin{cases} \hat{x}(t+1) = \hat{A}(t)\hat{x}(t) + \hat{B}(t)u(t), \\ \hat{y}(t) = C\hat{x}(t). \end{cases} \tag{4.2}$$

An adaptive residual generation algorithm for fault diagnosis normally involves both the state $\hat{x}(t)$ and parameters $\hat{A}(t)$ and $\hat{B}(t)$ estimation and can be considered as a combination of observer and identification based FDI approaches. Hence complementary advantages in both approaches can be gained.

For all adaptive methods, the main problem to be tackled is that fault effects may be compensated, as well as modeling errors and parameter variations. This makes the detection for incipient faults almost impossible, whilst for abrupt faults this can be acceptable. To overcome this problem, the effect of faults can be considered as a slow varying parameter which can be estimated together with other parameters. Under the assumption that parameters and faults are varying at different rates, two filters with different gains can be used. However, much research effort is still needed in the theory and application of adaptive residual generation methods.

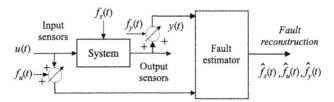

**FIGURE 4.2** Fault estimation scheme for fault tolerant control.

### 4.1.1 Integration of Fault Diagnosis and Control

A conventional feedback control design for complex systems may result in unsatisfactory performance in the event of malfunction in input–output sensors, actuators, and system components. A fault tolerant closed-loop control system is very attractive because it can tolerate faults whilst also maintaining desirable performance.

The conventional approach to the design of a fault-tolerant control includes different steps and separate modules: modeling or identification of the controlled system, design of the controller, FDI scheme, and a method for reconfiguring the control system. Identification and design of the controller can be performed separately or by using combined methods. Hence the FDI and controller are linked through the reconfiguration module. The fundamental problem with such a system lies in the identification stage, in the independent design of the control and FDI modules. Significant interactions occurring among these modules can be neglected. There is therefore a need for a research study into the interactions between system identification, control design, the FDI stage, and the fault-tolerant control design strategy.

Fault identification is the most important of all the fault diagnosis tasks. When a fault is estimated, detection and isolation can be easily achieved since the fault nature can improve the diagnosis process. However, the fault identification problem itself has not gained enough research attention.

Most fault diagnosis techniques, such as parameter identification, parity space and observer-based methods, cannot be directly used to identify faults in sensors and actuators. Very little research has been done to overcome the fault identification problem. The Kalman filter for statistical testing and fault identification was proposed in Patton et al. (1989a). However, the statistical testing methods can impose a high computational demand. A fault identification scheme solving a system inversion problem was proposed in Simani et al. (1998), Chen and Patton (2012), Simani et al. (1999d).

In the scheme depicted in Fig. 4.2 fault identification is performed by estimating the nonlinear relationship between residuals and fault magnitudes. This is possible because robust residuals should only contain fault information.

Such a nonlinear function approximation and estimation can be performed by using neural networks or an inversion of the transfer matrix between residuals and faults.

The central task in model-based fault detection is the residual generation. Most residual generation techniques are based on linear system models. For nonlinear systems, the traditional approach is to linearize the model around the system operating point. However, for systems with high nonlinearity and a wide dynamic operating range, the linearized approach fails to give satisfactory results.

One solution is to use a large number of linearized models corresponding to a range of operating points. This means that a large number of FDI schemes corresponding to each operating points is needed. Hence it is important to study residual generation techniques which tackle nonlinear dynamic systems directly. There are some research studies on the residual generation of nonlinear dynamic systems, for example, using nonlinear observers (Frank et al., 2000; De Persis and Isidori, 2001; Chen and Patton, 2012). There have been some attempts to use nonlinear observers to solve nonlinear system FDI problem (De Persis and Isidori, 2001; Chen and Patton, 2012), e.g., nonlinear unknown input observers, including adaptive observers and sliding mode observers. If the class of nonlinearities can be restricted, observers for bilinear systems were also proposed (Chen and Patton, 2012).

On the other hand, the analytical models, on which the nonlinear observer approaches are based, are not easy to obtain in practice. Sometimes it is impossible to model the system using an explicit mathematical model. To overcome this problem, it is desirable to find a universal approximate model which can be used to represent the real system with an arbitrary degree of accuracy. Different approaches were proposed and they are currently under investigation: neural networks, fuzzy models, and hybrid models.

As recalled in Chapter 3, neural networks are a powerful tool of handling nonlinear problems. One of the most important advantages of neural networks is their ability to implement nonlinear transformations for functional approximation problems (Schilling et al., 2001). Therefore neural networks can be used in a number of ways to tackle fault diagnosis problems for nonlinear dynamic systems. In early publications they were mainly exploited as fault classifier with steady state processes, whereas neural networks have been used as residual generators and for modeling nonlinear dynamic systems for FDI purposes (Chen and Patton, 2012).

Fuzzy models can be used both as residual classifier (Sneider and Frank, 1996) and as nonlinear system parametric model (Ying, 1994). In the second case the main idea is to build an FDI scheme based on fuzzy observers. Estimated outputs and residuals are computed as fuzzy fusion of local observer output and residuals. The main problem of this approach concerns the stability of the global observer. A linear matrix inequality method was proposed in Chen and Patton (2012) using Lyapunov theorem, but this solution can be quite conservative.

Hybrid models can describe the behavior of any nonlinear dynamic process if they are described as a composition of several local affine models selected according to the process operating conditions (Special Issue on Hybrid Control Systems, 1998). Instead of exploiting complicated nonlinear models obtained by modeling techniques, it is possible to describe the plant by a collection of affine models. Such a compound system requires the identification of the local models from data. Several works (Rovatti et al., 1998; Simani et al., 1999b, 1999a; Fantuzzi et al., 2003) addressed a method for the identification and the optimal selection of the local affine models from a sequence of noisy measurements acquired from the process. Application of these results to model-based fault diagnosis for wind turbine systems is another research area worth mentioning.

### 4.1.2 Nonlinear Adaptive Filters for Fault Estimation

As described in Section 3.2.4, the main point of proposed NLGA scheme for FDI consists in achieving the structural decoupling of critical disturbances and critical modeling errors. However, the nonlinear residual generators robustness can be improved by minimizing the effects of both noncritical disturbances and modeling errors, which are not decoupled, whilst maximizing the fault effects on the residual signals.

In order to apply the robustness improvement procedure presented in this section, the considered procedure is restricted to suitable scalar components of the $\bar{x}_1$-subsystem of Eq. (3.71). In particular, the vectors $\bar{x}_1$ and $\bar{y}_1$ are decomposed as

$$\bar{x}_1 = \begin{bmatrix} \bar{x}_{11} \\ \bar{x}_{1c} \end{bmatrix}, \qquad \bar{y}_1 = \begin{bmatrix} \bar{y}_{11} \\ \bar{y}_{1c} \end{bmatrix} \tag{4.3}$$

where $\bar{x}_{11} \in \Re$, $\bar{y}_{11} \in \Re$, and correspondingly it follows that

$$n_1(\cdot) = \begin{bmatrix} n_{11}(\cdot) \\ n_{1c}(\cdot) \end{bmatrix}, \qquad g_1(\cdot) = \begin{bmatrix} g_{11}(\cdot) \\ g_{1c}(\cdot) \end{bmatrix} \qquad \ell_1(\cdot) = \begin{bmatrix} \ell_{11}(\cdot) \\ \ell_{1c}(\cdot) \end{bmatrix}. \tag{4.4}$$

The following conditions are considered:

$$\bar{y}_{11} = h_{11}(\bar{x}_{11}), \qquad \bar{y}_{1c} = h_{1c}(\bar{x}_{1c}), \qquad \ell_{11}(\cdot) \neq 0, \tag{4.5}$$

where $h_{11}(\cdot)$ is a smooth map and $h_{1c}(\cdot)$ is an invertible smooth map.

It is important to highlight that if the constraints of Eq. (4.5) are satisfied, the decomposition of Eqs. (4.3) and (4.4) can always be applied to obtain the $\bar{x}_{11}$-subsystem in the form

$$\begin{cases} \dot{\bar{x}}_{11} = n_{11}(\bar{x}_{11}, \bar{y}_{1c}, \bar{y}_2) + g_{11}(\bar{x}_{11}, \bar{y}_{1c}, \bar{y}_2)c + \ell_{11}(\bar{x}_{11}, \bar{y}_{1c}, \bar{y}_2, \bar{x}_3)f, \\ \bar{y}_{11} = h_{11}(\bar{x}_{11}). \end{cases} \tag{4.6}$$

It can be shown that the conditions of Eq. (4.5) are satisfied for the considered application to wind turbines. Therefore the scalar $\bar{x}_{11}$-subsystem of Eq. (4.6) is referred to in place of the $\bar{x}_1$-subsystem of Eq. (3.71).

It can be noted that the tuning of the residual generator gains, in the framework of the $\bar{x}_{11}$-subsystem of Eq. (4.6), cannot be properly carried out. In fact, the critical disturbances are structurally decoupled, but the noncritical ones are not considered. For this reason, in order to achieve robustness of the residual generators, the tuning of the gains is performed by embedding the description of the noncritical disturbances in the $\bar{x}_{11}$-subsystem of Eqs. (4.7) as follows:

$$\begin{cases} \dot{\bar{x}}_{11} = n_{11}(\bar{x}_{11}, \bar{y}_{1c}, \bar{y}_2) + g_{11}(\bar{x}_{11}, \bar{y}_{1c}, \bar{y}_2)c + \ell_{11}(\bar{x}_{11}, \bar{y}_{1c}, \bar{y}_2, \bar{x}_3)f + e(\bar{x}_{11}, \bar{y}_{1c}, \bar{y}_2, \bar{x}_3)\zeta, \\ \bar{y}_{11} = \bar{x}_{11} + \nu, \end{cases} \tag{4.7}$$

where, according to the considered wind turbine applications, in order to simplify the treatment without loss of generality, the state variable $\bar{x}_{11}$ is supposed to be directly measured. Moreover, the variable $v \in \Re$ is the measurement noise on $\bar{x}_{11}$. Finally, the variable $\zeta \in \Re$ and the related scalar field $e(\cdot)$ represent the noncritical effects which have not been considered in the wind turbine model used for the NLGA scheme, as shown in Section 3.2.4.

The system of Eq. (4.8) is referred to as filter form, and it represents a generic scalar residual generator based on the subsystem of Eq. (4.7). It is worth noting that the residual generators designed in Chapter 5 belong to this class of systems as a particular case:

$$\begin{cases} \dot{\xi}_f = n_{11}(\bar{y}_{11}, \bar{y}_{1c}, \bar{y}_2) + g_{11}(\bar{y}_{11}, \bar{y}_{1c}, \bar{y}_2)c + k_f \left( \bar{y}_{11} - \xi_f \right), \\ r_f = \bar{y}_{11} - \xi_f, \end{cases} \tag{4.8}$$

where the gain $k_f$ has to be tuned in order to minimize the effects of the disturbances $\zeta$ and $v$, whilst maximizing the effects of the fault $f$ on the residual $r_f$.

In order to quantify the effects of both the disturbances and the faults on the residual, the estimation error can be defined in the form

$$\tilde{x}_f = \bar{x}_{11} - \xi_f, \tag{4.9}$$

allowing one to write the following equivalent residual model in the form

$$\begin{cases} \dot{\tilde{x}}_f = n_{11}(\bar{x}_{11}, \bar{y}_{1c}, \bar{y}_2) - n_{11}(\bar{y}_{11}, \bar{y}_{1c}, \bar{y}_2) + g_{11}(\bar{x}_{11}, \bar{y}_{1c}, \bar{y}_2)c - g_{11}(\bar{y}_{11}, \bar{y}_{1c}, \bar{y}_2)c + \\ \qquad + \ell_{11}(\bar{x}_{11}, \bar{y}_{1c}, \bar{y}_2, \bar{x}_3)f + e(\bar{x}_{11}, \bar{y}_{1c}, \bar{y}_2, \bar{x}_3)\zeta - k_f \tilde{x}_f - k_f v, \\ r_f = \tilde{x}_f + v. \end{cases} \tag{4.10}$$

As already remarked in Section 4.1, this problem formulation allows us to apply a mixed $\mathcal{H}_-/\mathcal{H}_\infty$ approach (Chen and Patton, 2012; Hou and Patton, 1996) for tuning the gain $k_f$. Therefore the system of Eq. (4.10) has to be linearized in the neighborhood of a steady-state working condition, with reference to the $\mathcal{H}_\infty$ optimization of nonlinear unknown input observers addressed in Amato et al. (2006) for the aerospace framework.

It is worth observing that the considered benchmark application is characterized by small excursions of the state, input and output variables with respect to their trim values $\bar{x}_{10}$, $\bar{x}_{30}$, $c_0$, $\bar{y}_{10}$, and $\bar{y}_{20}$, hence the robustness of the nonlinear residual generator is achieved. The linearization of the model of Eq. (4.10) has the form

$$\begin{cases} \dot{\tilde{x}}_f = -k_f \tilde{x}_f - k_f v + mf + \check{q}\check{\zeta}, \\ r_f = \tilde{x}_f + v, \end{cases} \tag{4.11}$$

where

$$\begin{cases} a' = \frac{\partial n_{11}(\cdot)}{\partial \bar{x}_{11}} \|_{(\bar{x}_{10}, \bar{y}_{20})}, \\ b = g_{11}(\cdot)\|_{(\bar{x}_{10}, \bar{y}_{20})}, \\ m = \ell_{11}(\cdot)\|_{(\bar{x}_{10}, \bar{y}_{20}, \bar{x}_{30})}, \\ q = e(\cdot)\|_{(\bar{x}_{10}, \bar{y}_{20}, \bar{x}_{30})}, \end{cases} \tag{4.12}$$

and

$$\check{q}\check{\zeta} = q\zeta - a'v. \tag{4.13}$$

It is worth noting that in place of the residual generators in the filter form of Eq. (4.8), the observer formulation in the following form is used:

$$\begin{cases} \dot{\xi}_o = n_{11}(\xi_o, \bar{y}_{1c}, \bar{y}_2) + g_{11}(\xi_o, \bar{y}_{1c}, \bar{y}_2)c + k_o \left( \bar{y}_{11} - \xi_o \right), \\ r_o = \bar{y}_{11} - \xi_o. \end{cases} \tag{4.14}$$

As previously remarked, the estimation error $\tilde{x}_o$ is introduced:

$$\tilde{x}_o = \bar{x}_{11} - \xi_o. \tag{4.15}$$

**TABLE 4.1** Observer and filter forms.

| | $\tilde{x}$ | $\varepsilon$ | $r$ | $a$ | $k$ | $e_{11}$ | $E_2$ |
|---|---|---|---|---|---|---|---|
| **General Form** | $\tilde{x}$ | $\varepsilon$ | $r$ | $a$ | $k$ | $e_{11}$ | $E_2$ |
| **Filter Form** | $\tilde{x}_f$ | $[\check{\zeta}\ v]^{\mathrm{T}}$ | $r_f$ | $0$ | $k_f$ | $\check{q}$ | $[0\ 1]$ |
| **Observer Form** | $\tilde{x}_o$ | $[\zeta\ v]^{\mathrm{T}}$ | $r_o$ | $a'$ | $k_o$ | $q$ | $[0\ 1]$ |

Hence

$$\begin{cases} \dot{\tilde{x}}_o = n_{11}(\bar{x}_{11}, \bar{y}_{1c}, \bar{y}_2) - n_{11}(\xi_o, \bar{y}_{1c}, \bar{y}_2) + g_{11}(\bar{x}_{11}, \bar{y}_{1c}, \bar{y}_2)\,c - g_{11}(\xi_o, \bar{y}_{1c}, \bar{y}_2)\,c + \\ \qquad + \ell_{11}(\bar{x}_{11}, \bar{y}_{1c}, \bar{y}_2, \bar{x}_3)\,f + e(\bar{x}_{11}, \bar{y}_{1c}, \bar{y}_2, \bar{x}_3)\,\zeta - k_o\,\tilde{x}_o - k_o\,v, \\ r_o = \tilde{x}_o + v. \end{cases} \quad (4.16)$$

In this way, by performing the linearization of the system of Eq. (4.16), we get

$$\begin{cases} \dot{\tilde{x}}_o = \left(a' - k_o\right)\tilde{x}_o - k_o\,v + mf + q\,\zeta, \\ r_o = \tilde{x}_o + v. \end{cases} \quad (4.17)$$

So both linearized models represented by Eqs. (4.11) and (4.17) of the residual generators in the filter and observer forms, respectively, can be represented in the general form

$$\begin{cases} \dot{\tilde{x}} = (a - k)\,\tilde{x} + (E_1 - kE_2)\,\varepsilon + m\,f, \\ r = \tilde{x} + E_2\,\varepsilon, \end{cases} \quad (4.18)$$

where $E_1 = [e_{11},\ 0]$, and by considering the relations reported in Table 4.1.

On the basis of Eq. (4.18) and Table 4.1, the mixed $\mathcal{H}_-/\mathcal{H}_\infty$ can be considered (Chen and Patton, 2012; Hou and Patton, 1996). Thus this approach is developed and applied for the robustness improvement of the residual generators, both in the filter and observer forms.

It is important to note that, since the considered NLGA residual generators are scalar, the $\mathcal{H}_-/\mathcal{H}_\infty$ procedure leads to a simple and straightforward analytical solution, which represents one of the main contributions of the NLGA scheme for FDI when applied to wind turbine systems, as shown in Chapter 5.

In the considered $\mathcal{H}_-/\mathcal{H}_\infty$ framework the norms $\mathcal{H}_\infty$ and $\mathcal{H}_-$ of a stable transfer function $G$ are defined in the form (Zhou et al., 1996b; Doyle et al., 1998):

$$\|G\|_\infty = \sup_{\omega \geq 0} \bar{\sigma}\left[G\left(j\,\omega\right)\right], \qquad \|G\|_- = \underline{\sigma}\left[G\left(j\,0\right)\right], \quad (4.19)$$

where $\bar{\sigma}$ and $\underline{\sigma}$ represent the maximum and the minimum singular value, respectively.

The problem of the trade-off between disturbance robustness and fault sensitivity is stated as follows. On the basis of the two scalars $\beta > 0$ and $\gamma > 0$, the set $\mathcal{K}$ in the following form can be defined:

$$\mathcal{K} = \left\{k \in \mathfrak{R} : (a - k) < 0,\ \|G_{r\varepsilon}\|_\infty < \gamma,\ \|G_{rf}\|_- > \beta\right\}, \quad (4.20)$$

where

$$G_{r\varepsilon}(s) = (s - a + k)^{-1}\,(E_1 - k\,E_2) + E_2 \quad (4.21)$$

and

$$G_{rf}(s) = (s - a + k)^{-1}\,m. \quad (4.22)$$

In order to obtain the analytical solution of this problem, it can be shown that $\forall k \in \mathfrak{R}$ with $(a - k) < 0$, it follows that

$$\|G_{r\varepsilon}\|_\infty^2 = \max\left\{1,\ \frac{\left(e_{11}^2 + a^2\right)}{(k - a)^2}\right\} \quad (4.23)$$

and

$$\sup_{\{k \in \mathfrak{R}: \, (a-k) < 0\}} \| G_{r\varepsilon} \|_{\infty} = +\infty. \tag{4.24}$$

In fact, from the relation of Eq. (4.22), the following relation holds:

$$G_{r\varepsilon}(s) = \left[ \begin{array}{cc} \frac{e_{11}}{s-a+k} & \frac{s-a}{s-a+k} \end{array} \right], \tag{4.25}$$

hence it is possible to write

$$\left\{ \bar{\sigma} \left[ G_{r\varepsilon}(j\omega) \right] \right\}^2 = \frac{e_{11}^2}{(k-a)^2 + \omega^2} + \frac{a^2 + \omega^2}{(k-a)^2 + \omega^2} = \frac{(e_{11}^2 + a^2) + \omega^2}{(k-a)^2 + \omega^2}, \tag{4.26}$$

so that

$$\| G_{r\varepsilon} \|_{\infty}^2 = \sup_{\xi \geq 0} \frac{(e_{11}^2 + a^2) + \xi}{(k-a)^2 + \xi}. \tag{4.27}$$

From the last expression it is straightforward to obtain the expressions of Eqs. (4.23) and (4.24).

On the other hand, the set defined by the relation

$$\mathscr{K}_{\gamma} = \{ k \in \mathfrak{R} : (a-k) < 0, \| G_{r\varepsilon} \|_{\infty} < \gamma, \gamma > 1 \} \tag{4.28}$$

is given by the form

$$k > \underline{k} \quad \text{with} \quad \underline{k} = a + \frac{\sqrt{e_{11}^2 + a^2}}{\gamma}. \tag{4.29}$$

In fact, by means of Eq. (4.22), it is possible to write

$$\frac{(e_{11}^2 + a^2)}{(k-a)^2} < \gamma^2 \tag{4.30}$$

that holds for

$$k > a + \frac{\sqrt{e_{11}^2 + a^2}}{\gamma}. \tag{4.31}$$

Moreover, if $\gamma > 1$, then the relation

$$\left\{ \| G_{rf} \|_{-} : \| G_{r\varepsilon} \|_{\infty} < \gamma \right\} \tag{4.32}$$

is described in the form

$$0 < \| G_{rf} \|_{-} < \beta_{\max}(\gamma) \quad \text{with} \quad \beta_{\max}(\gamma) = \frac{m\gamma}{\sqrt{e_{11}^2 + a^2}}. \tag{4.33}$$

In fact, from Eq. (4.22), it results that $G_{rf}(s) = m/(s-a+k)$. Moreover, assuming without loss of generality that $m > 0$, we obtain $\| G_{rf} \|_{-} = m/(k-a)$. By imposing that $\| G_{rf} \|_{-} > \beta$ with $\beta > 0$, the constraint $k < a + (m/\beta)$ holds. Then, by recalling the expression of Eq. (4.28), the maximum feasible value for $\beta$ fulfilling the constraint $\| G_{r\varepsilon} \|_{\infty} < \gamma$ is given by

$$\underline{k} = a + \frac{m}{\beta_{\max}(\gamma)}, \tag{4.34}$$

hence

$$\beta_{\max}(\gamma) = \frac{m}{\underline{k} - a} = \frac{m\gamma}{\sqrt{e_{11}^2 + a^2}}. \tag{4.35}$$

From the previous results, it can be shown that if $\gamma > 1$ and $\beta \in \left]0, \beta_{\max}(\gamma)\right[$, the set $\mathcal{K}$ fulfilling the constraints defined by the expression of Eq. (4.20) is given by

$$\mathcal{K} = \left\{ k \in \mathcal{R} : k \in \left]\underline{k}, \overline{k}\right[, \ \underline{k} = a + \frac{m}{\beta_{\max}(\gamma)}, \ \overline{k} = a + \frac{m}{\beta} \right\}. \tag{4.36}$$

It is worth noting that, if maximization of the performance index

$$J = \frac{\|G_{rf}\|_-}{\|G_{r\varepsilon}\|_\infty} \tag{4.37}$$

is considered, from Eq. (4.23) it follows that

$$\|G_{r\varepsilon}\|_\infty = \begin{cases} 1 & \text{if } k > \left(a + \sqrt{e_{11}^2 + a^2}\right), \\ \frac{\sqrt{e_{11}^2 + a^2}}{k-a} & \text{if } a < k \le \left(a + \sqrt{e_{11}^2 + a^2}\right), \end{cases} \tag{4.38}$$

hence

$$J = \begin{cases} \frac{m}{k-a} & \text{if } k > \left(a + \sqrt{e_{11}^2 + a^2}\right), \\ \frac{m}{\sqrt{e_{11}^2 + a^2}} & \text{if } a < k \le \left(a + \sqrt{e_{11}^2 + a^2}\right). \end{cases} \tag{4.39}$$

Moreover, from the expression of Eq. (4.39), it can be observed that

$$J = \frac{m}{k-a} < \frac{m}{\sqrt{e_{11}^2 + a^2}} \tag{4.40}$$

for $k > \left(a + \sqrt{e_{11}^2 + a^2}\right)$. In this way the maximum value of the performance index $J$ of Eq. (4.40) can be computed and described by

$$J_{\max} = \frac{m}{\sqrt{e_{11}^2 + a^2}} \qquad \forall k \in \mathcal{K}_J = \left\{ k \in \Re : a < k \le \left(a + \sqrt{e_{11}^2 + a^2}\right) \right\}. \tag{4.41}$$

The method considered in this section guarantees the maximum value of the performance index $J$, as well as the fulfilment of the constraints $\|G_{r\varepsilon}\|_\infty < \gamma$ and $\|G_{rf}\|_- > \beta$, when $\beta \ge m/\sqrt{e_{11}^2 + a^2}$. In fact, from $\beta \ge m/\sqrt{e_{11}^2 + a^2}$ it follows that

$$\|G_{rf}\|_- = \frac{m}{k-a} > \beta \ge \frac{m}{\sqrt{e_{11}^2 + a^2}} \tag{4.42}$$

and $k < \left(a + \sqrt{e_{11}^2 + a^2}\right)$.

Finally, from Eq. (4.33), it is always possible to determine a value for $\beta$ such that

$$\frac{m}{\sqrt{e_{11}^2 + a^2}} \le \beta \le \beta_{max}(\gamma) \tag{4.43}$$

when $\gamma > 1$.

On the basis of the expression of Eq. (4.36), the residual generator gain for FDI $k$ can be designed by means of the procedure, which consists of the following steps:

1. Choose $\gamma > 1$ to obtain a desired level of disturbance attenuation.
2. Compute $\beta_{\max}(\gamma)$, and choose $\beta \in \left]0, \beta_{\max}(\gamma)\right[$ for obtaining the desired level of fault sensitivity.

**3.** Choose $k \in ]\underline{k}, \overline{k}[$, with $\underline{k} = a + m/\beta_{\max}(\gamma)$ and $\overline{k} = a + m/\beta$.

**4.** Apply the fixed gain $k$ value to the $k_f$ of Eq. (4.8), or to $k_o$ of Eq. (4.14), if the NLGA residual generator is in the filter form or in the observer form, respectively.

Note that the NLGA residual generator, whose gain has been optimized according to the procedure described here, will be defined as NLGA Adaptive Filter (NLGA–AF) and it will be used for the estimation of the faults affecting the wind turbine systems considered in Chapter 5.

It is worth observing how the basic NLGA scheme based on residual signals as proposed in De Persis and Isidori (2001) is not able to provide fault size estimation. Different nonlinear geometric approaches providing the reconstruction of the fault signal can be found also, e.g., in Kabore et al. (2000), Kaboré and Wang (2001), in which the fault estimation method relies on the successive derivatives of the input and output signals. However, the drawback of this strategy is high sensitivity to measurement noise. Therefore, the original NLGA method has been modified in order to obtain an adaptive filtering algorithm, capable of reconstructing the fault signal. Moreover, the following section will show how the NLGA Adaptive Filter (NLGA–AF) scheme exploits the coordinate transformation detailed in Section 3.2.4 as a starting point for designing an adaptive filtering strategy for the FDI of sensors and actuators, as well as for estimating the magnitude of the considered faults affecting the considered wind turbine system.

In the following an adaptive nonlinear filter for the $\bar{x}_1$-subsystem, providing fault size estimation, is developed. Moreover, the asymptotic convergence of the estimate to the actual fault size is formally proven. It is worth noting that the NLGA–AF FDI scheme can be applied to the wind turbine system described in Chapter 2 only if the fault detectability condition presented in Section 3.2.4 holds, and the following new constraints required by the NLGA–AF for FDI are satisfied:

**1.** The $\bar{x}_1$-subsystem is independent of the $\bar{x}_3$ state components.

**2.** The fault is a step function of time, hence the parameter $f$ is a constant to be estimated.

**3.** There exists a proper scalar component $\bar{x}_{1s}$ of the state vector $\bar{x}_1$ such that the corresponding scalar component of the output vector is $\bar{y}_{1s} = \bar{x}_{1s}$ and the following relation holds:

$$\dot{\bar{y}}_{1s}(t) = M_1(t) \cdot f + M_2(t) \tag{4.44}$$

where $M_1(t) \neq 0, \forall t \geq 0$. Moreover, $M_1(t)$ and $M_2(t)$ can be computed for each time instant, since they are functions of input and output measurements. The relation of Eq. (4.44) describes the general form of the wind turbine system under diagnosis described in Chapter 2.

With reference to the system model of Eq. (4.44), the design of an adaptive filter is required for providing an estimate $\hat{f}(t)$, which asymptotically converges to the magnitude of the actual fault $f$. The proposed adaptive filter is based on the least-squares algorithm with forgetting factor (Ioannou and Sun, 2012) and described by the adaptation law in the form

$$\begin{cases} \dot{P} = \beta P - \frac{1}{N^2} P^2 \check{M}_1^2 & P(0) = P_0 > 0, \\ \dot{\hat{f}} = P \varepsilon \check{M}_1 & \hat{f}(0) = 0, \end{cases} \tag{4.45}$$

with the expressions of Eq. (4.46) representing the output estimate and the corresponding normalized estimation error:

$$\begin{cases} \hat{\bar{y}}_{1s} = \check{M}_1 \hat{f} + \check{M}_2 + \lambda \check{\bar{y}}_{1s}, \\ \varepsilon = \frac{1}{N^2} \left( \bar{y}_{1s} - \hat{\bar{y}}_{1s} \right), \end{cases} \tag{4.46}$$

where all the involved variables of the adaptive filter are scalar. In particular, $\lambda > 0$ is a parameter related to the bandwidth of the filter, $\beta \geq 0$ is the forgetting factor, and $N^2 = 1 + \check{M}_1^2$ is the normalization factor of the least-squares algorithm.

Moreover, the proposed adaptive filter adopts the signals $\check{M}_1, \check{M}_2, \check{\bar{y}}_{1s}$, which are obtained by means of low-pass filtering the signals $M_1, M_2, \bar{y}_{1s}$ defined in the form

$$\begin{cases} \dot{\check{M}}_1 = -\lambda \check{M}_1 + M_1 & \check{M}_1(0) = 0, \\ \dot{\check{M}}_2 = -\lambda \check{M}_2 + M_2 & \check{M}_2(0) = 0, \\ \dot{\check{\bar{y}}}_{1s} = -\lambda \check{\bar{y}}_{1s} + \bar{y}_{1s} & \check{\bar{y}}_{1s}(0) = 0. \end{cases} \tag{4.47}$$

The considered adaptive filter is described by Eqs. (4.45)–(4.47). The asymptotic relation between the normalized output estimation error $\varepsilon(t)$ and the fault estimation error $f - \hat{f}(t)$ has the form

$$\lim_{t \to \infty} \varepsilon(t) = \lim_{t \to \infty} \frac{\check{M}_1(t)}{N^2(t)} \left( f - \hat{f}(t) \right). \tag{4.48}$$

The design of this filter can be achieved by considering the auxiliary system

$$\begin{cases} \dot{y}'_1 = -\lambda y'_1 + \dot{\bar{y}}_{1s}, & y'_1(0) = 0, \\ \dot{y}'_2 = -\lambda y'_2 + \lambda \bar{y}_{1s}, & y'_2(0) = 0, \\ y' = y'_1 + y'_2. \end{cases} \tag{4.49}$$

In this way it is easy to show that the solution is given by

$$\begin{aligned} y'(t) &= \int_0^t e^{-\lambda(t-\tau)} \dot{\bar{y}}_{1s}(\tau) d\tau + \int_0^t e^{-\lambda(t-\tau)} \lambda \bar{y}_{1s}(\tau) d\tau \\ &= \int_0^t e^{-\lambda(t-\tau)} \left( M_1(\tau) f + M_2(\tau) \right) d\tau + \lambda \check{\bar{y}}_{1s} \\ &= \check{M}_1(t) f + \check{M}_2(t) + \lambda \check{\bar{y}}_{1s}(t). \end{aligned} \tag{4.50}$$

The function $V$ is considered in the form

$$V = \frac{1}{2} \left( y' - \bar{y}_{1s} \right)^2, \tag{4.51}$$

which is trivially positive definite and radially unbounded. Moreover, its first time derivative can be computed as

$$\begin{aligned} \dot{V} &= (y' - \bar{y}_{1s})(\dot{y}'_1 + \dot{y}'_2 - \dot{\bar{y}}_{1s}) \\ &= (y' - \bar{y}_{1s})(-\lambda y'_1 - \lambda y'_2 + \lambda \bar{y}_{1s}) \\ &= -\lambda \left( y' - \bar{y}_{1s} \right)^2. \end{aligned} \tag{4.52}$$

Since $\dot{V}$ is trivially negative definite $\forall\, y' \neq \bar{y}_{1s}$, $V$ is a Lyapunov function, so that $y'(t)$ globally asymptotically tends to the output function $\bar{y}_{1s}(t)$. Moreover, from Eq. (4.50), the following relation holds:

$$\lim_{t \to \infty} \bar{y}_{1s}(t) = \check{M}_1(t) f + \check{M}_2(t) + \lambda \check{\bar{y}}_{1s}(t). \tag{4.53}$$

From Eqs. (4.46) and (4.53), the asymptotic behavior of the normalized output estimation error $\varepsilon(t)$ can be straightforwardly obtained in the form

$$\begin{aligned} \lim_{t \to \infty} \varepsilon(t) &= \lim_{t \to \infty} \frac{1}{N^2(t)} \left( \bar{y}_{1s}(t) - \check{M}_1(t) \hat{f}(t) - \check{M}_2(t) - \lambda \check{\bar{y}}_{1s}(t) \right) \\ &= \lim_{t \to \infty} \frac{1}{N^2(t)} \left( \check{M}_1(t) f - \check{M}_1(t) \hat{f}(t) \right). \end{aligned} \tag{4.54}$$

Therefore the adaptive filter described by Eqs. (4.45)–(4.47) represents a solution of the problem of the fault estimation for FDI purposes, so that $\hat{f}(t)$ provides an asymptotically convergent estimation of the magnitude of the step fault $f$.

The function $W$ is considered in the form

$$W = \frac{1}{2} \left( \hat{f} - f \right)^2, \tag{4.55}$$

which is trivially positive definite and radially unbounded. Moreover, its first time derivative is

$$\dot{W} = \left( \hat{f} - f \right) \left( P \varepsilon \check{M}_1 - 0 \right). \tag{4.56}$$

It is worth noting that the smoothness of the involved functions allows us to apply the asymptotic approximation of Eq. (4.48) to the expression of Eq. (4.56). In fact, $\exists\, t_\star > 0$ so that the sign of $\dot{W}(t)$, $\forall\, t \geq t_\star$ is not affected by the asymptotic approximation of Eq. (4.48).

Hence it follows that

$$\dot{W}(t) = -P(t)\frac{\check{M}_1^2(t)}{N^2(t)}\left(\hat{f}(t) - f\right)^2 \qquad \forall\, t \geq t_\star, \tag{4.57}$$

which is negative definite $\forall\, \hat{f} \neq f$. In fact, $\check{M}_1(t)$ is a low-pass filtered signal $M_1(t)$, which is a smooth function and always not null by hypothesis.

Moreover, $N^2(t) = 1 + \check{M}_1^2(t) > 0$ and

$$P(t) = \left(e^{-\beta t}P_0^{-1} + \int_0^t e^{-\beta(t-\tau)}\frac{\check{M}_1^2(\tau)}{N^2(\tau)}\,d\tau\right)^{-1} > 0. \tag{4.58}$$

Therefore $W$ is a Lyapunov function and $\hat{f}(t)$ globally asymptotically tends to $f$.

## 4.2  WIND TURBINE CONTROL STRATEGIES

As already highlighted in Chapter 2, wind turbines are complex dynamic systems forced by gravity and stochastic wind disturbances, which are affected by gravitational, centrifugal, and gyroscopic loads. Their aerodynamics are nonlinear, whilst their rotors are subject to complicated turbulent wind inflow fields driving fatigue loading. Therefore wind turbine modeling and control is a challenging task (Johnson et al., 2006a).

Accurate models have to contain many degrees of freedom in order to capture the most important dynamic effects. Moreover, the rotation of the turbine adds further complexity to the dynamics modeling. In general, off-the-shelf commercial software usually is not adequate for wind turbine dynamics modeling, and special dynamic simulation codes are required. It is clear that the design of control algorithms for wind turbines has to take into account these complexities. The main goal of the controller can consist of maintaining safe turbine operation, achieving prescribed control performances, and managing possible fault conditions, as shown, e.g., in Sami and Patton (2012), Odgaard and Stoustrup (2012a), Rotondo et al. (2012).

Today's wind turbines employ different control actuation and strategies to achieve the required goals and performances. Some turbines perform the regulation action through passive control methods, such as in fixed-pitch, stall control machines. In these machines, the blades are designed so that the power is limited above rated wind speed through the blade stall. Thus no pitch mechanism is needed (Zhang et al., 2004). In this case the rotational speed control is proposed, thus avoiding the inaccuracy of measuring the wind speed. Rotors with adjustable pitch are often used in constant-speed machines, in order to provide turbine power control better than the one achievable with blade stall (Sakamoto et al., 2004). In order to maximize the power output below the wind speed, the rotational speed of the turbine must vary with wind speed. Blade pitch control is used above rated wind speed in order to limit power (Johnson et al., 2006a). Another control strategy for large commercial wind turbines can employ yaw regulation to orient the machine into the wind. A yaw error signal from a nacelle-mounted wind direction sensor is used to calculate a control error. In this situation the yaw motor is used when the yaw error exceeds a certain amount, as shown, e.g., in Zhao and Stol (2007).

Other data-driven approaches can be based on schemes relying on the direct fuzzy identification of the controller model. As the wind turbine mathematical model is partially known and nonlinear, fuzzy identification represents an alternative for developing experimental models from input–output data. In contrast to pure nonlinear identification methods, fuzzy identification is capable of deriving nonlinear models without detailed system assumptions. Therefore this approach yields the controller models using the data acquired from the plant under investigation. These fuzzy controllers are described by a collection of local affine systems of the type of Takagi–Sugeno (TS) fuzzy prototypes (Babuška, 1998), whose parameters are obtained by identification procedures. In this way the fuzzy controllers adjust both the wind turbine blade pitch angle and the generator torque of the wind turbine benchmark.

Note that, with respect to Galdi et al. (2008), an offline identification approach for fuzzy prototypes is exploited here, without the need of any further optimization procedure, thus enhancing the real-time application. Moreover, in Simani (2012b, 2012a) a different solution to the design of the fuzzy regulators was presented. In particular, Simani (2012b, 2012a) presented the development of a fuzzy regulator, which is mainly based on the fuzzy identification of the wind turbine system; after this step the fuzzy regulator is derived without any further identification procedure, but the PI fuzzy controller parameters are analytically computed from the identified process; the PI fuzzy controller parameters are thus derived using a suboptimal design procedure, i.e., using a fuzzy combination of the local PI controller parameters; the fuzzy membership functions of the PI fuzzy regulators are the same as those of the identified fuzzy model of the wind turbine system. On the other hand, in the controller, the design relies on the direct identification of the fuzzy controllers, whose parameters are directly identified, and not derived from the fuzzy models of the wind turbine process. Also the fuzzy membership functions of these controllers are directly identified, and they are different from the membership functions of the identified fuzzy models of the wind turbine system.

Other controller design methods can be also used that exploit adaptive schemes, which were not addressed in Simani (2012b, 2012a). With reference to this adaptive control method, it considers the application of model online identification mechanisms in connection with model-based adaptive control design. This control method belongs to the field of adaptive control, see, e.g., Bobál et al. (2005). Online parametric model identification schemes represent an alternative for developing experimental prototypes. Therefore this approach exploits the implementation of controllers based on adaptive identification schemes used for the online estimation of the controlled process. Recursive identification approaches extended to the adaptive case can make use of exponential forgetting algorithms, as proposed, e.g., in Linden et al. (2012). In this way a system identification scheme is exploited for the online estimation of the parameters of time-varying systems (Bobál et al., 2005).

### 4.2.1 Fuzzy Modeling for Control

This section recalls the approach exploited for obtaining the fuzzy description of the wind turbine controller, with the proposed controller model estimation.

The approach suggested in this section employs fuzzy clustering techniques to partition the available data into subsets characterized by linear behaviors. Relationships between clusters and linear regression are exploited, thus allowing for the combination of fuzzy logic techniques with system identification tools. In addition, an implementation in the Matlab® Toolbox of the Fuzzy Modeling and IDentification (FMID) technique presented in the following is available (Babuška, 2000). In this study TS fuzzy models are exploited (Babuška, 1998), as they are able to provide the mathematical description of the nonlinear system. The switching and the scheduling between the submodels is achieved through a smooth function of the system state, the behavior of which is defined using fuzzy set theory.

As already highlighted in Chapter 3 with reference to the FDI scheme design, the fuzzy modeling and identification algorithm is based on a two-step procedure, in which at first the operating regions are determined using the data clustering technique, and in particular, the Gustafson–Kessel (GK) fuzzy clustering, since it is already available in Babuška (2000). Then, in the second stage, the estimation of the controller parameters is achieved using the identification algorithms already available in the literature (Babuška, 1998), which can be seen as a generalization of the classical least-squares approach.

The TS fuzzy models have the form

$$y(k+1) = \frac{\sum_{i=1}^{M} \mu_i\left(x(k)\right) y_i}{\sum_{i=1}^{M} \mu_i\left(x(k)\right)}, \tag{4.59}$$

where $y_i = a_i x + b_i$, with $a_i$ being the parameter vector (regressand) and $b_i$ the scalar offset. $x = x(k)$ represents the regressor vector, which can contain delayed samples of $u(k)$ and $y(k)$ acquired for the wind turbine system.

The *antecedent* fuzzy sets $\mu_i$ are extracted from the fuzzy partition matrix (Babuška, 1998). The *consequent parameters* $a_i$ and $b_i$ are estimated from the data (Babuška, 1998). This identification scheme exploited for the estimation of the TS model parameters has been integrated into the FMID toolbox for Matlab®. This approach is usually preferred when the TS model should serve as predictor, as it computes the consequent parameters via the Frisch scheme, in particular developed for the Errors-In-Variables (EIV) descriptions (Linden et al., 2012).

Once a reasonably accurate fuzzy description of the considered wind turbine system becomes available, it is used offline to directly estimate the nonlinear fuzzy controllers. As already remarked, the recalled control design methodology relies on the *model inverse control* principle (Babuška, 1998), but is solved within the fuzzy identification framework recalled here. It is well-known that for stable fuzzy systems, whose inverted dynamics are stable, a nonlinear controller can be simply designed by inverting the fuzzy model. Moreover, in the ideal situation of no modeling errors and disturbances, this controller provides perfect tracking with zero steady-state errors. However, in practice, one has to deal with both modeling errors and disturbances, which can be tackled with an arbitrary degree of accuracy by exploiting the fuzzy modeling approach recalled in this section for directly *identifying* the inverse model controller.

Moreover, since the fuzzy model derived as recalled in Chapter 3 for FDI purpose is able to predict any system behavior, it will be used for the direct identification of the inverse model controller. An optimal control strategy is thus obtained by minimizing a cost function which includes the difference between the desired and controller outputs, and a penalty

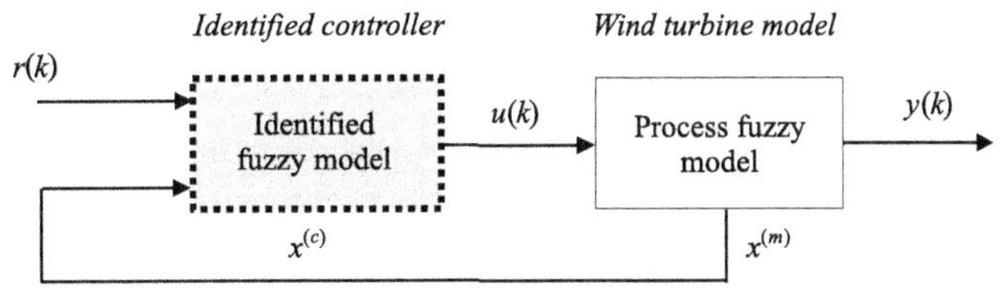

**FIGURE 4.3** The process fuzzy model and the controller based on its inverse.

on the system stability. Constraints on the complete system stability are included as a part of the optimization problem. Generally, a nonconvex optimization problem must be solved at each control sample, which hampers the direct and practical application of the approach. However, to solve this problem, the optimization scheme described in Babuška (1998), which is based on a parametrized search technique, is applied at a higher level to formulate the control objectives and constraints.

The remainder of this section tries to explain how the fuzzy controller can be directly identified, again by exploiting a fuzzy identification scheme, inspired by the inverse model control. The nonlinear controller based on a fuzzy inverse model takes into account the process nonlinearities, including the inherent saturation (level) constraints of the control inputs and other process variables, and is capable of controlling a nonlinear system in the entire operating range.

First, it is assumed that the rule-based model of Eq. (4.59) has been identified for describing the continuous-time behavior of the wind turbine system described in Chapter 2 in the discrete-time form

$$y(k+1) = f(x(k), u(k)) \tag{4.60}$$

and in particular the TS prototype

$$y(k+1) = \frac{\sum_{i=1}^{M} \mu_i^{(m)}\left(x^{(m)}(k)\right)\left(a_i^{(m)} x^{(m)}(k) + b_i^{(m)}\right)}{\sum_{i=1}^{M} \mu_i^{(m)}\left(x^{(m)}(k)\right)}. \tag{4.61}$$

The inputs of the model are the current state $x(k) = [y(k), \ldots, y(k-n+1), u(k-1), \ldots, u(k-n+1)]^T$ and the current input $u(k)$. The output is a prediction of the system's output at the next sample $y(k+1)$. In Eq. (4.61) the estimated membership functions $\mu_i^{(m)}$, the state $x^{(m)}$, and the parameters $a_i^{(m)}$, $b_i^{(m)}$ of the controlled system are denoted by the superscript $(m)$.

The objective of the control algorithm is to compute the control input $u(k)$, such that the system output at the next sampling instant is equal to the desired (reference) output $r(k+1)$. In principle, this can be achieved by inverting the model of the process. Given the generic current state $x^c(k)$ and the reference $r(k)$, the control input is given by

$$u(k+1) = f^{-1}\left(x^c(k), r(k)\right). \tag{4.62}$$

Generally, it is difficult to find the *analytical inverse* function $f^{-1}$. Therefore, the method exploited here uses the identified fuzzy TS of the process under investigation of Eq. (4.61) for providing the particular state $x^{(m)}(k)$ at each time step $k$. From this mapping, the inverse mapping $u(k+1) = f^{-1}\left(x^{(c)}(k), r(k)\right)$ is easily identified as a prototype in the form of Eq. (4.59), if the controlled system is stable, and in particular in the form

$$u(k+1) = \frac{\sum_{i=1}^{M} \mu_i^{(c)}\left(x^{(c)}(k)\right)\left(a_i^{(c)} x^{(c)}(k) + b_i^{(c)}\right)}{\sum_{i=1}^{M} \mu_i^{(c)}\left(x^{(c)}(k)\right)}, \tag{4.63}$$

where the inputs of the identified controller model are the state $x^{(c)}(k) = [x^{(m)}(k), r(k-1), \ldots, r(k-n+1)]^T$ and the current reference $r(k)$. In Eq. (4.63) the estimated membership functions $\mu_i^{(c)}$ and the parameters $a_i^{(c)}$, $b_i^{(c)}$ of the identified controller model are denoted now by the superscript $(c)$.

As shown in Fig. 4.3, the series connection of the controller and the identified inverse model should yield an identity mapping:

$$y(k+1) = f\left(x^{(m)}(k), u(k)\right) = f\left(x^{(m)}(k), f^{-1}\left(x^{(c)}(k), r(k)\right)\right) = r(k+1) \tag{4.64}$$

when $u(k)$ exists such that $r(k+1) = f\left(x^{(m)}(k), u(k)\right)$. However, due to modeling errors, noise, and disturbances, by means of the fuzzy identification procedure recalled here, the difference $|r(k+1) - f\left(x^{(m)}(k), u(k)\right)|$ is made arbitrarily small by an appropriate choice of the identification parameters, i.e., the fuzzy membership functions $\mu_i^{(c)}$, the number of clusters $M$, and the regressands $a_i^{(c)}, b_i^{(c)}$.

As sketched in Fig. 4.3, the process fuzzy model is used for the recursive prediction of the state vector $x^{(m)}(k)$. Therefore, the state of the fuzzy controller $x^{(c)}(k)$ is updated using the process model state $x^{(m)}(k)$ and the reference input $r(k)$. Apart from the computation of the membership degrees $\mu_i^{(c)}$, both the process model and the controller are estimated using standard matrix operations and linear interpolations, which makes the algorithm suitable for real-time implementation, as it will be shown, e.g., in Chapter 5.

Note however that, with reference to the fuzzy control strategy proposed here, model–reality mismatch can cause differences in the behavior of the process and of the model, which results in an error between the reference and the process output. A mechanism to compensate this error can be exploited, e.g., via online adaptation of both the process and controller models of Eqs. (4.61) and (4.63), respectively.

Classical adaptive algorithms adapt only the consequent parameters, while the antecedent membership functions remain unchanged. However, if a good initial model is not available, the membership function parameters can be adapted using some nonlinear optimization techniques (Babuška, 1998). In addition, the fuzzy clustering scheme recalled here could be repeated regularly, after gathering new data or online. However, when industrial applications require short sampling times, extensive computations are not allowed in practice. Therefore, Section 4.2.2 motivates the adaptive strategy based on *linear* models, whose parameters are estimated online and exploited for the controller parameter estimation.

### 4.2.2 Recursive Identification for Adaptive Control

This section describes the recursive approach exploited for obtaining the mathematical description of the wind turbine system, which is used for the design of the adaptive control strategy. A recursive implementation of the batch Frisch scheme algorithm recalled in Section 4.2.1 is proposed here to identify dynamic EIV systems. For the update of the estimated model parameters, a recursive bias-compensating strategy is also exploited. Thus, a recursive Frisch scheme identification approach is extended to enhance its online applicability. It is shown that, by incorporating adaptivity via the introduction of exponential forgetting, the algorithm is able to compensate for systematic errors (Linden et al., 2012) arising from the original method. Therefore, this adaptive recursive Frisch scheme is able to deal with linear time-varying (LTV) systems, and it is used in connection with the design of an adaptive control scheme, shown in the following.

Thus the recalled scheme is proposed for the online identification of the process modeled by the following transfer function $G(z)$:

$$G(z) = \frac{A(z^{-1})}{B(z^{-1})} = \frac{b_1 z^{-1} + \cdots + b_{n_b} z^{-n_b}}{1 + a_1 z^{-1} + \cdots + a_{n_a} z^{-n_a}}, \tag{4.65}$$

where $a_i, b_i, n_a$, and $n_b$ represent the unknown parameters and the structure of the model, defining the polynomials $A(z^{-1})$ and $B(z^{-1})$, whilst $z$ is the discrete-time complex variable.

The parameter vector describing the linear relationship is given by

$$\theta = \begin{bmatrix} a_1 \ldots a_{n_a} \ b_1 \ldots b_{n_b} \end{bmatrix}^T \tag{4.66}$$

and its extended version is denoted by

$$\bar{\theta} = \begin{bmatrix} 1 \ \theta^T \end{bmatrix}^T. \tag{4.67}$$

Hence an alternative expression for the considered difference equation is given by

$$\psi^T(k)\bar{\theta} = 0 \tag{4.68}$$

where

$$\psi(k) = \begin{bmatrix} -y(k) \ -y(k-1) \ \ldots \ -y(k-n_a) \ u(k-1) \ \ldots \ u(k-n_b) \end{bmatrix}^T \tag{4.69}$$

is the extended regressor vector.

The Frisch scheme provides estimates for the measurement errors affecting the input and output signals $u(k)$ and $y(k)$, i.e., $\sigma_u$ and $\sigma_y$, and $\theta$ for a linear time-invariant dynamical system. Moreover, it can also be utilized to determine the polynomial orders $n_a$ and $n_b$. However, since this work is oriented to the design of an adaptive controller, the polynomial orders $n_a$ and $n_b$ are assumed to be fixed in advance.

From the Frisch scheme method, the following expression is considered:

$$\left( \Sigma_\psi - \Sigma_{\tilde{\psi}} \right) \bar{\theta} = 0, \tag{4.70}$$

where the noise covariance matrix is given by

$$\Sigma_{\tilde{\psi}} = \begin{bmatrix} \sigma_y \, I_{n_a+1} & 0 \\ 0 & \sigma_u \, I_{n_b} \end{bmatrix}, \tag{4.71}$$

which is approximated by the sample covariance matrix over $N$ samples, namely

$$\Sigma_{\tilde{\psi}} \approx \frac{1}{N} \sum_{k=1}^{N} \psi(k) \, \psi^T(k). \tag{4.72}$$

Thus the Frisch scheme aims to find suitable noise variances $\sigma_u$ and $\sigma_y$ such that $\left( \Sigma_\psi - \Sigma_{\tilde{\psi}} \right)$ is singular positive semidefinite, i.e., it is rank-one deficient, and the corresponding overdetermined system of Eq. (4.70) is solvable, with $\bar{\theta}$ denoting the solution.

If the following residuals are computed:

$$\varepsilon\left(k, \bar{\theta}\right) = A(z^{-1}) \, y(k) - B(z^{-1}) \, u(k), \tag{4.73}$$

their sample autocovariance is determined by

$$r_\varepsilon(h, N) = \frac{1}{N} \sum_{l=1}^{N} \varepsilon\left(l, \bar{\theta}\right) \varepsilon\left(l+h, \bar{\theta}\right), \tag{4.74}$$

where $h$ here denotes a time-shift.

In this adaptive control application it is essential to obtain online estimates of the model parameters $\theta(k)$ of Eq. (4.65), while the process generating the data is running. In fact, this compensation strategy relies on adaptive control, where the control action at time step $k$ relies on a current estimate of the plant model, which is estimated using data up to the sample $k$. Therefore the Frisch scheme relying on the batch of Eqs. (4.70), (4.72), and (4.74) has to be modified in a recursive algorithm.

Therefore the sums in Eqs. (4.72) and (4.74) are required to be updated recursively. The update of the covariance matrix is straightforward:

$$\Sigma_{\tilde{\psi}}(k) = \frac{k-1}{k} \Sigma_{\tilde{\psi}}(k-1) + \frac{1}{k} \psi(k) \, \psi^T(k), \tag{4.75}$$

but the autocovariance $r_\varepsilon(h, k)$ cannot be computed recursively when $1 \leq l \leq k$. However, if the following approximation holds:

$$\varepsilon\left(l, \bar{\theta}(k)\right) \approx \varepsilon\left(l, \bar{\theta}(l)\right) \quad \text{for } l < k, \tag{4.76}$$

i.e., at time step $k$ only the residual $\varepsilon\left(k, \bar{\theta}(k)\right)$ is calculated using the data stored in $\psi(k)$, together with the current parameter estimate $\bar{\theta}(k)$. This allows for recursive calculation of the autocovariance:

$$r_\varepsilon(h, k) = \frac{k-1}{k} r_\varepsilon(k, k-1) + \frac{1}{k} \varepsilon\left(k, \bar{\theta}(k)\right) \varepsilon\left(k+h, \bar{\theta}(k)\right), \tag{4.77}$$

which needs only $\varepsilon\left(k+h, \bar{\theta}(k)\right)$ at each recursion step.

Note that the initial values for $\theta(0)$, $\Sigma_{\tilde{\psi}}(0)$, and $r_\varepsilon(0, h)$ are initialized using, e.g., the Frisch scheme batch procedure (Linden et al., 2012).

The previous recursive estimation algorithm has been modified for tracking variation of system properties, i.e., for coping with time-varying systems. This can be achieved by reducing the importance of older observations, while placing more emphasis on the more recent data and realized using a forgetting factor. The recursive algorithm described by Eqs. (4.75) and (4.77) can show numerical problems for large values of $k$. One benefit of using adaptivity within the recursive Frisch scheme is that the storage of $k$ is avoided. The error arising due to the approximation of the residuals in Eq. (4.76) can be also reduced.

These comments lead to the use of an *exponential forgetting* by means of the new sample covariance matrix and auto-covariance defined as

$$
\begin{cases}
H_{\Sigma_{\tilde{\psi}}}(k) = \omega(\delta)\, \Sigma_{\tilde{\psi}}(k), \\
h_{\varepsilon}(h,\, k) = \omega(\delta)\, r_{\varepsilon}(h,\, k),
\end{cases}
\tag{4.78}
$$

where $\omega(\delta)$ is a scaling factor that coincides with $k$ when no adaptation is introduced. In this way the updated expressions have the form

$$
\begin{cases}
H_{\Sigma_{\tilde{\psi}}}(k) = (1-\delta)\, H_{\Sigma_{\tilde{\psi}}}(k-1) + \delta\, \psi(k)\, \psi^T(k), \\
h_{\varepsilon}(h,\, k) = (1-\delta)\, h_{\varepsilon}(h,\, k-1) + \delta\, \varepsilon\left(k,\, \bar{\theta}(k)\right) \varepsilon\left(k+h,\, \bar{\theta}(k)\right),
\end{cases}
\tag{4.79}
$$

with $0 < \delta < 1$ representing the forgetting factor. Thus the adaptive Frisch scheme algorithm uses Eq. (4.79) on the basis of the following steps:

1. Initialize $\theta(0)$, $\Sigma_{\tilde{\psi}}(0)$, and $r_{\varepsilon}(0,\, h)$ with $h \leq n_a$;
2. At each recursion step, by means of $r_{\varepsilon}(h,\, k)$, compute the noise variances $\sigma_u$ and $\sigma_y$;
3. At each recursion step, determine $\bar{\theta}(k)$ by solving Eq. (4.70) via (4.79).

The vector $\theta(k)$ represents the model parameter estimates computed at step $k$.

The online identification method described here was implemented in the Matlab® and Simulink® environments and integrated in the Simulink® toolbox (Bobál et al., 2005). Note that the initial values of the parameters $\theta(0)$, $\Sigma_{\tilde{\psi}}(0)$, and $r_{\varepsilon}(0,\, h)$ are properly selected.

Finally, once the time-varying parameters $\theta(k)$ of the discrete-time linear model approximating the nonlinear process of Eq. (2.19) have been computed at each time step $k$, the adaptive controller is derived as recalled in the remainder of this section.

In particular, with reference to the wind turbine system considered in Chapter 2, adaptive controllers for second order models ($n_a = n = 2$) are exploited. Note that the design of digital self-tuning controllers do not use any fuzzy modeling presented in Section 4.2.1.

With reference to Eq. (4.65), with $n_a = n_b = n = 2$, the transfer function of the time-varying controlled system has the form

$$
G(z) = \frac{b_1 z^{-1} + b_2 z^{-2}}{1 + a_1 z^{-1} + a_2 z^{-2}}
\tag{4.80}
$$

whose parameters, estimated using the online identification approach already recalled, are

$$
\theta(k) = \left[\hat{a}_1,\, \hat{a}_2,\, \hat{b}_1,\, \hat{b}_2\right]^T.
\tag{4.81}
$$

The subscript $k$ for model and controller parameters is dropped in order to simplify equations and formulas.

It is worth noting here the motivation for selecting *second order* online identified models of Eq. (4.80). This choice seems valid and appropriate. In fact, as described, for example, in Odgaard et al. (2009b), Sloth et al. (2011), Odgaard and Stoustrup (2012b), a linearized model for the system of Eq. (2.19) could be described as a fourth order model when considering only the two controlled inputs and the two monitored outputs. Therefore, the approximation of the behavior of the wind turbine process from each one of its two outputs as a two-input one-output second order model description motivates the choice of the online identification of second order model of Eq. (4.80). Moreover, each of the two outputs of the wind turbine model is tuned via a PI adaptive controller, whose parameters were computed from the adaptive second-order parametric model.

The control law corresponding to a discrete-time PI adaptive controller using the forward Euler's discretization method has the form (Franklin et al., 1998)

$$u(k) = K_p \left[ e(k) - e(k-1) + \frac{T_s}{T_I} e(k) \right] + u(k-1), \tag{4.82}$$

where $e(k)$ is the tracking error, i.e., $e(k) = r(k) - y(k)$, with $r(k)$ being the set-point or reference signal, $T_s$ the sampling time. The (time-varying) controller variables $K_p$ and $T_I$ are now computed from the time-varying model parameters $\theta(k)$.

The controller parameters $K_p$ and $T_I$ are computed using the following relations: $K_p = 0.6 K_{P_u}$ and $T_I = 0.5 T_u$, and are described in Bobál et al. (2005). The variables $K_{P_u}$ and $T_u$ are now the (time-varying) critical gain and the critical period of oscillations, respectively. Also these variables are functions of the time-varying model parameters $\theta(k)$.

In particular, when considering a second order *stable* model described by its (time-varying) parameters $\hat{a}_2$, $\hat{a}_1$, $\hat{b}_2$, and $\hat{b}_1$, the required variables $K_{P_u}$ and $T_u$ can be computed at each time step $k$ from the following relations:

$$\begin{cases} K_{P1} = \frac{1 - a_2}{b_2}, \\ K_{P2} = \frac{a_1 - a_2 - 1}{b_2 - b_1}, \\ d = (b_1 K_{P1} + a_1)^2 + 4 (b_2 K_{P1} + a_2), \\ \omega_u = \frac{1}{T_s} \arccos(-\frac{b_1 K_{P1} + a_1}{2}). \end{cases} \tag{4.83}$$

The critical gain and the critical period depend on the value of $d$, and in particular,

$$\begin{cases} T_u = \frac{2\pi}{\omega_u}, & K_{P_u} = K_{P1}, & \text{if } d < 0, \\ T_u = 2 T_s, & K_{P_u} = K_{P1}, & \text{if } d = 0, \\ T_u = 2 T_s, & K_{P_u} = K_{P2}, & \text{if } d > 0, \end{cases} \tag{4.84}$$

The choice of the tuning methodology for the PI regulator parameters described above was motivated by its direct implementation in the Simulink® toolbox (Bobál et al., 2005).

It is worth noting that, instead of using the Ziegler–Nichols relations, the digital PI controller design can be based on the pole assignment method directly performed in the discrete time domain. In this case a controller based on the assignment of poles in a closed feedback control loop is derived to stabilize the closed loop, while the characteristic polynomial should have previously determined poles. Apart from the stability requirement, proper poles configuration can lead to obtaining desired closed loop response (e.g., the maximum overshoot, damping, etc.). The general approach to this method is based on the algebraic theory. Therefore, a suitable PI controller design ensures the required control loop dynamic behavior by choosing the characteristic polynomial in connection with various regulator and controller model structures (Bobál et al., 2005). In particular, this is the case if the online identified plant is described by the second order discrete-time model $G(z)$ of Eq. (4.80), with known polynomials $A(z^{-1})$ and $B(z^{-1})$, whilst $G_R(z) = \frac{Q(z^{-1})}{P(z^{-1})}$ is the closed-loop transfer function. The characteristic polynomial $D(z^{-1}) = A(z^{-1}) P(z^{-1}) + B(z^{-1}) Q(z^{-1})$ of the closed-loop transfer function depends on the PI controller parameters $K_p$ and $T_I$ of Eq. (4.82).

The characteristic polynomial can be defined by various methods. The most frequently used are those, for example, which meet requirements of the response of a discrete second-order plant, the dead-beat control, or quadratic optimal control. However, in particular when controlling the process under investigation, there is the requirement for a control response with limited overshoot. Then the design leads to the characteristic polynomial in the form (Bobál et al., 1995)

$$D(z^{-1}) = (z - \alpha) \left[ z - (\alpha + j\,\omega) \right] \left[ z - (\alpha - j\,\omega) \right]. \tag{4.85}$$

In this case, the characteristic polynomial $D(z^{-1})$ has a real pole $z_1 = \alpha$ placed inside the unit circle ($0 \leq \alpha < 1$), and a couple of conjugate complex poles $z_{2,3} = \alpha \pm j\,\omega$, with $\alpha^2 + \omega^2 < 1$. These parameters are used to change the speed of the control response and the size of the changes in the controller output at the same time; different overshoots can be also achieved. The parameters $K_p$ and $T_I$ of the adaptive PI controller can be easily computed via a linear equation system depending on the parameters of the online identified controlled system $b_1, b_2, a_1, a_2$, as well as the desired poles and their position in the $z$ complex plane.

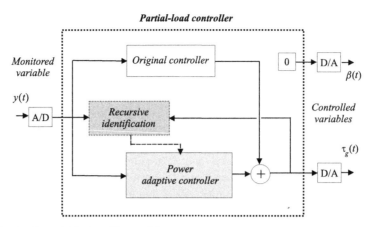

**FIGURE 4.4** Adaptive controller for the partial load working condition.

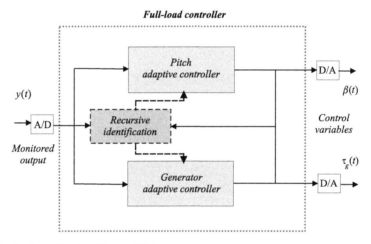

**FIGURE 4.5** Adaptive controller for the full load working conditions.

The proposed control scheme employs 3 adaptive PI regulators, which are sketched in Figs. 4.4 and 4.5. In particular, when the wind turbine works in partial load (Region 1), the corresponding adaptive controller is depicted in Fig. 4.4. These working conditions and the baseline controller for the wind turbine were already recalled in Chapter 2.

Fig. 4.4 shows that, in partial load working conditions, the torque controller of the benchmark is compensated via the adaptive regulator of Eq. (4.82). In this case the adaptive PI controller exploits the output variable $y(t)$, namely $\omega_g(t)$, whilst the actuated input $u(t)$ is only $\tau_g(t)$, since the blade pitching is used in these working conditions. The reference signal $r(k)$ is the sampled signal $\omega_r(t)$ exploited by the adaptive controller.

On the other hand, when the wind turbine is working under full load, the proposed control strategy is shown in Fig. 4.5.

Regarding Fig. 4.5, one adaptive PI controller is exploited for the regulation of the blade pitch angle, $u(t) = \omega_g(t)$, and uses the constant reference signal $r(k) = \omega_{nom}$. The second PI adaptive controller generates the signal $u(t) = \tau_g(t)$, with $y(t) = P_g(t)$, and $r(k)$ is the constant generated power reference $P_r$.

In the remainder of this section, some comments are highlighted regarding the proposed self-tuning control methodology. The first point concerns the assumption that Eqs. (4.83) and (4.84) hold for the stable model of Eq. (4.80). However, when the model identification procedure is considered, e.g., in Linden et al. (2012), in general it is assumed that the system under investigation, and that generating the identification data, has the form of (4.65) and is stable. The identification experiment provided stable linear models describing the nonlinear system of the wind turbine. However, for the *linear case*, staring from a linear model, it is known that the spectral factorization result leads to stable models, when the input–output sequences are quasi-stationary. Moreover, with reference to the practical problem of the identification of stable linear models from the observed inputs and outputs, for a finite number of data points, the estimated system is not guaranteed to be stable (even when the true linear system is known to be stable). Therefore, the stability constraints may be included in the identification phase, as, e.g., in Chui and Maciejowski (1996).

On the other hand, when unstable models have to be controlled, gain and phase margin requirements must be included, as described, e.g., in Ho and Xu (1998), and the controller parameters can be computed using the $w$-plane design. In this case, it can be shown that the model of Eq. (4.80) is transformed into its equivalent description

$$\frac{T_s\,(b_1\,T_s - b_2\,T_s)\,w^2 + (-2\,b_1\,T_s + 2\,b_2\,T_s + T_s\,(2\,b_1 + 2\,b_2))\,w - 4\,b_1 - 4\,b_2}{(a_1\,T_s^2 - T_s^2 - a_2\,T_s^2)\,w^2 + (4\,a_2\,T_s - 4\,T_s)\,w - 4 - 4\,a_2 - 4\,a_1} \approx \frac{k}{w\,\tau_o - 1}\,e^{-w\,L}. \qquad (4.86)$$

Assuming the validity of the approximation of Eq. (4.86), which neglects the fast dynamic stable modes, and considering only the unstable pole $\tau_o$ with the effective delay $L$ (see, e.g., the approximation in Skogestad, 2003 recalled in the following), the PI controller parameters are computed via

$$\begin{cases} K_p = \frac{\delta\,\tau_o}{A_m\,k}, \\[2mm] T_I = \frac{1}{\frac{\pi}{2}\,\delta - \delta^2\,L - \frac{1}{\tau_o}}, \\[2mm] \delta = \frac{A_m\,\phi_m + \frac{\pi}{2}\,A_m\,(A_m - 1)}{(A_m^2 - 1)\,L}. \end{cases} \qquad (4.87)$$

The relations of Eq. (4.87) provide the parameters $K_p$ and $T_I$ of the PI regulator that allow us to obtain the gain and phase margins ($A_m$, $\phi_m$) for the identified (unstable) model of Eq. (4.86). Note that, even if approximations are used to derive the tuning formulas of (4.87), it can be seen that the achievable gain and phase margins can be quite close to those specified, and in general within a 5% of maximal error.

The further important issue concerns the possible nonminimum phase behavior of the considered wind turbine described in Chapter 2. In this case, the self-tuning design procedure represented by Eqs. (4.83) and (4.84) would lead to poor performances. However, it is worth noting that, according to Chapter 2, there are two possible regions of turbine operation, namely the high and low wind speed regions. High-speed operation is frequently bounded by the speed limit of the machine. Conversely, regulation in the low-speed region is usually not restricted by speed constraints. However, the system has possible nonminimum phase dynamics in this low-speed region. Moreover, this behavior in the low-speed region can be observed for very low tip-speed ratio, i.e., when $\lambda$ is near zero. However, the condition $\lambda \approx 0$ is not always satisfied, also for low wind speed. Therefore the benchmark under consideration can present nonminimum features only for working conditions different from those considered in Chapter 2.

However, alternative design methodologies that take into account nonminimum phase or time delay behavior can be found, e.g., in Skogestad (2003). Considering again Eq. (4.86), Taylor approximation of a time delay transfer function leads to a verification that an inverse response time constant $T_o^{inv}$ (negative numerator time constant in Eq. (4.86)) may be approximated as a time delay:

$$(-T_o^{inv}\,w + 1) \approx e^{-T_o^{inv}\,w}. \qquad (4.88)$$

This represents the deteriorating effect of an inverse response similar to that of a time delay. In the same way a lag time constant $\tau_s$ may be approximated again as a time delay:

$$\frac{1}{\tau_s\,w + 1} \approx e^{-\tau_s\,w}. \qquad (4.89)$$

Furthermore, considering the product of Eqs. (4.88) and (4.89), it follows that the effective delay $L$ of Eq. (4.86) can be taken as the sum of the contribution from the various approximated terms $T_o^{inv}$ and $\tau_s$. In addition, for digital implementation with sampling period $T_s$, $L$ can consider also the contribution of the D/A converter $T_s/2$. Therefore the approximation of Eq. (4.86) is valid, and an approximate first-order time delay *stable* model is obtained as already shown by Eq. (4.86):

$$\frac{k}{w\,\tau_o + 1}\,e^{-w\,L}, \qquad (4.90)$$

where $k$ is the model gain, $\tau_o$ the dominant lag time constant, and $L$ the effective time delay. Under these assumptions, the following rules for the tuning of the PI controller can be exploited for first-order time delay (nonminimum phase) models:

$$\begin{cases} K_p = \frac{1}{k}\,\frac{\tau_o}{\tau_c + L}, \\[2mm] T_I = \min\{\tau_o,\,4\,(\tau_c + L)\}, \end{cases} \qquad (4.91)$$

where $\tau_c$ is the desired first-order closed-loop response time, and it represents the only tuning parameter. Small values of $\tau_c$ lead to response fast speed, whilst large values for stable and robust behavior. A good rule would suggest setting $\tau_c = L$ for fast response with good robustness properties (Skogestad, 2003).

The next remark is related with the control design rules that, according to Eqs. (4.83) and (4.84), seem dependent on the sampling time $T_s$, and for small sample times very aggressive control actions could be obtained. However, regarding the sampling time, the point that the data are acquired with a given sampling time and that the implementation of the control strategy is performed at a given frequency $1/T_s$ does not imply that also the model identification must be accomplished at the same rates. This remark is particular valid for the problem under investigation, where the model–reality mismatch variations are slower than the measurement errors affecting the wind turbine benchmark, as described in Odgaard et al. (2008). In fact, the dynamics of the main source of uncertainty and that representing the model–reality mismatch are slower than the disturbance effects, such as the wind signal or the other measurement errors. Therefore, under these considerations, the sampling time $T_s$ used in Eqs. (4.83) and (4.84) does not represent a key issue here, as it can be smaller than the data sampling time.

Another consideration regards the need of a PI controller, which should be quite evident, and, for example, a simple proportional compensation would not be sufficient. In fact, the simple proportional adaptive controller does not allow for asymptotic tracking and/or disturbance rejection (i.e., the steady state accuracy when the time tends to infinity). On the other hand, there are disadvantages of the PI controller. In fact, it is well known that PI controllers applied to systems with input saturation may yield possible overflow of the integral control action – the integrator or integral windup. Integral windup for PI control may severely deteriorate control performance, e.g., large overshoots and/or oscillations in the regulated output may occur. However, also in this case, a solution is possible. In fact, integral control action with *antiwindup strategy* can be used (Visioli, 2003), and in particular, the conditional integration scheme, which is quite easy and straightforward to implement. In this situation, the conditional integration scheme considers that the PI integral action is properly compensated depending on certain conditions of the control error, as already implemented by the PI controller block-set of the Simulink® environment (Visioli, 2003).

Finally, a further interesting problem connected with the issue above regards the bumpless transfer for adaptive control. In more detail, in the adaptive control setting, the controller output can have undesired transients (bumps), if a current online controller and a new controller to be switched have different outputs at the switching instant. To attenuate these bumps associated with controller switching, a variety of bumpless transfer methods have been suggested. As in this case, the plant is not precisely known at the outset, and the goal of adaptive control is to change the controllers to improve performance as plant data begins to reveal some information about the plant. Thus in adaptive switching control an exact plant is generally unavailable at the time of switching. Moreover, in the adaptive application considered in this monograph where the true plant model may only be poorly known at controller switching times, it may be preferable to employ a bumpless transfer technique that does not depend on a precise knowledge of the true plant model. Therefore the bumpless transfer method presented in Cheong and Safonov (2008), Jun and Safonov (1999) based on the slow decomposition of the controller is considered, as described in Chapter 2. It consists of a method that can be implemented without precise knowledge of the true plant at switching times. In particular, by appropriately reinitializing the states of the PI modes at switching times, this method ensures that not only will the controller output be continuous, but also that it will avoid fast transient bumps after switching.

### 4.2.3 Sustainable Control

In general, wind turbines in the megawatt size range are expensive, and hence their availability and reliability must be high in order to maximize the energy production. This issue could be particularly important for offshore installations, where operation and maintenance (O&M) services have to be minimized, since they represent one of the main factors of the energy cost. The capital cost, as well as the wind turbine foundation and installation, determines the basic term in the cost of the produced energy, which constitutes the energy "fixed cost." The O&M represent a "variable cost" that can increase the energy cost up to about 30%. At the same time industrial systems have become more complex and expensive, with less tolerance for performance degradation, productivity decrease, and safety hazards.

This leads also to an ever increasing requirement on reliability and safety of control systems subjected to process abnormalities and component faults. As a result, it is extremely important to consider the FDD or the FDI tasks, as well as the achievement of fault-tolerant features, for minimizing possible performance degradation and avoiding dangerous situations. With the advent of computerized control, communication networks, and information techniques, it is possible to develop novel real-time monitoring and fault-tolerant design techniques for industrial processes, but this also brings challenges.

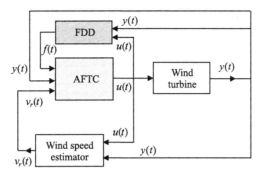

**FIGURE 4.6** General scheme of the Active FTC (AFTC) strategy.

In the last years many works have been proposed on wind turbine FDI/FDD, and the most relevant are, e.g., in Gong and Qiao (2013a), Freire et al. (2013). On the other hand, regarding the FTC problem for wind turbines, it was recently analyzed with reference to an offshore wind turbine benchmark, e.g., in Odgaard et al. (2013).

In general, FTC methods are classified into two types, i.e., Passive Fault Tolerant Control (PFTC) scheme and Active Fault Tolerant Control (AFTC) scheme (Mahmoud et al., 2003). In PFTC, controllers are fixed and are designed to be robust against a class of presumed faults. In contrast to PFTC, AFTC reacts to the system component failures actively by reconfiguring control actions so that the stability and acceptable performance of the entire system can be maintained. In particular for wind turbines, FTC designs were considered and compared in Odgaard et al. (2013). These processes are nonlinear dynamic systems, whose aerodynamics are nonlinear and unsteady, whilst their rotors are subject to complicated turbulent wind inflow fields driving fatigue loading. Therefore, the so-called wind turbine "sustainable" control represents a complex and challenging task (Odgaard and Stoustrup, 2013b).

Therefore the purpose of this section is to outline the basic solutions to sustainable control design, which are capable of handling faults affecting the controlled wind turbine. For example, changing dynamics of the pitch system due a fault cannot be accommodated by signal correction. Therefore, it should be considered in the controller design, to guarantee stability and a satisfactory performance. Among the possible causes for changed dynamics of the pitch system, they can be due to a change in the air content of the hydraulic system oil. This fault is considered since it is the most likely to occur, and since the reference controller becomes unstable when the hydraulic oil has high air content, as highlighted in Chapter 2. Another issue arises when the generator speed measurement is unavailable, and the controller should rely on the measurement of the rotor speed, which is contaminated with much more noise than the generator speed measurement. This makes it necessary to reconfigure the controller to obtain a reasonable performance of the control system.

Section 4.3 outlines the main differences between active and passive fault-tolerant control systems and suggests how they are applied to the considered wind turbine system.

## 4.3 FAULT TOLERANT CONTROL ARCHITECTURES

In order to outline and compare the controllers developed using active and passive fault-tolerant design approaches, they should be derived using the same procedures in the fault-free case. In this way any differences in their performance or design complexity would be caused only by the fault tolerance approach, rather than the underlying controller solutions.

Furthermore, the controllers should manage the parameter-varying nature of the wind turbine along its nominal operating trajectory caused by the aerodynamic nonlinearities. Usually, in order to comply with these requirements, the controllers are designed, for example, using LPV modeling or fuzzy descriptions (Bianchi et al., 2007; Galdi et al., 2008).

The two fault-tolerant control solutions have different structures as shown in Figs. 4.6 and 4.7. Note that only the AFTC relies on a fault diagnosis algorithm (FDD). This represents the main difference between the two control schemes.

The main point between AFTC and PFTC schemes is that an active fault-tolerant controller relies on a fault diagnosis system, which provides information about the faults $f$ to the controller. In the considered case the fault diagnosis system FDD contains the estimation of the unknown input (fault) affecting the system under control. The knowledge of the fault $f$ allows the AFTC to reconfigure the current state of the system.

On the other hand, the FDD is able to improve the controller performance in fault-free conditions, since it can compensate, e.g., the modeling errors, uncertainty, and disturbances. On the other hand, the PFTC scheme does not rely on a fault diagnosis algorithm, but is designed to be robust towards any possible faults. This is accomplished by designing a controller

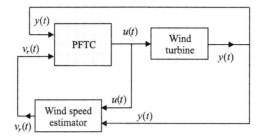

**FIGURE 4.7** General layout of the passive FTC (PFTC) structure.

that is optimized for the fault-free situation, while satisfying some graceful degradation requirements in the faulty cases. However, with respect to the robust control design, the PFTC strategy provides reliable controllers that guarantee the same performance with no risk of false FDI or reconfigurations.

In general, the methods used in the fault-tolerant controller designs should rely on output feedback, since only part of the state vector is measured. Additionally, they should take the measurement noise into account. Moreover, the design methods should be suited for nonlinear systems or linear systems with varying parameters. The latest proposed solutions for the derivation of both active and the passive fault-tolerant controllers rely on LPV and fuzzy or neural network descriptions, to which the fault-tolerance properties are added, since these frameworks methods are able to provide stability and guaranteed performance with respect to parameter variations, uncertainty, and disturbances.

Additionally, LPV and fuzzy or neural controller design methods are well-established in multiple applications including wind turbines (Bianchi et al., 2007; Galdi et al., 2008). To add fault-tolerance to the common LPV and fuzzy or neural controller formulation, different approaches can be exploited. For example, the AFTC scheme can use the parameters of both the LPV and fuzzy structures estimated by the FDD module for scheduling the controllers (Sloth et al., 2010; Sami and Patton, 2012).

On the other hand, different approaches can be used to obtain fault-tolerance in the PFTC methods. For this purpose, the design methods described in Bianchi et al. (2007), Galdi et al. (2008) can be modified to cope with parametric uncertainties, as addressed, e.g., in Puig (2010), Rodrigues et al. (2007). Alternatively, other methods could have been used such as Niemann and Stoustrup (2005), which preserves the nominal performance. Generally, these approaches rely on solving some optimization problems where a controller is calculated subjected to maximizing the disturbance attenuation. These problems are formulated as LMI (Chen et al., 1997).

### 4.3.1 Controller Compensation and Active Fault Tolerance

The key point of the controller compensation and active fault tolerance issues regards the fault estimation task, which is also the most important of all the fault diagnosis phases. In fact, when a fault is estimated, both the detection and the isolation phases can be easily achieved, since the fault nature can improve the diagnosis process. However, the fault identification problem itself has not gained enough research attention.

Most fault diagnosis techniques, such as parameter identification, parity space, and observer-based methods addressed in Chapter 3, in general cannot be directly used to identify faults in sensors and actuators.

Very little research has been done to overcome the fault identification problem. The Kalman filter for statistical testing and fault identification was proposed, e.g., in Patton et al. (1989a). However, the statistical testing methods can impose a high computational demand. A fault identification scheme solving a system inversion problem was proposed, e.g., in Simani and Patton (2002), Simani and Fantuzzi (2002) and applied to power systems.

In the scheme represented in Fig. 4.8, the fault identification task is performed by estimating the nonlinear relationship between residuals and fault magnitudes. This is possible because the generated robust residuals must contain only fault information, and do not depend on the system under diagnosis.

The fault estimation module depicted in Fig. 4.8 requires a nonlinear function approximation that can be performed by using data-driven or model-based approaches as highlighted in Chapter 3.

Another important fault identification strategy can be achieved via a purely nonlinear scheme, which provides the fault detection, the isolation, and the fault size estimation. As already remarked, this FDD method is based on the NLGA principle developed in De Persis and Isidori (2004a) and described in Chapter 3. By means of this methodology, disturbance decoupled adaptive nonlinear filters providing the fault reconstruction are developed. It is worth observing that the original NLGA FDD scheme based on residual signals cannot provide fault size estimation. The achieved results in fault-free and

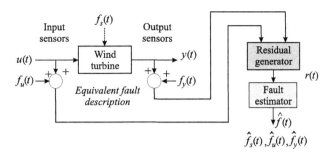

**FIGURE 4.8**  The fault estimation scheme for FTC.

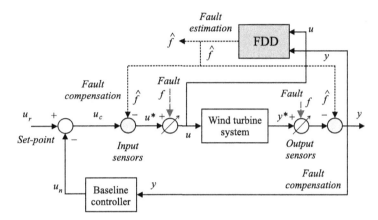

**FIGURE 4.9**  Active FTC strategy relying on fault compensation.

faulty conditions that will be illustrated in Chapter 5 highlight the enhancement of the control requirements, the asymptotic fault accommodation, and the control objective recovery.

Finally, regarding the AFTC strategies for fault compensation that are summarized in this chapter, the possible logic scheme of the integrated fault tolerant approach is represented in Fig. 4.9.

With reference to Fig. 4.9, the following nomenclature and symbols are used:

$u_r$, reference inputs (e.g., the reference set-point);
$u$, actuated inputs;
$u^*$, unmeasurable inputs;
$u_c$, controlled inputs;
$u_n$, the feedback signals from the baseline controller;
$y$, controlled outputs (e.g., the wind turbine monitored outputs);
$y^*$, unmeasurable outputs;
$f$, generic equivalent faults;
$\hat{f}$, estimated faults.

Therefore the logic scheme depicted in Fig. 4.9 shows how the AFTC strategy has been implemented by integrating the fault diagnosis module (FDD) with the existing control system. From the controlled input and output signals, the FDD module provides the correct estimate $\hat{f}$ of the $f$ actuator fault, which is injected to the control loop, for compensating the effect of the actuator fault. After this correction, the current controller provides the exact tracking of the reference signal $u_r$. Note that this signal can be generated by a further compensation block, as suggested in Chapter 2. It can be shown that the feedback of the estimated fault $\hat{f}$ improves the identification of the fault signal $f$ itself, by reducing also the estimation error and possible bias due to the model–system mismatch.

Further results recalled in this monograph will state the achieved performance of this integrated FDD and AFTC strategy. However, the enhancement of the control requirements, the asymptotic fault accommodation, and the control objective recovery, that in this book are verified in simulation, can require further studies and investigations when applied to real wind turbine installations, as remarked in Chapter 7.

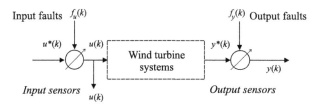

**FIGURE 4.10** The equivalent additive input and output sensor faults affecting the wind turbine system.

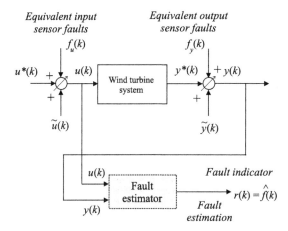

**FIGURE 4.11** A general residual generation scheme providing the fault estimates.

Finally, these last sections suggested the possible development of advanced fault tolerant control schemes. The methodologies were based on fault detection and diagnosis procedures relying on adaptive filters designed via model-based and data-driven approaches. The controller reconfiguration can exploit a further control loop, depending on the online estimate of the fault signal. One of the advantages of this strategy is that, for example, a structure of logic-based switching controller recalled in Chapter 2 is not modified. The adaptive fault tolerant control schemes are applied to different wind turbine systems in different working conditions, in the presence of faults, disturbances, measurement noise, and modeling errors.

## 4.4 FAULT TOLERANT CONTROL ORIENTED FAULT DIAGNOSIS

In the following the discrete-time monitored systems, i.e., the wind turbine and the wind farm, are assumed to be affected by equivalent and additive faults on the input and output sensor measurements, which are able to properly describe actuator, system, and sensor faults affecting the considered systems, as represented in Fig. 4.10, in the forms

$$\begin{cases} \mathbf{u}(k) = \mathbf{u}^*(k) + \mathbf{f}_u(k), \\ \mathbf{y}(k) = \mathbf{y}^*(k) + \mathbf{f}_y(k), \end{cases} \tag{4.92}$$

where $\mathbf{u}^*(k), \mathbf{y}^*(k)$ are the actual unmeasurable variables, $\mathbf{u}(k), \mathbf{y}(k)$ represent the sensor acquisitions, affected by both the measurement noise and the faults. As already remarked in Chapter 3, $\mathbf{f}_u(k)$ and $\mathbf{f}_y(k)$ are additive signals, which assume values different from zero only in the presence of faults.

Fig. 4.10 shows a general scheme with the faults affecting the system under diagnosis, i.e., the wind turbine or the wind farm, as additive signals on the input (actuator) and output measurements.

Among the different approaches to generate the residual signals, already recalled in Chapter 3, the solutions proposed in this chapter exploit data-driven or model-based prototypes, which are able provide an online estimation of the faulty signals. Hence, as shown in Fig. 4.11, the residual signals $\mathbf{r}$ are provided by means of these filters by using the measured inputs $\mathbf{u}(k)$ and outputs $\mathbf{y}$ from the systems under diagnosis:

$$\mathbf{r}(k) = \hat{\mathbf{f}}(k), \tag{4.93}$$

which are the estimated faults $\hat{\mathbf{f}}(k)$ representing the equivalent faults $\mathbf{f}_u(k)$ and $\mathbf{f}_y(k)$ of Eq. (4.92).

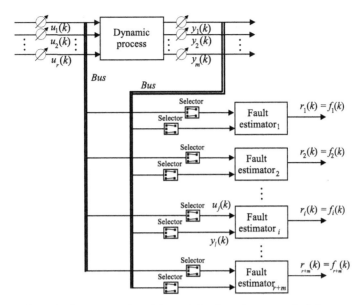

**FIGURE 4.12**  A general estimator scheme for the reconstruction of the *equivalent* input $\mathbf{f}_u$ or output $\mathbf{f}_y$ fault.

Fig. 4.11 highlights the residual generation scheme that is achieved by the proper processing of the acquired measurements, as described in Chapter 3. As already remarked, the fault diagnosis process involves, as a first step, the fault detection task. It is performed here by using a proper thresholding logic operating on the residuals after their elaboration into a proper evaluation function:

$$\mathbf{r}_e(k) = F\left(\mathbf{r}(k)\right), \tag{4.94}$$

where the proposed function $F(\cdot)$ can be the identity function, in case of the fault diagnosis solutions considered in this monograph, or suitable moving average or statistical operations on the residual signals, as explained in Chapter 3. Then the occurrence of the $i$th fault can be detected according to a simple thresholding logic described as

$$\begin{cases} \bar{r}_{e_i} - \delta\sigma_{r_i} \leq r_{e_i} \leq \bar{r}_{e_i} + \delta\sigma_{r_i} & \text{in fault-free situations,} \\ r_{e_i} < \bar{r}_{e_i} - \delta\sigma_{r_i} \text{ or } r_{e_i} > \bar{r}_{e_i} + \delta\sigma_{r_i} & \text{in faulty cases,} \end{cases} \tag{4.95}$$

where the $i$th item $r_{e_i}$ of the residual vector $\mathbf{r}_e$ is considered a random variable, whose unknown mean $\bar{r}_{e_i}$ and variance $\sigma_{r_i}^2$ can be estimated in fault-free conditions, after the acquisition of $N$ samples, as described by the relations

$$\begin{cases} \bar{r}_{e_i} = \frac{1}{N}\sum_{k=1}^{N} r_{e_i}(k), \\ \sigma_{r_i}^2 = \frac{1}{N}\sum_{k=1}^{N} (r_{e_i}(k) - \bar{r}_{e_i})^2. \end{cases} \tag{4.96}$$

Note that he tolerance parameter $\delta \geq 2$ has to be properly tuned in order to separate the fault-free from the faulty conditions. The $\delta$ value determines the trade-off between the false alarm rate and the fault detection probability. A common choice of $\delta$ relies on the three-sigma rule, otherwise extensive simulations can be performed to optimize the $\delta$ value.

Consequently to the fault detection, the fault isolation task is easily achieved by means of a bank of estimators. As described by Eq. (4.92), the faults are considered as equivalent signals that affect the input measurements, i.e., $\mathbf{f}_u$, or the output measurements, i.e., $\mathbf{f}_y$.

Under this assumption, by following the scheme of the generalized estimator configuration of Fig. 4.12, in order to uniquely isolate one of the input or output faults, by considering that multiple faults cannot occur, a bank of Multiple-Input Single-Output (MISO) fault estimators is used. In general, the number of these estimators is equal to the number of faults that have to be diagnosed, i.e., equal to the number of input and output measurements, $r + m$. Therefore, in general the $i$th fault estimator that reconstructs the fault $\hat{f}(k) = r_i(k)$ is driven by the components of the input and output signals $\mathbf{u}(k)$ and $\mathbf{y}(k)$ that are sensitive to the specific fault $f_i(t)$. Therefore, it should be clear that the design of these fault estimators is enhanced by the so-called Failure Mode & Effect Analysis (FMEA) described in Chapter 2. For each fault case, the failure

**TABLE 4.2** Fault signatures for FDI.

|        | $u_1$ | $u_2$ | ... | $u_r$ | $y_1$ | $y_2$ | ... | $y_r$ |
|--------|-------|-------|-----|-------|-------|-------|-----|-------|
| $r_1$  | 0     | 1     | ... | 0     | 0     | 0     | ... | 0     |
| $r_2$  | 1     | 0     | ... | 0     | 0     | 0     | ... | 0     |
| $\vdots$ |     |       | $\ddots$ |  |       |       | ... | $\vdots$ |
| $r_i$  | 0     | 0     | ... | 1     | 0     | 1     | ... | 0     |
| $\vdots$ |     |       | ... |       |       |       | $\ddots$ | $\vdots$ |
| $r_{r+m}$ | 0  | 0     | ... | 0     | 0     | 0     | ... | 1     |

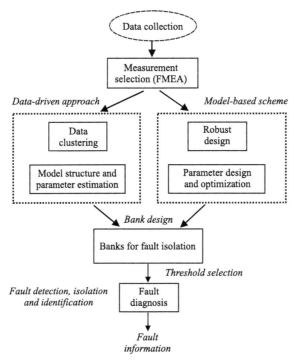

**FIGURE 4.13**   Fault diagnosis tasks oriented to the design of the FTC scheme.

modes and their resulting effects on the rest of the system are analyzed, and in particular the most sensitive input and output measurements to that specific fault situation are identified. In this way, it is possible to derive the dynamic relationships between the input–output measurements and the faults, as represented by the estimator bank of Fig. 4.12.

Fig. 4.12 shows this generalized fault estimator scheme, where the fault estimators are driven only by the input–output signals selected via the FMEA tool recalled in Chapter 2 and analyzed for the wind turbine systems considered in this monograph, so that the relative residual $r_i(k) = \hat{f}_i(k)$ is insensitive only to the fault affecting those inputs and outputs defined by the selector blocks. It is worth noting that multiple faults occurring at the same time cannot be correctly isolated using this configuration.

The capabilities of the adopted fault diagnosis module can be summarized by means of the so-called *fault signature* matrix, depicted in Table 4.2, where each entry that is characterized by a value equal to "1" means that the considered residual (i.e., the equivalent fault) is sensitive to the actual fault effect ("0" otherwise), under the hypothesis above mentioned.

As already remarked, the FMEA tool (Stamatis, 2003b) recalled in Chapter 2, which has to be executed before the design of the fault estimators, suggests how to select the input–output configuration for the fault estimator blocks. Then the design of the fault diagnosis block can be performed, as recalled in Chapter 3. Finally, the threshold test logic of Eq. (4.95) allows achieving the fault diagnosis tasks.

A summary of the complete design flow is shown in Fig. 4.13.

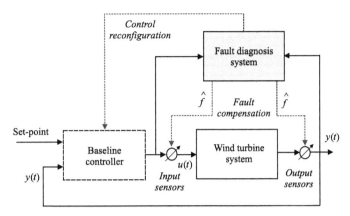

**FIGURE 4.14**   FTC strategy scheme comprising both the fault diagnosis and control compensation/reconfiguration tasks.

### 4.4.1   Fault Tolerant Control for Wind Turbine Systems

The structure of the FTC systems considered in this chapter, and applied to both the wind turbine and the wind farm benchmarks described in Chapter 2, is mainly based on fault diagnosis modules that provide the online fault estimations. The result of a proper fault identification allows for the compensation of the faulty measurement signals, before their access to the controller, so that the proper reference signal can be send to the turbine system, without the modification of the preexistent controller. Fig. 4.14 shows the overall FTC strategy.

Fig. 4.14 shows a general scheme of the FTC strategies proposed in this chapter: the fault diagnosis module can provide the online fault estimates that are used to compensate the faulty input–output signals, so that the controller can force the system to track the desired reference. On the other hand, the same fault diagnosis module can be used also for controller reconfiguration.

Therefore the fault estimates $\hat{\mathbf{f}}$ in Fig. 4.14, i.e., the signals $\hat{\mathbf{f}}_u$, $\hat{\mathbf{f}}_y$ of Eq. (4.92), are exploited for the compensation of both the input and the output measurements used by the system controller. In particular, the actuator signal coming from the controller is compensated by $\hat{\mathbf{f}}_u$, while $\hat{\mathbf{f}}_y$ corrects the output measurement acquired from the monitored system. After the fault compensation, the controller can track the nominal power reference signals. It is worth noting that, thanks to this fault estimation feedback, the controller could be easily designed considering the fault-free system condition.

Further investigations regarding the stability analysis of the overall FTC module are highlighted in Chapter 5, where it is shown that the variables of the models remain bounded in a set, which assures control performance, even in the presence of faults. Moreover, these faults do not modify the system structure, hence the global stability is guaranteed. However, whilst the fault effect is eliminated in steady-state conditions, during the transient the compensation can be improperly handled, and the stability properties should be considered.

Some concluding remarks can be finally drawn here. This chapter provided some theoretical results for the diagnosis and the compensation of faults in the actuators, systems, and sensors of wind turbine plants through the use of different fault diagnosis and control accommodation schemes. These strategies were designed from the linear and nonlinear input–output descriptions of the system under diagnosis, and the disturbance decoupling was obtained. Procedures for optimizing the fault sensitivity and dynamic response were also suggested.

An important aspect of the strategies based on linear residual generators is the simplicity of the technique used to generate these residuals when compared with different schemes. The algorithmic simplicity is a very important aspect when considering the need for verification and validation of demonstrable schemes for the viable application of these strategies to ream systems. The more complex the computations required to implement the scheme, the higher the cost and complexity in terms of verification and validation.

On the other hand, nonlinear methodologies can rely on a design scheme based on the structural decoupling of the disturbances obtained by means of proper coordinate transformations in the state and output spaces. To apply the nonlinear theory, simplified models of the system under diagnosis can be required. The mixed $\mathscr{H}_-/\mathscr{H}_\infty$ optimization of the trade-off between fault sensitivity, disturbances, and modeling errors is now well understood in the theoretical work and is a promising area for application study to energy conversion systems.

The nonlinear fault diagnosis and fault compensation strategies were based also on adaptive filters scheme. In addition to a proper detection and isolation, these methods provided also a fault size estimation. This feature is not usual for a fault detection and isolation method and can be very useful during online automatic control system reconfiguration, in order to recover from a faulty operating condition. Compared with similar methods proposed in the literature, the nonlinear adaptive fault diagnosis and accommodation techniques described here have the advantage of being applicable to more general classes of nonlinear systems and less sensitive to measurement noise, since they do not use input/output signal derivatives.

Suitable filtering algorithms for stochastic systems were also recalled. The knowledge regarding the noise process acting on the system under diagnosis can be exploited by the fault diagnosis and accommodation designs, hence the proposed schemes provided possible solutions to nonlinear system diagnosis with non-Gaussian noise and disturbances.

The main advantage of nonlinear fault diagnosis and compensation techniques with disturbance decoupling features is represented by the fact that they take into account directly the model nonlinearity and the system reality–model mismatch.

The fault compensation techniques that were outlined in this chapter will be applied to high fidelity simulators of wind turbines, which is able to take into account disturbances and measurement errors acting on the system under investigation. Moreover, the robustness characteristics and the achievable performances of the fault tolerant control approaches described have been carefully considered and investigated.

The effectiveness and the reliability of the proposed fault tolerant control schemes were shown by simulations and a comparison with widely used data-driven and model-based schemes with disturbance decoupling. The robustness properties of the designed fault estimators to model uncertainty, disturbances, and measurements noise will be analyzed via extensive simulations, including the use of Monte Carlo simulation experiments to tune the fault diagnosis and compensation module parameters.

Finally, the need to bridge the design gap between fault diagnosis and recovery mechanisms, i.e., the sustainable control schemes is obvious and properly analyzed in this chapter. Moreover, fault diagnosis and fault tolerant control strategies can be properly designed and combined as shown in this chapter.

## 4.5 SUMMARY

This chapter introduced the fault diagnosis and the fault tolerant schemes that will be adopted in the simulations of the wind turbine systems considered in this monograph. Firstly, in the light of the design of the fault diagnosis module already proposed and oriented to the design of the control reconfiguration and accommodation system, the effective design of the fault estimators for fault tolerant control is summarized. Then, by exploiting the fault diagnosis scheme, interesting by-products consisting of the fault detection and isolation tasks are achieved. Finally, possible implementations of fault tolerant controller schemes are shown, which represent the key issue of the sustainable control design for safety-critical systems, such as the offshore wind turbine installations. In general, the most effective fault tolerant control schemes rely on the online estimation and compensation of the system faults, modeled as equivalent input and output sensor faults, which are able to effectively describe any fault conditions affecting the considered wind turbine systems.

# Chapter 5

# Application Results

## 5.1 INTRODUCTION

This chapter shows the simulations related to the considered benchmark systems, in which the considered solutions for the fault diagnosis and the fault tolerant control solutions proposed in Chapters 3 and 4 have been analyzed, implemented, and discussed. Firstly, the focus is placed on the single wind turbine benchmark presented in Chapter 2, both the data-driven and model-based solutions are considered and validated by means of a Monte Carlo analysis. Then their performances are investigated with respect to other fault diagnosis methods, commonly adopted in the related literature recalled in Chapter 1. Furthermore, the tracking capability of the fault tolerant controllers, based on the fault accommodation strategy, are also considered.

Afterwards the wind farm benchmark system is considered as addressed in Chapter 2. Similarly to the analysis carried out for the single turbine system, the fault diagnosis and fault tolerant control solutions available in the related literature are analyzed and evaluated with respect to other commonly adopted methodologies.

Finally, in order to assess the considered systems in a more realistic framework, with reference to the wind turbine benchmark, Hardware-In-the-Loop (HIL) tests supported by the laboratory facilities have been considered, by means of an industrial computer interacting with the on-board electronics.

## 5.2 WIND TURBINE MODEL APPLICATION

In the following, with reference to the wind turbine benchmark model of Chapter 2, all the simulations are driven by the same wind mean speed sequence, depicted in Fig. 5.1.

It comes from the realistic acquisition of wind speed data from a wind turbine, as described in Chapter 6. It represents a good coverage of typical operating conditions, as it ranges from 5 to 20 m/s, with a few spikes at 25 m/s. The other wind speed components are represented by uniform random variables, as addressed in Chapters 2 and 6.

The simulations last for 4400 s, during which only one fault may occur. The discrete-time benchmark model runs at a sampling frequency of 100 Hz, so that $N = 440,000$ samples per simulation are acquired.

With reference to the different scenarios described in Chapter 2, Table 5.1 recalls the shape and the commencing times of the fault signals affecting the wind turbine system. They are modeled as input (actuator) or output (sensor) additive fault, based on the FMEA results of Chapter 2.

This section shows the results achieved from the application of the FDI/FDD and FTC schemes to the wind turbine system described in Chapter 2. Firstly, the so-called FMEA tool is described, as it can lead to an effective design of the FDI/FDD schemes. Then the FDD strategies are presented, highlighting their features for achieving the FDI task. Finally,

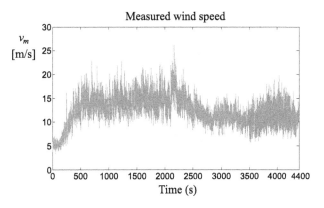

**FIGURE 5.1** The measured wind speed signal driving the simulations of the wind turbine benchmark.

Fault Diagnosis and Sustainable Control of Wind Turbines. DOI: 10.1016/B978-0-12-812984-5.00005-5

**TABLE 5.1** Fault scenarios for the wind turbine simulator.

| Fault # | Fault Location | Fault Model | Timing (s) |
|---------|----------------|-------------|------------|
| 1 | actuator | step | 2000–2100 |
| 2 | actuator | step | 2300–2400 |
| 3 | actuator | step | 2600–2700 |
| 4 | actuator | step | 1500–1600 |
| 5 | actuator | step | 1000–1100 |
| 6 | sensor | step | 2900–3000 |
| 7 | sensor | trapezoidal | 3500–3600 |
| 8 | sensor | step | 3800–3900 |
| 9 | sensor | step | 4100–4300 |

**TABLE 5.2** The FMEA results for the wind turbine benchmark.

| Fault # | 1 | 2 | 3 | 4 | 5 | 6 | 7 | 8 | 9 |
|---------|---|---|---|---|---|---|---|---|---|
| Signal | $\beta_{1,m1}$ | $\beta_{2,m2}$ | $\beta_{3,m1}$ | $\omega_{r,m1}$ | $\omega_{r,m2}$ | $\beta_{2,m1}$ | $\beta_{3,m2}$ | $\tau_{g,m}$ | $\omega_{g,m1}$ |
| RMSE % | 11.29 | 0.98 | 2.48 | 1.44 | 1.45 | 0.80 | 0.73 | 0.84 | 0.77 |

the results of the FTC methodologies are shown. In general, they rely on the online accommodation of the system faults, or the controller reconfiguration, modeled as described in Chapters 3 and 4.

As remarked in Chapter 2, by following the guidelines reported in Stamatis (2003b), the Failure Mode & Effect Analysis (FMEA) has been performed on the wind turbine simulator. The FMEA relies on the sensitivity analysis summarized in the following, aimed at determining the most sensitive measurements with respect to the simulated fault conditions.

In practice the monitored fault signals have been injected into the wind turbine benchmark simulator, assuming that only a single fault may occur in the considered plant. Then the Relative Mean Square Errors (RMSEs) between the fault-free and faulty measured signals are computed, so that, for each fault, the most sensitive signal can be selected. The results of the FMEA are shown in Table 5.2 for the wind turbine benchmark.

In particular, the FMEA has been performed on the basis of a selection algorithm that is achieved by introducing the normalized sensitivity function $N_x$, defined as

$$N_x = \frac{S_x}{S_x^*},\tag{5.1}$$

where

$$S_x = \frac{\|x_f(k) - x_n(k)\|_2}{\|x_n(k)\|_2^2},\tag{5.2}$$

$$S_x^* = \max \frac{\|x_f(k) - x_n(k)\|_2}{\|x_n[k]\|_2^2}.\tag{5.3}$$

Its value represents the effect of the considered fault case with respect to a certain measured discrete-time signal $x(k)$ (the integer $k$ is the sample number). The subscripts "$f$" and "$n$" indicate the faulty and the fault-free case, respectively. Therefore the measurements most affected by the considered faults imply a value of $N_x$ equal to 1. Otherwise, a smaller value of $N_x$, i.e., closer to zero, denotes a signal $x$ not affected by the fault. The signals characterized by the maximum values of $N_x$ are selected as the most sensitive measurements and they are considered in the design of the FDI blocks.

The results of the FMEA sensitivity procedure are summarized in Table 5.3 for the wind turbine simulator. The selected signals for each fault included in the wind turbine system are divided as inputs and outputs.

As a result, the fault diagnosis blocks that have to be designed can implement the reduced fault models instead of the overall system model of Eq. (2.47) with a noteworthy simplification of the inner structure, thus providing a decrease in the computational effort.

In order to recall how faults may affect the measurements acquired from a wind turbine simulator, comparisons between the faulty and the fault-free sensitivity functions of Eq. (5.1) are depicted in Figs. 5.2, 5.3, and 5.4. They regard the most sensitive signals of the FMEA tests and the related analysis reported in Chapter 2.

**TABLE 5.3** The fault sensitivity for the wind turbine benchmark.

| Fault # | Input Fault Sensitivity | Output Fault Sensitivity |
|---|---|---|
| 1 | $\beta_{1,m1}$, $\beta_{1,m2}$ | $\omega_{g,m2}$ |
| 2 | $\beta_{1,m2}$, $\beta_{2,m2}$ | $\omega_{g,m2}$ |
| 3 | $\beta_{1,m2}$, $\beta_{3,m1}$ | $\omega_{g,m2}$ |
| 4 | $\beta_{1,m2}$ | $\omega_{g,m2}$, $\omega_{r,m1}$ |
| 5 | $\beta_{1,m2}$ | $\omega_{g,m2}$, $\omega_{r,m2}$ |
| 6 | $\beta_{1,m2}$, $\beta_{2,m1}$ | $\omega_{g,m2}$ |
| 7 | $\beta_{1,m2}$, $\beta_{3,m2}$ | $\omega_{g,m2}$ |
| 8 | $\beta_{1,m2}$, $\tau_{g,m}$ | $\omega_{g,m2}$ |
| 9 | $\beta_{1,m2}$ | $\omega_{g,m1}$, $\omega_{g,m2}$ |

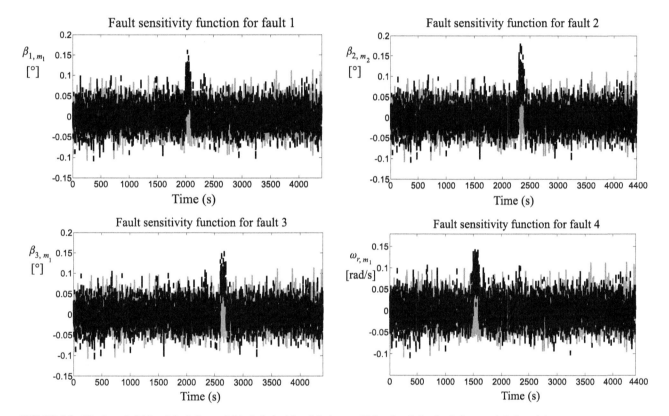

**FIGURE 5.2**  The (grey bold lines) fault-free and (black dashed lines) faulty sensitivity signals for the fault cases 1, 2, 3, and 4.

As an example, the normalized sensitivity functions for the fault cases 1, 2, 3, and 4 are considered in Fig. 5.2. On the other hand, the normalized sensitivity functions for the fault cases 5, 6, 7, and 8 are reported in Fig. 5.3. Finally, the normalized sensitivity functions for the fault case 9 are depicted in Fig. 5.4.

## 5.2.1  Data-Driven Fault Diagnosis Examples

The data-driven FDI methodologies recalled in Chapter 3 were applied to the wind turbine simulator. The considered process has $r = 2$ inputs that represent the aerodynamic torque $\tau_{aero}(t)$ and the reference signal $\tau_{ref}(t)$ measurements, whilst its $m = 2$ outputs are the rotor angular speed $\omega_r(t)$ and the generator torque $\tau_{gen}(t)$ measurements. The available data from the measured inputs and outputs were $440 \times 10^3$ samples from normal operating records acquired with a sampling rate of 100 Hz.

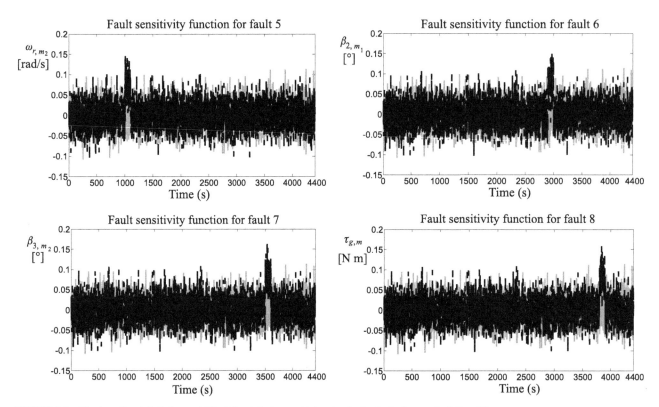

**FIGURE 5.3** The (grey lines) fault-free and (black lines) faulty sensitivity signals for the fault cases 5, 6, 7 and 8.

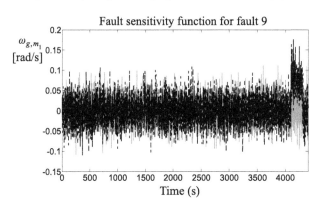

**FIGURE 5.4** The (grey lines) fault-free and (black lines) faulty sensitivity signals fault case 9.

Because of the underlying physical mechanisms of the wind turbine dynamics and because of the switching control logic described in Odgaard et al. (2009a), Odgaard and Stoustrup (2012b), the wind turbine system has a nonlinear steady state, as well as dynamic characteristics.

As an example, according to Chapter 3, the fuzzy approach to FDI is described here. Therefore the GK clustering algorithm (Babuška, 1998) with 4 clusters and $n = 2$ shifts was applied to the sampled data $\{\tau_{ref}(t), \tau_{aero}(t), \omega_r(t)\}$. On the other hand, 4 clusters and $n = 3$ shifts were considered for clustering of the sampled data $\{\tau_{ref}(t), \tau_{aero}(t), \tau_{gen}(t)\}$.

After clustering, the residual generator parameters were estimated using the identification method presented in Chapter 3. In this way, the $i$th output $y(t)$ of the wind turbine ($i = 1, \ldots, m$ and $m = 2$) continuous-time model is approximated by a TS fuzzy MISO discrete-time prototype with 2 inputs.

The mean square errors of the output estimations $r(t)$, under no-fault conditions, are 0.2549 for the first output, and 0.0125 for the second one. The fitting capabilities of the estimated fuzzy models can be expressed also in terms of the so-called Variance Accounted For (VAF) index (Babuška, 1998). In particular, the VAF value for first output is bigger than 90%, whilst for the second one it is bigger than 99%. Hence, the fuzzy multiple description seems to approximate the

**TABLE 5.4** Capabilities of the data-driven residual generators for FDI.

| Output # | VAF % |
| --- | --- |
| 1 | 90 |
| 2 | 99 |

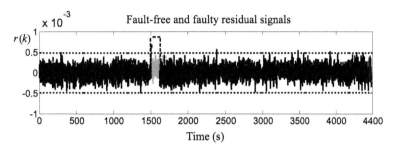

**FIGURE 5.5**  Residuals $r(k)$ with the FDI thresholds for the $\omega_r(t)$ sensor fault.

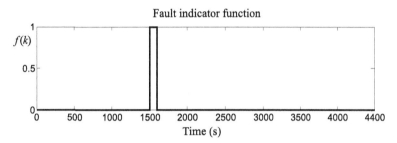

**FIGURE 5.6**  The fault indicator function for the $\omega_r(t)$ sensor.

process under diagnosis quite accurately. Using these fuzzy prototypes, the data-driven approach recalled in Chapter 3 for fault diagnosis is exploited, and applied to the benchmark wind turbine process. Table 5.4 summarizes the achieved results.

As an example, the results shown in the following were obtained by considering a fault $f_y(t)$ affecting the $\omega_r(t)$ sensor, whose measurement gets stuck at the constant value 1.4 rad/s for 100 s, and commencing at the instant $t = 1500$ s. On the other hand, a different fault, $f_u(t)$, corresponding to a converter fault situation is considered. This fault is active for 100 s in the period between 3800 and 3900 s. This fault is due to a constant offset of 100 in the converter torque control.

As described in Chapter 2, the controller in this wind turbine simulation model runs in two modes: power optimization (speed controlled by converter torque) and speed control (speed controlled by pitching blades) (Odgaard et al., 2009a). A wind speed in the range from approximately 5 to 15 m/s is simulated. This wind speed scenario is used to cover the relevant wind speed region of power optimization including turbulence.

The considered faults cause alteration of the signals $u(t)$ and $y(t)$, and therefore of the residuals $r(t)$ given by the predictive model recalled in Chapter 3. Residuals indicate fault occurrence according to the considered FDI logic, whether their values are lower or higher than the thresholds fixed in fault-free conditions.

As an example, Fig. 5.5 represents the fault-free (grey continuous line) and the faulty (black dashed line) residuals $r(t)$ related to the $\omega_r(t)$ sensor fault.

The fault detection thresholds used for FDI are represented as dotted constant lines in Fig. 5.5. Their values were properly settled by selecting a proper margin that leads to minimizing the false alarm and missed fault rates, while maximizing the correct detection and isolation rates. In these conditions, the fault is correctly detected when the corresponding residual signals exceed the thresholds by a suitable number of consecutive samples that are evaluated by a fault indicator function implemented as described in Chapter 6. This function is computed by means of a software counter defined in the Simulink® environment, whose output is zero in fault-free conditions, whilst having value 1 between 1500 and 1600 s. The value of this function is depicted in Fig. 5.6.

On the other hand, Fig. 5.7 represents the fault-free (grey continuous line) and the faulty (black dashed line) residuals $r(t)$ regarding the $\tau_{gen}(t)$ converter fault, $f_u(t)$.

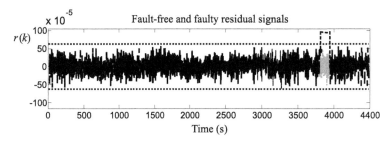

**FIGURE 5.7** Residuals $r(k)$ regarding the $\tau_{gen}(t)$ converter fault.

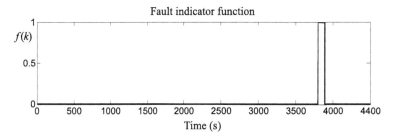

**FIGURE 5.8** The fault indicator function for the $\tau_{gen}(t)$ converter.

The fault detection thresholds represented as dotted constant lines in Fig. 5.7 were obtained with a suitable margin to minimize the false alarm and missed fault rates, while maximizing the correct detection and isolation rates. Also in these conditions, the fault is correctly detected when the residuals exceed the thresholds by a suitable number of consecutive samples, defined by the fault indicator function implemented in the Simulink® environment as described in Chapter 6. This function is 1 between 3800 and 3900 s and it is depicted in Fig. 5.8.

Finally, it is worth noting that the developed FDI strategy based on a data-driven solution allows the detection and isolation of realistic faults using uncertain measurements acquired from the wind turbine simulator benchmark described in Chapter 2. Further examples and comparisons will be shown in Section 5.2.3.

### 5.2.2 Model-Based Fault Diagnosis Examples

A multiple model-based methodology was applied to the wind turbine for the FDI of the actuators and the sensors of the benchmark system, as described in Chapter 3. In particular, the wind turbine process was described as an LPV state-space model with $r = 3$ inputs, i.e., the reference signal $\tau_{ref}(t)$, the wind speed $v_{hub}(t)$, and the pitch angle $\beta_i(t)$ measurements ($i = 1, 2, 3$). The single output considered for FDI purposes corresponds to the rotor angular speed $\omega_r(t)$.

The available data from the measured inputs and outputs were $440 \times 10^3$ samples from normal operating records acquired with a sampling rate of 100 Hz. On the basis of the state-space LPV model, an observer for the LPV model was designed and used for residual generation, as described in Chapter 3.

Note that, because of the underlying physical mechanisms described in Chapter 2, and due to the switching control logic described in Odgaard et al. (2009a), the wind turbine system has a nonlinear steady state, as well as dynamic characteristics. Therefore the considered model-based LPV strategy allowed achieving interesting solutions, even if purely nonlinear methodologies are analyzed in the following.

Two series of data were acquired from the benchmark process. The first one was used for the identification of the LPV model of the wind turbine, and the second one was exploited for its validation. According to the algorithm recalled in Chapter 3 for the selection of the LPV model order, the initial value of $n = 1$ and $\epsilon = 10^{-7}$ were fixed. Under these assumptions, as stated in Chapters 3 and 4, the partition of the input–output domain into suitable regions was performed. The partition of the domain was obtained by exploiting a data clustering algorithm addressed in Babuška (1998). The partition of the domain for the wind turbine with $k = 1$ has been achieved by considering the Cartesian product of the input–output intervals.

Five regions were considered for applying the LPV model and performing the identification task. Five local affine models were therefore estimated. In this case, the state-space vector is $x(t) = [y(t), u^T(t)]^T$, and the data belonging to the input–output domain have been thus clustered into 5 sets.

**TABLE 5.5** Capabilities of the model-based LPV residual generators for FDI.

| Data Set | VAF % |
| --- | --- |
| Estimation | 97.97 |
| Validation | 89.15 |

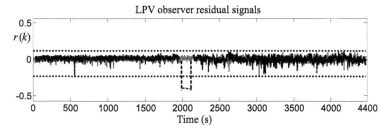

**FIGURE 5.9**   Residuals $r(k)$ for the $\beta_1(t)$ sensor fault.

The mean square errors of the output estimation $r(k)$, under no-fault conditions, are 0.0043 with respect to the estimation data and 0.0044 for the validation data set. The fitting capabilities of the estimated LPV model are expressed again in terms of the VAF index (Babuška, 1998). In particular, the VAF value computed for the identification data is 97.97%, whilst it is 89.15% for the validation data. Hence the fuzzy multiple description seems to approximate the process under diagnosis quite accurately. Table 5.5 summarizes the achieved results.

It is worth noting how the identified model represents a trade-off between simulation accuracy (also depending on the available data in each region) and structure complexity. Using this LPV description, the model-based approach considered in this section for fault diagnosis is exploited, and applied to the benchmark wind turbine process.

The following simulation results were obtained by considering a fault $f_u(t)$ affecting the $\beta_1(t)$ sensor, whose measurement gets stuck at the constant value 5 degrees for 100 s, and commencing at the instant $t = 2000$ s. On the other hand, a second fault case, $f_u(t)$, corresponding to the $\beta_3(t)$ sensor stuck at the constant value of 10 degrees is considered. This fault is active for 100 s in the period between 2600 and 2700 s.

Again, as recalled in Chapter 2, the controller in this wind turbine simulation model works mainly in two conditions: power optimization (speed controlled by converter torque) and speed control (speed controlled by pitching blades). A wind speed in the range from approximately 5 to 15 m/s is simulated. This wind speed scenario is used to cover the relevant wind speed region of power optimization including turbulence.

The considered faults cause alteration of the signals $u(t)$, and therefore of the residuals $r(t)$ given by the observer residuals. Its residuals indicate fault occurrence according to the considered FDI logic, whether their values are lower or higher than the thresholds fixed in fault-free conditions.

As an example, Fig. 5.9 represents the fault-free (grey continuous line) and the faulty (black dashed line) residuals $r(t)$ related to the $\beta_1(t)$ pitch sensor fault.

The fault detection thresholds of the residual evaluation logic are represented as dotted constant lines in Fig. 5.9. Their values were properly settled by selecting suitable margins, which minimize the false alarm and missed fault rates, while maximizing the correct detection and isolation rates. In these conditions, in practice, the fault is correctly detected when the corresponding residual signals exceed the thresholds by a sufficient number of consecutive samples, as indicated by the fault indicator function depicted in Fig. 5.10. This function is zero in fault-free conditions, and has value 1 between 2000 and 2100 s.

On the other hand, Fig. 5.11 represents the fault-free (grey continuous line) and the faulty (black dashed line) residuals $r(k)$ related to the $\beta_3(t)$ pitch sensor fault.

The fault detection thresholds, represented as dotted constant lines in Fig. 5.11, were obtained with suitable margins to minimize the false alarm and missed fault rates, while maximizing the correct detection and isolation rates. Also in these conditions, the fault is correctly detected when the residuals exceed the thresholds by a proper number of consecutive samples, as indicated by the fault indicator function depicted in Fig. 5.12. This function is equal to 1 between 2600 and 2700 s.

It is worth noting that the model-based strategy based on LPV prototypes allows the detection and isolation of realistic faults using uncertain measurements acquired from the wind turbine simulator.

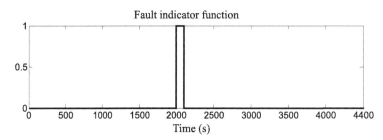

**FIGURE 5.10** Fault indicator function for the $\beta_1(t)$ sensor FDI.

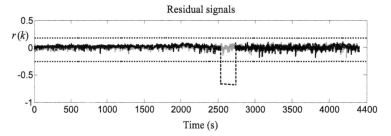

**FIGURE 5.11** Residuals $r(k)$ for the $\beta_3(t)$ sensor fault.

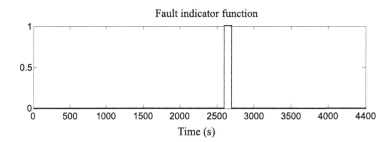

**FIGURE 5.12** Fault indicator function for the $\beta_3(t)$ sensor FDI.

### 5.2.3 Fault Diagnosis Comparative Results

Some examples of the fault diagnosis of the wind turbine benchmark model via fuzzy models have been already discussed in Section 5.2.1. In particular, the fuzzy $c$-means clustering exploited $n_C = 4$ clusters and $o = 3$ as delay on input and output regressors. The algorithm generated the membership function points that were fitted through Gaussian membership functions.

Afterwards the parameters of the TS fuzzy model of Eq. (3.92) were identified for each cluster, following the procedure explained in Chapter 3. As a result, the TS models could be implemented, and the 9 fault estimators were built and organized into 2 generalized observer schemes, in order to accomplish the fault diagnosis and identification task.

The modeling capabilities of the fuzzy TS models were evaluated in terms of $RMSE\%$, where the error was calculated as the difference between the measured and the estimated signals, for each output provided by the fuzzy estimators, in nominal fault-free conditions. Therefore Table 5.6 shows the achieved modeling performances of the 9 designed fault estimators. Furthermore, these reconstruction errors are directly exploited as residual signals and they are compared with the geometric threshold logic recalled in Chapter 3, optimally selected in order to achieve the optimization of the fault diagnosis performance indices, e.g., the missed fault and the false alarm rate, defined in the following. Thus Table 5.7 reports the adopted $\delta$ value for the threshold logic of each fault estimator.

The simulation results described here were achieved by considering the equivalent actuator and sensor faults, $f_u$ and $f_y$, respectively, of the scenario recalled in Chapter 2. These faults change the monitored input and output signals, **u** and **y**, affecting the residuals $r_i$ (with $i = 1, \ldots, 9$) generated by the fuzzy fault estimators. These residuals allow the achievement of the fault detection task, as they are significantly above the threshold bounds only when the relative fault is active. These residuals generated in faulty conditions by the fuzzy estimators were compared with the corresponding fault-free residuals. Therefore, fixed thresholds easily led to a solution of the fault detection task.

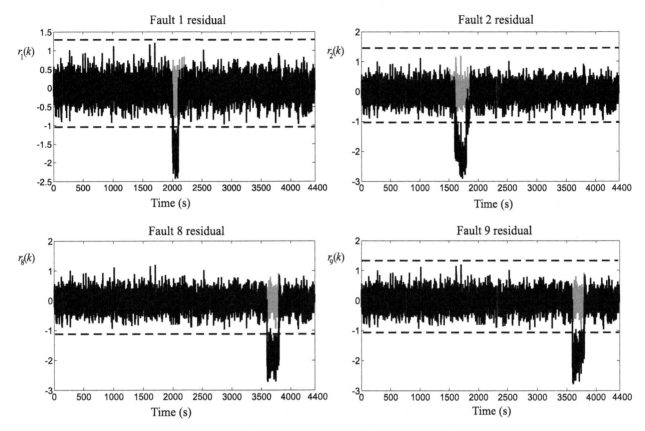

**FIGURE 5.13** Fault-free (grey line) and faulty (black continuous line) residuals for the fault cases 1, 2, 8 and 9.

**TABLE 5.6** RMSE % fuzzy and NN estimator performance.

| Fault Case # | 1 | 2 | 3 | 4 | 5 | 6 | 7 | 8 | 9 |
|---|---|---|---|---|---|---|---|---|---|
| Fuzzy RMSE % | 1.6 | 2.3 | 2.1 | 2.0 | 1.9 | 2.1 | 1.7 | 2.1 | 1.9 |
| NN RMSE % | 0.9 | 0.8 | 0.9 | 1.2 | 1.1 | 1.2 | 0.9 | 0.8 | 1.3 |

As an example for the fuzzy residual generators, these residuals are depicted in Fig. 5.13. They clearly show the achievement of the fault detection task, as they are significantly above the threshold bounds only when the relative fault is active.

Fig. 5.13 shows the residual signals generated in faulty conditions by the fuzzy estimators (black continuous line) compared with the fault-free residuals (grey line). The fixed thresholds are depicted with dotted lines. The considered residuals regard the fault cases 1, 2, 8, and 9. It is worth noting that in fault-free conditions the estimated residual signals $r_i(t)$ are not zero due to both the model–reality mismatch and the measurement errors.

With reference to the FDI example using NN for FDI, 9 open-loop NARX NN described in Chapter 3 have been designed to estimate the nonlinear behavior between the acquired measurements and the considered fault cases. The selected architecture of the networks involves two layers, namely the hidden layer and the output layer. The number of neurons in the hidden layer has been fixed to $n_h = 16$. Finally, a number of $d_u = d_y = 4$ has been chosen for the input–output delays. Similarly to the fuzzy models, the neural networks modeling capabilities have been tested in terms of RMSE and the results are reported in Table 5.6 in fault-free conditions.

Again the fault detection task is achieved by comparing the residual with a fixed optimized threshold, in this case, in contrast to the direct approach exploited by the fuzzy estimators, the residuals are filtered by an evaluation function ahead of the threshold comparison. This evaluation function can be either a Mobile Average (MA) or a Mobile Variance (MV), with a properly tuned window size ($n$), as reported in Table 5.7.

Also in this case meaningful results were achieved, as shown in Section 5.2.1, residual signals for actuator and sensor faults, together with the relative thresholds, were generated.

**TABLE 5.7** Residual evaluation logic parameters.

| Fault # | 1 | 2 | 3 | 4 | 5 | 6 | 7 | 8 | 9 |
|---|---|---|---|---|---|---|---|---|---|
| $\delta$ | 2.7 | 3.8 | 3.9 | 4.1 | 2.9 | 3.8 | 3.6 | 2.9 | 4.1 |
| Filter | MV (20) | MA (60) | MV (20) | MA (45) | MA (50) | MV (60) | MV (70) | MA (50) | MA (50) |

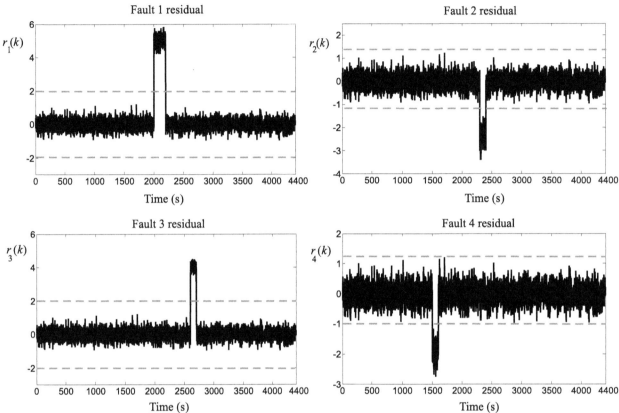

**FIGURE 5.14** Faulty residuals (continuous line) and fixed thresholds (dashed line) for the fault cases 1, 2, 3, and 4.

Fig. 5.14 shows some meaningful residual signal for actuator faults, together with the relative thresholds, while Fig. 5.15 shows the estimated signals regarding the sensor faults. Further details on validation and comparative results are described in the following.

As an example, Fig. 5.14 shows the residuals $r_i(k)$ generated in faulty conditions by neural network estimators (continuous line) compared the fixed thresholds (dashed line). The considered residuals concern the actuator faults 1, 2, 3, and 4.

On the other hand, Fig. 5.15 shows the residuals $r_i(k)$ generated by the NN residual generators (continuous line) compared the fixed thresholds (dashed line). The considered residuals concerns the sensor fault cases 6, 7, 8, and 9.

### 5.2.4 Performance and Robustness Analysis

The evaluation of the performances of the considered FDI/FDD strategies is based on the computation of the following indices:

- **False Alarm Rate** (FAR), which is the ratio between the number of wrongly detected faults and the number of simulated faults;
- **Missed Fault Rate** (MFR), which is the ratio between the total number of missed faults and the number of simulated faults;
- **True FDI Rate** (TFR), which is the ratio between the number of correctly detected faults and the number of simulated faults (complementary to MFR);
- **Mean FDI Delay** (MFD), which is the delay time between the fault occurrence and the fault detection.

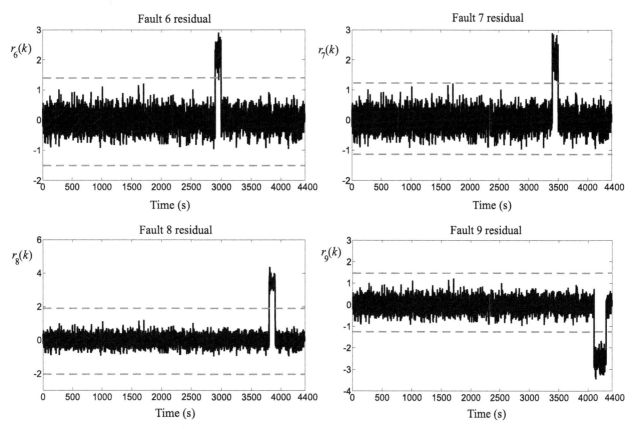

**FIGURE 5.15** Faulty residuals (continuous line) and fixed thresholds (dashed line) for the faults 6, 7, 8 and 9.

**TABLE 5.8** Monte Carlo sensitivity analysis.

| Parameter | Value | Standard Deviation % |
|---|---|---|
| $\rho$ | $1.225$ kg/m$^3$ | $\pm 25$ |
| $J$ | $7.794 \times 10^6$ kg/m$^3$ | $\pm 35$ |
| $C_p$ | $C_{p0}$ | $\pm 35$ |

A proper Monte Carlo analysis has been performed in order to compute these indices and to test the robustness of the considered FDI schemes. Indeed, the Monte Carlo tool is useful at this stage, as the efficacy of the diagnosis depends on both the model approximation capabilities and the measurement errors.

In particular, 1000 Monte Carlo runs have been performed, during which realistic wind turbine uncertainties have been considered by modeling some meaningful variables as Gaussian stochastic processes around the nominal values and with standard deviations corresponding to the realistic minimal and maximal error values of Table 5.8.

In addition to the considered fuzzy and neural network fault estimators, the performance indices of other fault diagnosis schemes are analyzed, as recalled in Chapter 3.

The first alternative approach considered here uses a Support Vector machine based on a Gaussian Kernel (GKSV) developed in Laouti et al. (2011). The scheme defines a vector of features for each fault, which contains relevant signals obtained directly from measurements, filtered measurements, or their combinations. These vectors are subsequently projected onto the kernel of the Support Vector Machine (SVM), which provides suitable residuals for all of the defined faults. Data with and without faults were used for learning the model for the FDI of the specific faults.

The second scheme consists of an Estimation-Based (EB) solution shown in Zhang et al. (2011). In particular, a fault detection estimator is designed to detect a fault, and an additional bank of estimators is derived to isolate them. The method was designed on the basis of a system linear model and used fixed thresholds. Each estimator for fault isolation was computed on the basis of the particular fault scenario under consideration.

**TABLE 5.9** FDI result comparison: first fault cases.

| Fault # | Index | GKSV | EB | UDC | COK | GFM | Fuzzy | NN |
|---------|-------|------|-----|-----|-----|-----|-------|-----|
| 1 | FAR | 0.001 | 0.001 | 0.001 | 0.001 | 0.001 | 0.001 | 0.001 |
|   | MFR | 0.002 | 0.003 | 0.002 | 0.003 | 0.002 | 0.001 | 0.001 |
|   | TFR | 0.978 | 0.977 | 0.987 | 0.977 | 0.982 | 0.999 | 0.999 |
|   | MFD (s) | 0.03 | 0.03 | 0.04 | 10.32 | 0.05 | 0.02 | 0.01 |
| 2 | FAR | 0.234 | 0.224 | 0.123 | 0.003 | 0.235 | 0.001 | 0.228 |
|   | MFR | 0.343 | 0.333 | 0.232 | 0.029 | 0.532 | 0.003 | 0.001 |
|   | TFR | 0.657 | 0.667 | 0.768 | 0.971 | 0.468 | 0.997 | 0.999 |
|   | MFD (s) | 47.24 | 44.65 | 69.03 | 19.32 | 13.74 | 0.08 | 0.08 |
| 3 | FAR | 0.004 | 0.141 | 0.123 | 0.056 | 0.135 | 0.003 | 0.001 |
|   | MFR | 0.006 | 0.132 | 0.241 | 0.128 | 0.232 | 0.008 | 0.001 |
|   | TFR | 0.974 | 0.868 | 0.769 | 0.872 | 0.768 | 0.992 | 0.999 |
|   | MFD (s) | 0.05 | 0.54 | 0.05 | 19.32 | 0.74 | 0.02 | 0.008 |

**TABLE 5.10** FDI result comparison: second fault set.

| Fault # | Index | GKSV | EB | UDC | COK | GFM | Fuzzy | NN |
|---------|-------|------|-----|-----|-----|-----|-------|-----|
| 4 | FAR | 0.006 | 0.005 | 0.123 | 0.056 | 0.236 | 0.004 | 0.001 |
|   | MFR | 0.005 | 0.006 | 0.113 | 0.128 | 0.333 | 0.004 | 0.001 |
|   | TFR | 0.975 | 0.994 | 0.887 | 0.872 | 0.667 | 0.996 | 0.999 |
|   | MFD (s) | 0.15 | 0.33 | 0.04 | 19.32 | 17.64 | 0.02 | 0.69 |
| 5 | FAR | 0.178 | 0.004 | 0.234 | 0.256 | 0.236 | 0.002 | 0.001 |
|   | MFR | 0.223 | 0.005 | 0.254 | 0.329 | 0.242 | 0.003 | 0.001 |
|   | TFR | 0.777 | 0.995 | 0.746 | 0.671 | 0.758 | 0.997 | 0.998 |
|   | MFD (s) | 25.95 | 0.07 | 0.04 | 31.32 | 9.49 | 0.03 | 0.001 |
| 6 | FAR | 0.897 | 0.173 | 0.334 | 0.156 | 0.096 | 0.042 | 0.001 |
|   | MFR | 0.987 | 0.234 | 0.257 | 0.129 | 0.042 | 0.033 | 0.001 |
|   | TFR | 0.013 | 0.766 | 0.743 | 0.871 | 0.958 | 0.967 | 0.999 |
|   | MFD (s) | 95.95 | 11.37 | 12.94 | 34.02 | 9.49 | 3.03 | 0.001 |

The third method relying on Up–Down Counters (UDC) was addressed in Ozdemir et al. (2011). These tools are commonly used in the aerospace framework, and they provide a different approach to the decision logic applied to the FDI residuals. Indeed, the decision to declare the fault occurrence involves discrete-time dynamics and is not simply a function of the current residual value.

The fourth approach Combines Observer and Kalman filter (COK) methods (Chen et al., 2011). It relies on an observer used as a residual generator for diagnosing the faults of the drive-train, in which the wind speed is considered a disturbance. This diagnosis observer was designed to decouple the disturbance and simultaneously achieve optimal residual generation in a statistical sense. For the other two subsystems of the wind turbine, a Kalman filter-based approach was applied. The residual evaluation task used a generalized likelihood ratio test, and cumulative variance indices were applied. For fault isolation purposes, a bank of residual generators was exploited. Sensor and system faults were thus isolated via a decision table.

Finally, the fifth method is a General Fault Model (GFM) scheme, which is a method of automatic design (Svärd et al., 2011). The FDI strategy consists of three main steps. In the first step, a large set of potential residual generators was designed. In the second step, the most suitable residual generators to be included in the final FDI system were selected. In the third step, tests for the selected set of residual generators were performed, which were based on comparisons of the estimated probability distributions of the residuals, evaluated with fault-free and faulty data.

The comparative analysis results are reported in Tables 5.9, 5.10, and 5.11, according to the fault typology and topology. In particular, different model-based and data-driven approaches to the FDI of the wind turbine benchmark model are compared.

**TABLE 5.11** FDI result comparison: third fault set.

| Fault # | Index | GKSV | EB | UDC | COK | GFM | Fuzzy | NN |
|---|---|---|---|---|---|---|---|---|
| 7 | FAR | 0.899 | 0.044 | 0.134 | 0.134 | 0.123 | 0.047 | 0.676 |
|   | MFR | 0.899 | 0.035 | 0.121 | 0.101 | 0.098 | 0.023 | 0.001 |
|   | TFR | 0.101 | 0.965 | 0.879 | 0.899 | 0.902 | 0.977 | 0.999 |
|   | MFD (s) | 99.95 | 26.17 | 13.93 | 35.01 | 29.79 | 5.07 | 6.87 |
| 8 | FAR | 0.004 | 0.045 | 0.144 | 0.109 | 0.099 | 0.003 | 0.466 |
|   | MFR | 0.007 | 0.011 | 0.101 | 0.032 | 0.124 | 0.002 | 0.001 |
|   | TFR | 0.993 | 0.989 | 0.899 | 0.968 | 0.876 | 0.998 | 0.999 |
|   | MFD (s) | 0.07 | 0.08 | 0.09 | 0.06 | 8.94 | 0.05 | 0.20 |
| 9 | FAR | N/A | N/A | N/A | N/A | N/A | 0.134 | 0.002 |
|   | MFR | N/A | N/A | N/A | N/A | N/A | 0.165 | 0.001 |
|   | TFR | N/A | N/A | N/A | N/A | N/A | 0.835 | 0.998 |
|   | MFD (s) | N/A | N/A | N/A | N/A | N/A | 0.30 | 0.001 |

The results summarized in Tables 5.9, 5.10, and 5.11 serve to highlight the efficacy of the considered FDI solutions. In details, both data-driven approaches seem to work better than other approaches, and they have a noteworthy performance level considering the mean delay time, which is significantly lower than 10 s for all the fault cases. Also false alarm and missed fault rate are often lower than those of other approaches, particularly NN features an almost null missed fault rate for all the considered faults. However, for both model-based and data-driven FDI designs, optimization stages are required, for example, for the selection of the optimal thresholds. Furthermore, GKSV involves delays bigger than 25 s, with false alarms and missed fault rate up to 35%. EB has comparable performance with respect to GKSV in terms of false alarm, true detection, and missed fault rate, but with a quicker detection. UDC often involves high false alarm rates, higher than 12% for all the detectable faults. COK and GFM have similar performances, with delay time higher than 10 s, false alarm and missed fault bigger than 10%. Fault 9 concerns the drive-train. This fault is difficult to detect at wind turbine level, therefore it is investigated also in the context of the wind farm benchmark. However, data-driven approaches to FDI can detect it, with a minimum delay but with a lower True FDI Rate, with respect to the other fault cases.

## 5.3 ADVANCED CONTROL DESIGNS FOR WIND TURBINES

This section describes the control designs relying on data-driven and model-based solutions, as presented in Chapter 2.

Regarding the data-driven approach using the fuzzy modeling method, the GK clustering algorithm discussed in Chapter 4, $M = 3$ clusters and $n = 2$ shifts were applied to estimate and validate sampled data sets $\{P_g(k), \omega_g(k), \beta(k)\}$, with $k = 1, 2, \ldots, N$ and $N = 440 \times 10^3$. On the other hand, $M = 3$ and $n = 2$ were considered for achieving a suitable clustering of the sampled data sets $\{P_g(k), \omega_g(k), \tau_g(k)\}$. After clustering the TS model parameters for each output $P_g(k)$ and $\omega_g(k)$ were estimated. Therefore the two outputs $y(t)$ of the wind turbine model described in Chapter 2 are approximated by 2 TS fuzzy prototypes of Eq. (4.59). The relative mean square errors of the output estimations are 0.0254 for the first output and 0.0125 for the second one.

The data-driven fuzzy model estimation procedure was implemented in order to guarantee the identification of stable TS prototypes. This point was realized in Matlab® by modifying the optimization procedure of the FMID toolbox (Babuška, 2000) so that the identified parameters $n$, $\mu_i^{(m)}$, $a_i^{(m)}$, and $b_i^{(m)}$ provide stable systems.

The fitting capabilities of the estimated fuzzy models can be expressed also in terms of the VAF index (Babuška, 1998). In particular, the VAF value for first output was bigger than 90%, whilst it was bigger than 99% for the second one. Hence the fuzzy TS models seem to approximate the wind turbine process quite accurately.

Regarding the data-driven fuzzy controller estimation recalled in Chapter 4, the experimental setup employs $m = 2$ (Multiple-Input Single-Output) MISO fuzzy regulators used for the control of the blade pitch angles $\beta(t)$ and the generator control torque $\tau_g(t)$, respectively, which have been identified according to the scheme of Chapter 4.

Also in this case the GK clustering algorithm was applied again for the estimation of the two fuzzy inverse model regulators. $M = 3$ clusters and $n = 3$ shifts were applied to estimate and validate sampled data sets $\{\beta(k), P_g(k), \omega_g(k)\}$. On the other hand, $M = 3$ and $n = 3$ were considered again for achieving a description of the second fuzzy inverse model regulator via the clustering of the data $\{\tau_g(k), P_g(k), \omega_g(k)\}$.

**TABLE 5.12** $NSSE\%$ values of the data-driven controllers.

| Working Condition | $NSSE\%$ |
|---|---|
| Partial load | 36.36 |
| Full load | 16.57 |

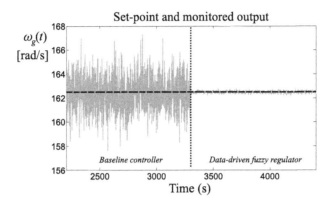

**FIGURE 5.16** Generator speed (bold gray line) $\omega_g(t)$ compared with its set-point (dashed black line) $\omega_{nom}$.

The data-driven fuzzy model estimation procedure was implemented in order to guarantee the identification of stable systems. This feature was realized in Matlab® by modifying the optimization procedure of the FMID toolbox (Babuška, 2000) so that the identified parameters $n$, $\mu_i^{(c)}$, $a_i^{(c)}$, and $b_i^{(c)}$ of the fuzzy controllers led to stable closed-loop simulations.

The controller capabilities have been assessed in simulation, and Normalized Sum of Squared tracking Error ($NSSE$) values (in percents), defined by

$$NSSE\% = 100 \sqrt{\frac{\sum_{k=1}^{N}(r(k) - y(k))^2}{\sum_{k=1}^{N} r^2(k)}}, \tag{5.4}$$

have been computed for the controller solutions addressed in Chapter 4.

With reference to Table 5.12, in partial load (Region 1), the performance is represented by the comparison between the power produced by the generator, $y(k) = P_g$, with respect to the theoretical maximum power output, $r(k) = P_{g,max} = \tau_{aero}(k)\,\omega_r(k)$, given the instant wind speed. Thus, the tracking error corresponds to $e(k) = r(k) - y(k) = P_{g,max} - P_g(k)$. On the other hand, in full load (Region 2), the performance depends on the generator speed, $y(k) = \omega_g(k)$, with respect to the nominal one, $r(k) = \omega_{nom}$. Therefore, the tracking error is $e(k) = \omega_{nom} - \omega_g(k)$.

According to the values summarized in Table 5.12, the capabilities of the considered data-driven controllers appear quite interesting, as shown also in the example of Fig. 5.16.

In particular, Fig. 5.16 shows the generator speed $\omega_g(t)$ versus its desired value $\omega_{nom}$ in full load (Region 2). When the data-driven controllers are exploited, the tracking error is smaller than the case of the baseline benchmark controller recalled in Chapter 2.

As highlighted in Fig. 5.16, during the time interval $2200 < t < 3300$ s the baseline regulator is working. In contrast, when the data-driven controllers are exploited in the control loop, during the interval $3300 < t < 4400$ s, the tracking error is lower.

With reference to the model-based control design, the wind turbine benchmark behavior was approximated using 2 time-varying MISO discrete-time prototypes as in Eq. (4.80) with 2 inputs.

In order to highlight the approximation capabilities of this model-based solution, Table 5.13 reports the so-called estimation or reconstruction errors, again with respect to the VAF index (Babuška, 1998). Values near 100% indicate that the reconstructed outputs are very close to the corresponding measured ones, $y(k)$. These models are compared with the efficacy of *third order models*, thus highlighting that the second order models identified online are highly appropriate for describing the behavior of the wind turbine process outputs.

Fig. 5.17 shows the comparison between the measured outputs $P_g(k)$ and $\omega_g(k)$, and the reconstructed ones in both the partial and full load working conditions.

**TABLE 5.13** Reconstruction errors as $VAF\%$ values.

| Process Output | Model Order | Performance % |
|---|---|---|
| $P_g$ | 2 | 98.20 |
| $\omega_g$ | 2 | 98.13 |
| $P_g$ | 3 | 98.21 |
| $\omega_g$ | 3 | 98.14 |

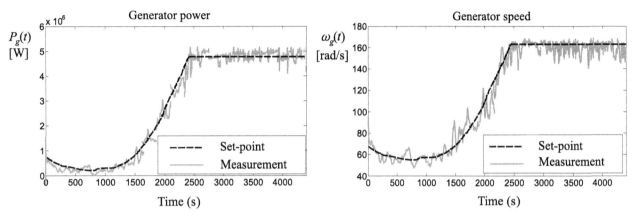

**FIGURE 5.17**  Tracking performances in terms of $P_g(k)$ and $\omega_g(k)$ outputs.

**TABLE 5.14** Adaptation algorithm initial values.

| Parameter | $\bar{\theta}(0)$ | | $\Sigma_{\bar{\psi}}(0)$ | $\delta$ |
|---|---|---|---|---|
| Value | $[0.1, 0.15, 0.20, 0.25\,0.30, 0.35]^T$ | | $10^{-1} I_7$ | 0.995 |

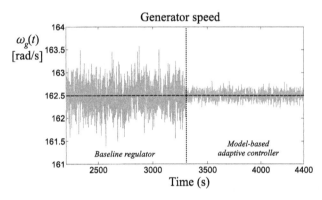

**FIGURE 5.18**  $\omega_g(t)$ tracking capabilities in full load conditions.

Using these second order models identified recursively, according to Chapter 3, the parameters of the adaptive controllers were computed online. In particular, for each output, the adaptive regulator parameters of Eqs. (4.82) were computed analytically at each time step $k$.

Simulations were performed under the same conditions of the data-driven controllers, and three adaptive regulators were used for the control of the blade pitch angles, and the generator control torque, in partial and full load working conditions.

As an example, the initial values for the parameters of the online estimation algorithm are listed in Table 5.14.

According to the simulation results presented in this section, good tracking capabilities of the model-based adaptive controllers seem to be reached, and they are better than the data-driven regulators, as shown in the example of Fig. 5.18.

In particular, Fig. 5.18 shows the set-point $\omega_g(t)$ versus its desired value $\omega_{nom}$ in full load working conditions (Region 2). The model-based adaptive regulators have replaced the benchmark controller at $t \geq 260$ s.

**TABLE 5.15** Adaptive controllers in partial and full load conditions tuned using third order models.

| Working Condition | $NSSE\%$ Value |
|---|---|
| Partial load | 23.81 |
| Full load | 12.26 |

**TABLE 5.16** Realistic wind turbine uncertainty.

| Model Parameter | Error % |
|---|---|
| $\beta(t)$ | 11 |
| $\omega_g(t)$ | 18 |
| $\tau_g(t)$ | 21 |
| $P_g(t)$ | 20 |
| Pitch 2nd order model natural frequency & damping ratio | 49 |
| Drive train model efficiency | 5 |
| Converter 1st order model time constant | 50 |

Also for this situation Fig. 5.18 shows how the model-based adaptive controller solution is able to achieve good performance.

It is worth noting that, as already highlighted in Table 5.13, the recursive identification of third order prototypes used for the design of the model-based regulators leads to quite limited improvements of the control performances. As an example, the results achieved by third order models are summarized in Table 5.15 in terms of $NSSE\%$ values.

It is worth observing that the previous results were obtained using proper advanced control solutions and implementations described in Chapters 2 and 6. In addition to the antiwindup method, a proper switching mechanism is needed to achieve a bumpless control transfer between adaptive regulators.

In the following, further results are reported. They regard the performance evaluation of the wind turbine control schemes with respect to modeling errors and measurement uncertainty. In particular, different simulations have been considered by exploiting the wind turbine benchmark, and a Matlab® Monte Carlo analysis. In fact, the Monte Carlo tool is useful at this stage as the control strategy performances depend on the error magnitude due to the model–reality mismatch, as well as on input–output measurement errors.

In particular, the nonlinear wind turbine simulator described in Chapter 2 developed in Matlab® and Simulink® environments is able to vary the statistical properties of the signals used for modeling possible process parameter uncertainty and measurement errors. Under this assumption, Table 5.16 recalls the nominal values of the considered wind turbine variables with respect to their simulated but realistic uncertainty (Odgaard et al., 2009a).

It is worth noting that the sensitivity analysis performed in this section represents one of the key issues of the monograph. In fact, when considering parametric uncertainty in connection with low order systems, the considered solutions remain valid. Therefore the relevant fact here is that the wind turbine system can be approximated with very good accuracy by LPV models. On the other hand, due to the wind turbine model approximations described in Chapter 2, also the stiffness assumptions are valid for the considered simulation benchmark. It may be true that these underlying stiffness assumptions typically hold up to some frequency, and the robustness analysis must take into account that the controller does not rely on high-frequency behavior. That is, the bandwidth of the controller is limited. These considerations could have been important in full load working conditions, as addressed in Bianchi et al. (2007).

On the other hand, for the high wind speeds, i.e., in full load operation, the desired operation of the wind turbine is to keep the rotor speed and the generated power at constant values, as described in Chapter 2. The main idea is to use the pitch system to control the efficiency of the aerodynamics, while applying the rated generator torque. However, in order to improve tracking of the power reference and cancel steady-state errors on the output power, a power controller is also introduced. The structure of the controllers operating above the rated wind speed was already shown in Chapter 2.

As described in Chapter 2, the wind speed can be considered as the disturbance input to the system. However, higher frequency components (e.g., the resonant frequency of the drive-train) can affect the measured generator speed. Therefore the measured generator speed is band-stop filtered before it is fed to the controller, to remove possible high frequency from the measurement. This solution is also found in other wind turbine control schemes to mitigate the effects of drive-train

**TABLE 5.17** Controller performance comparisons.

| Controller Type | Partial Load $NSSE\%$ | Full Load $NSSE\%$ |
|---|---|---|
| Data-driven fuzzy | 37.19 | 17.94 |
| Model-based adaptive | 24.52 | 13.72 |

oscillations, see, e.g., Sloth et al. (2011). The applied notch filters can be described as second order models that are discretized using, e.g., zeroth-order hold. Clearly, the notch frequency of the filter must be close to the resonant frequency to be damped. In this way, the speed controller is implemented as a PI controller that is able to track the speed reference $\omega_{nom}$ and cancel possible steady-state errors on the generator speed $\omega_g(t)$. On the other hand, the power controller is implemented in order to cancel possible steady-state errors on the output power $P_g(t)$ with respect to the power reference $P_r$. This suggests again introducing an integral control for the power controller, thus leading to a second PI regulator.

Therefore, due these motivations, the Monte Carlo analysis for robustness evaluation is adequate and it was performed by describing the wind turbine uncertain variables as Gaussian stochastic processes, with zero-mean and standard deviations corresponding to the maximal error values in Table 5.16. Moreover, as remarked in Chapter 2, it is assumed that the power coefficient $C_p$-map is uncertain, as described in Chapter 2.

In Table 5.17 the performances of the data-driven and model-based regulators are compared. Therefore, the average values of the $NSSE\%$ index were computed and evaluated with 1000 Monte Carlo runs, in both partial and full load conditions.

In particular, Table 5.17 summarizes the values of the considered performance index $NSSE\%$, with reference to the possible combinations of the random parameters described in Table 5.16. Table 5.17 shows that the considered control schemes, and in particular the model-based adaptive solution, allowing to maintain good control performances even in the presence of considerable error and uncertainty effects.

At this point, this section compares the previous model-based and data-driven methods with respect to different approaches recalled in Chapter 3, and in particular relying on the sliding mode, the neural controller, and LPV gain scheduling design, already proposed for wind turbines, and applied to the considered benchmark.

Sliding mode control is designed on the basis of a linearized model of the wind turbine process (Edwards and Spurgeon, 1998). In this case, the design procedure is based on the selection of an appropriate switching manifold, and then on the determination of a control law, including a discontinuous term, that ensures the sliding motion in this manifold. This control strategy can manage disturbances and modeling errors, which represent the effect of both the linearization and measurement errors.

The second control scheme relies on the Neural Network (NN) tool (Korbicz et al., 2004). The NN controller was on a 2-input 2-output time-delayed three-layer quasi-static Multiple-Layer Perceptron (MLP) NN with 5 neurons in the input layer, 10 neurons in the hidden layer, and 2 neurons in the output layer. The NN has been trained in order to provide the optimal reference tracking on the basis of the training patterns and target sequences.

The third approach uses an LMI-based method for synthesizing the LPV controller, presented in Sloth et al. (2010) and recalled in Chapter 3. The considered wind turbine model has varying parameters caused by the nonlinearity of the aerodynamic model along the nominal operating trajectory and due to the model uncertainty. In particular, the instantaneous partial derivatives of the aerodynamic torque are part of the linearized model and change along the nominal operating trajectory. These changes were approximated using an affine description in the wind speed.

In order to provide a brief but clear insight into the above mentioned techniques, the comparisons were performed under the same previous working conditions, and based on the $NSSE\%$ index, when the realistic wind turbine uncertainty of Table 5.16 was considered. It is worth noting that the schemes implemented via the sliding mode, LPV, or neural controllers do not exploit any adaptation mechanism. In fact, the sliding mode control strategy is able to decouple the uncertainty via the sliding motion, whilst neural networks or the LPV controller were designed to passively tolerate disturbances and modeling errors.

Table 5.18 summarizes the results obtained by comparing the control techniques considered in this section. It can be seen how the different schemes are able to tolerate uncertainty and errors.

As an example, Fig. 5.19 shows the achieved results obtained with the sliding mode controller. In particular, the tracking capabilities with respect to the 2 controlled variables $P_g(t)$ and $\omega_g(t)$ are depicted, in full load conditions.

In Fig. 5.19 the reference values are represented in black dashed lines, whilst the controlled signals are gray solid lines. With reference to the sliding mode controller, it appears that acceptable control performances are not achieved.

**TABLE 5.18** Wind turbine control technique comparison.

| Controller Type | Partial Load $NSSE\%$ | Full Load $NSSE\%$ |
|---|---|---|
| Data-driven fuzzy | 37.19 | 17.94 |
| Model-based LPV | 24.52 | 13.72 |
| Sliding mode | 28.52 | 27.72 |
| NN controller | 37.49 | 37.94 |
| Gain scheduling | 23.92 | 13.02 |

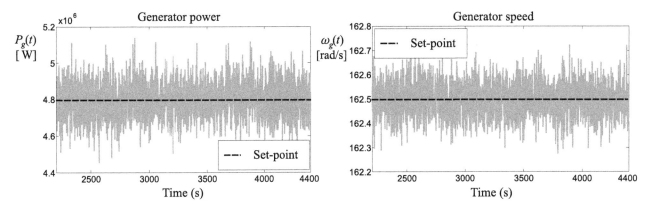

**FIGURE 5.19** Generator power $P_g(t)$ and speed $\omega_g(t)$ with the sliding mode controller.

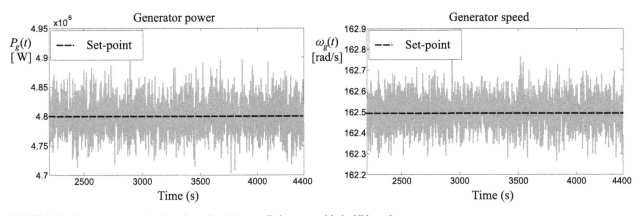

**FIGURE 5.20** Generator power $P_g(t)$ and speed $\omega_g(t)$ controlled outputs with the NN regulator.

As further example, Fig. 5.20 reports the results with the NN controller recalled in Chapter 4. As summarized in Table 5.18, the average case of the $NSSE\%$ index for $P_g(t)$ (with respect to $P_r$ in partial load) and $\omega_g(t)$ (with respect to $\omega_{nom}$ in full load) during the Monte Carlo analysis are shown.

Fig. 5.20 depicts the reference values in black dashed lines, whilst the controlled signals are gray solid lines. Also in this case, with reference to the NN controller, it appears that good control performances are not obtained.

As the final example, Fig. 5.21 shows the obtained results of the Monte Carlo analysis with the LPV controller recalled in Chapter 4. As reported again in Table 5.18, the average case for the controller outputs $P_g(t)$ and $\omega_g(t)$, with respect to their set-points $P_r$ and $\omega_{nom}$, respectively, are depicted.

Fig. 5.21 depicts the reference values in black dashed lines, whilst the controlled signals are gray solid lines. The LPV controller seems quite effective, even if its performances remain quite limited.

The comparisons among the different controllers summarized in Table 5.18 show that the scheme using the LPV strategy allows achieving acceptable performances in terms of tracking error. With reference to the sliding mode solution, the control input energy required by the sliding mode controller is bigger than for the other cases. Moreover, the sliding mode controller can increase the computational time considerably with respect to the other solution, without any gain scheduling, whilst the LPV control strategy may require larger computational effort at the design stage.

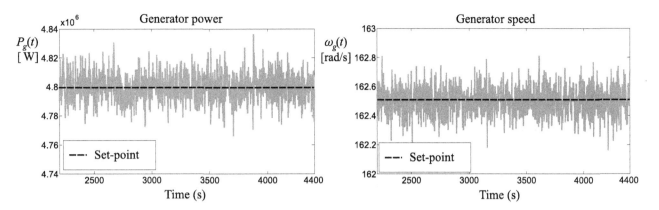

**FIGURE 5.21**   Generator power $P_g(t)$ and speed $\omega_g(t)$ outputs with the LPV controller.

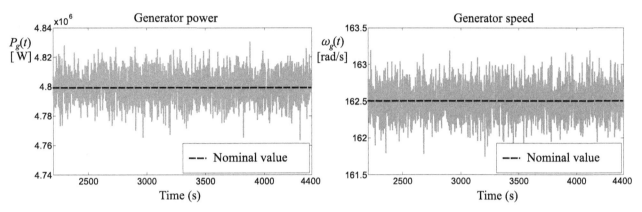

**FIGURE 5.22**   Sensitivity analysis in full load conditions with the data-driven fuzzy regulators.

A few further comments can be drawn here. When the modeling of the dynamic system can be perfectly obtained, in general model-based control strategies are preferred. On the other hand, when modeling errors and uncertainty are present, alternative control schemes relying on adaptation mechanisms, or passive robust control methods, showed interesting robustness properties in the presence of unmodeled disturbances, modeling mismatches, and measurement errors. With reference to data-driven methodologies, and in particular to the neural controller, in the case of a controlled system with modeling errors, the offline learning can lead to quite good results. Other explicit disturbance decoupling techniques can take advantage of their robustness capabilities, but with quite complicated and not straightforward design procedures. The NN-based scheme relies on the learning accumulated from offline simulations, but the training stage can be computationally heavy. Regarding the considered methods using adaptive, or fuzzy tools, they seem rather simple and straightforward, even if optimization stages can be required.

Also the stability properties of the overall control strategies were checked by means of a Monte Carlo campaign based on the wind turbine benchmark. In fact, the Monte Carlo analysis represents the only method for estimating the efficacy of the developed control schemes when applied to the monitored process. It is worth noting that Johnson et al. (2004) provided an analytical demonstration of the stability of an adaptive control scheme for wind turbines.

All simulations were performed by considering noise signals modeled as Gaussian processes, according to the standard deviations reported in Table 5.16. Different wind sequences were generated by the wind turbine benchmark simulator. Moreover, the initial conditions of the dynamic models recalled in Chapter 2 (i.e., the drive-train, the generator/converter, and the pitch system) were changed randomly. Therefore the random wind signal $v(t)$, the parameters of Table 5.16, and the dynamic model initial conditions allowed obtaining different sequences of the wind turbine signals $\beta(t)$, $\tau_g(t)$, $\lambda(t)$, $\omega_g(t)$, and $P_g(t)$ for each Monte Carlo simulation.

As an example of a single Monte Carlo run, Fig. 5.22 highlights that the main wind turbine model variables, such as the generator torque $\tau_g(t)$, the tip-speed ratio $\lambda(t)$, and the generator power $P_g(t)$ remain bounded around the reference values, proving the overall system stability in simulation, even in the presence of disturbances and uncertainty. These results refer to the case of full load operation with the data-driven fuzzy controllers.

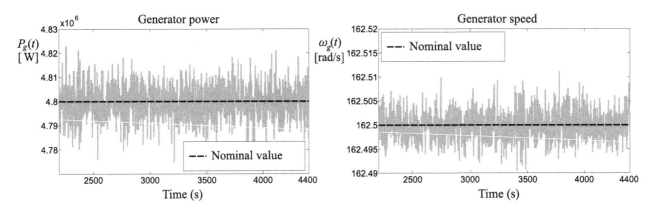

**FIGURE 5.23** Simulations in full load with the adaptive controllers.

**TABLE 5.19** HIL results with the fuzzy controller: $NSSE\%$ values.

| Control Algorithm | Partial Load $NSSE\%$ | Full Load $NSSE\%$ |
|---|---|---|
| Data-driven fuzzy | 38.34 | 18.35 |
| Model-based adaptive | 28.52 | 14.72 |
| Sliding mode | 27.55 | 16.34 |
| NN controller | 37.67 | 21.76 |
| Gain scheduling | 29.76 | 18.23 |

As a further Monte Carlo example run, the results achieved with the model-based adaptive controller in full load are summarized in Fig. 5.23.

Fig. 5.23 depicts the perturbed signals representing the generator speed $\omega_g(t)$ and power $P_g(t)$ versus their nominal values. Also in this case the main wind turbine variables remain bounded around the reference values, thus assessing the overall system stability in simulation, even in the presence of modeling errors and disturbances.

Finally, in order to evaluate the potentials of the considered control algorithms also in real-time applications and analyze their features for on-board implementations, the remainder of this section presents the results of the HIL tests. These experimental results serve to validate the considered control algorithms considering almost real conditions that the wind turbine may experience during its working situations. For this purpose HIL test-bed described in Chapter 6 was exploited, in order to verify the control algorithms in an almost real-time condition.

Therefore the results achieved from one test are summarized in Table 5.19 for the considered controller solutions.

Table 5.19 illustrates that there are some deviations between the achieved results, but consistent with those from the Monte Carlo analysis. Although there are some deviations between the simulation and experimental results, these deviations are not critical and they seem accurate enough for future wind turbine real applications.

### 5.3.1 Sustainable Control Design

This section presents some results obtained from the application of the FTC strategies proposed in Chapter 4 to the wind turbine simulator. In general, in order to design robust FTC schemes, the information provided by the fault estimation modules addressed in Chapter 3 is required.

In general, as recalled in Chapter 4, fault tolerant control methods are classified into two types, namely, Passive FTC Schemes (PFTCS) and Active FTC Schemes (AFTCS) In PFTCS, controllers are fixed and are designed to be robust against a class of presumed faults. This approach needs neither FDD schemes nor controller reconfiguration, but it has limited fault tolerance capabilities. In contrast to PFTCS, AFTCS react to the system component failures actively by reconfiguring control actions so that the stability and acceptable performance of the entire system can be maintained. To design a successful AFTCS, it relies heavily on real-time FDD schemes to provide the most up-to-date information about the true status of the system. Therefore, the main goal in a fault-tolerant control system is to design a controller with a suitable structure to achieve stability and satisfactory performance, not only when all control components are functioning normally, but also in cases when there are faults in sensors, actuators, or other system components.

Regarding the AFTCS design, in Chapter 4 it was argued that, in AFTCS, good FDD is needed. We claimed that, for the system to react properly to a fault, timely and accurate detection and location of the fault is needed. The most researched area in fault diagnosis is the residual generation approach using dynamic observers or filters. Plant–model mismatches can cause false alarms or, even worse, missed faults. Robustness issues in FDD are therefore very important, as remarked in Chapter 3.

The data-driven and model-based FTC strategies addressed in this monograph are based on the application of fuzzy modeling and identification, as well as on model-based control design, that have gained increasing attention in both theory and application. The main point concerns the considered mathematical model, which is a possible description of system behavior; therefore, accurate modeling for a complex nonlinear system can be very difficult to achieve in practice. Sometimes for nonlinear systems it can be impossible to describe them via analytical equations. Moreover, very often the system structure, parameters, and measurements are not precisely known. Thus parametric model identification and purely nonlinear models represent an alternative for developing experimental models of complex systems, as addressed in Chapter 2. However, in contrast to pure nonlinear identification methods, where detailed knowledge about the model's structure is required, data-driven approaches, such as fuzzy systems and neural networks, are capable of deriving nonlinear models directly from measured input–output data without detailed system assumptions.

Because of these considerations, one of the strategies considered in this monograph suggests using data-driven methodologies, and in particular the fuzzy system theory, since it seems to be a natural tool to handle uncertain conditions and measurements. Thus, instead of considering complicated nonlinear models obtained by modeling techniques, as in the model-based schemes, it is suggested to describe the plant under investigation by a collection of local affine systems of the type of TS fuzzy prototypes, whose parameters are obtained by identification procedures.

With reference to the alternative passive strategy considered in this section, it is based on the application of model identification mechanisms in connection with model-based adaptive control design. The strategy recalled in Chapter 4 belongs to the field of adaptive control, which has undergone significant development in recent years. As already remarked, the aim of this approach is to solve the problem of controller design, for instance, where the characteristics of the process under investigation are not sufficiently known, or change over time. Since a mathematical model is a description of system behavior, accurate modeling for a complex system is very difficult to achieve in practice.

Online or adaptive parametric model identification schemes represent an alternative for developing experimental models for complex systems, such as wind turbine systems. In contrast to traditional pure nonlinear identification methods also addressed in Chapter 4, where detailed knowledge about the model structure is required, recursive and online identification of parametric linear models can be capable of providing good description of nonlinear models directly from measured input–output data, without detailed system assumptions.

Because of these considerations, this section shows the implementation an alternative data-driven FTC solution based on an adaptive identification scheme, used for the online estimation of the controlled process, which can be affected by both uncertainty and faults. While the process time-varying parameters have been estimated, which can be the effect of both uncertainty and faults, the time-varying controller parameters are computed online, in order to maintain the required control performances. Thus, instead of exploiting complicated analytical nonlinear models, it is suggested to describe the plant under investigation by a parameter-varying linear model, whose variables are obtained by an online identification procedure. The interesting feature of this scheme is that can be seen as a powerful tool for applications where complete information about faults and model uncertainty are not available to the designer.

In particular, Section 5.3.3 shows the results achieved with the application of model-based FTC designs to the wind turbine simulator, as recalled in Chapter 4, whilst Section 5.3.2 considers different data-driven FTC designs for the same benchmark.

## 5.3.2    Data-Driven Fault Tolerant Control Examples

This section describes the simulations with some of the data-driven (passive) methods considered in Chapter 4 for the fuzzy modeling technique oriented to the design of the fault tolerant controller relying on the multiple-model TS approach. As already remarked, because of the underlying physical mechanisms involved in the wind turbine process and because of the switching control logic described in Chapter 2, the considered plant has nonlinear characteristics.

The GK clustering algorithm discussed in Chapter 4 with $M = 3$ clusters and $n = 2$ shifts was applied to the fault-free and faulty sampled data $\{P_g(t), \omega_g(t), \beta_r(t)\}$. On the other hand, $M = 3$ and $n = 2$ were considered for achieving a suitable clustering of the fault-free and faulty sampled data $\{P_g(t), \omega_g(t), \tau_g(t)\}$. After clustering the TS model parameters for each output were estimated. Therefore, the $i$th output $y(t)$ of the wind turbine ($i = 1, \ldots, m$ and $m = 2$) continuous-time model is approximated by a Takagi–Sugeno fuzzy MISO discrete-time prototype with $r = 2$ inputs. The relative mean square errors of the output estimates under no-fault conditions are 0.0254 for the first output and 0.0125 for the second one. The fitting capabilities of the estimated fuzzy models can be expressed also in terms of VAF index. In particular, the VAF value for first output is bigger than 90%, whilst it is bigger than 99% for the second one. Hence the fuzzy multiple models seem to approximate the process under diagnosis quite accurately.

Note that the GK fuzzy clustering algorithm is applied to two different data sets. The first set consists of the validation data, namely the fault-free sequences, whilst the second one contains the estimation data, i.e., the faulty sequences. Therefore the optimal number of clusters $M$, the estimated MFs, and the set of the optimal parameters $a_i$ and $b_i$ are determined in order to minimize the so-called model–reality mismatch. In this way the estimated TS prototype is able to provide the optimal fitting of both the fault-free and the faulty conditions.

Using this identified TS fuzzy prototype, the model-based approach for determining the FTC scheme is exploited and applied to the actual wind turbine benchmark. According to the methodology for deriving the fuzzy controller described in Chapter 4, the parameters of the fuzzy PI controllers have been computed. In particular, as the identified TS models for each output consists of a fuzzy collection of 3 MISO second order ($n = 2$) models, the regulator parameters are computed analytically.

In more detail, by considering a second order local model described by its identified parameters $a_i = [\alpha_2^{(i)}, \alpha_1^{(i)}]^T$ and $b_i[\delta_2^{(i)}, \delta_1^{(i)}]^T$, the so-called critical gain $K_0^{(i)}$ and critical period of oscillations $T_0^{(i)}$ required by the Ziegler–Nichols method (Ziegler and Nichols, 1942) are computed as follows (Bobál et al., 2005):

$$
\begin{cases}
K_0^{(i)} = \dfrac{\alpha_1^{(i)} - \alpha_2^{(i)} - 1}{\delta_2^{(i)} - \delta_1^{(i)}}, \\[2ex]
T_0^{(i)} = \dfrac{2\pi T_s}{\arccos \gamma^{(i)}} \quad \text{with} \quad \gamma^{(i)} = \dfrac{\alpha_2^{(i)} \delta_1^{(i)} - \alpha_1^{(i)} \delta_2^{(i)}}{2 \delta_2^{(i)}}.
\end{cases}
\tag{5.5}
$$

These relations are thus used for calculating the parameters $K_P^{(i)}$ and $K_I^{(i)}$ of the (local) $i$th PI controller:

$$
K_P^{(i)} = 0.6\, K_0^{(i)} \left(1 - \frac{T_s}{T_0^{(i)}}\right), \quad K_I^{(i)} = \frac{1.2\, K_0^{(i)}}{K_P^{(i)}\, T_0^{(i)}},
\tag{5.6}
$$

where $T_s$ is the sampling time.

Also in this case, it is worth remarking the strategy applied for achieving the required *passive* FTC characteristics. With reference to fuzzy regulator, the PI controller parameters $K_P^{(j)}$ and $K_I^{(j)}$ are tuned via the Ziegler–Nichols rules, applied to the identified local linear TS submodels, and by considering the *validation* data set, i.e., the faulty sequences. Therefore the optimal controller performances with respect to set-point variations are enhanced for the *faulty* working conditions. In this way, if both the TS model identification and fuzzy regulator tuning procedures are properly preformed, the gain scheduling mechanism of the fuzzy regulator parameters leads to acceptable *passive* fault tolerance properties. However, it is true that, due to the considered *passive* FTC scheme, when fault conditions different for the ones summarized in Chapter 2 are considered, acceptable characteristics of the designed fuzzy controller might not be achieved.

In the following the considered FTC fuzzy PI controllers and the original switching strategy described in Chapter 2 have been implemented and compared in the Matlab® and Simulink® environments. The simulation setup employs 2 MISO fuzzy PI regulators used for the control of the blade pitch angles and the generator control torque, respectively. As an example, by using the previous relations for the design of the fuzzy PI regulator, the following tuned parameter sets have been computed for the pitch angle control:

$$
\begin{cases}
\{K_P^{(1)}, \ldots, K_P^{(3)}\} = \{4.3,\, 4.1,\, 4.2\}, \\[1.5ex]
\{K_I^{(1)}, \ldots, K_I^{(3)}\} = \{1.2,\, 1.4,\, 1.5\}.
\end{cases}
\tag{5.7}
$$

In order to compare the advantages of the considered fuzzy PI strategy, the obtained results are also compared with the ones achieved by using the original switching wind turbine benchmark regulator recalled in Chapter 2.

The controller capabilities have been assessed in simulation by considering different fault-free and faulty data sequences. In Table 5.20, the $NSSE\%$ values are computed for both the controllers, and for different data sequences. It is worth noting that in *partial load operation*, the performance is represented by the comparison between the power produced by the generator $P_g$ with respect to the theoretical maximum power output $P_r$, given the instant wind speed. On the other hand, in *full load operation* the performance depends on the generator speed $\omega_g$ with respect to the nominal one, $\omega_{nom}$.

**TABLE 5.20** Controller performance in partial and full load operations.

| Working Condition | Baseline Controller $NSSE\%$ | Fuzzy PI $NSSE\%$ |
|---|---|---|
| Partial load fault-free | 39.34 | 36.36 |
| Partial load faulty | 42.19 | 37.17 |
| Full load fault-free | 19.53 | 16.57 |
| Full load faulty | 21.01 | 17.85 |

**TABLE 5.21** Fuzzy FTC performance.

| Working Condition | Partial Load $NSSE\%$ | Full Load $NSSE\%$ |
|---|---|---|
| Fault-free | 36.37 | 16.57 |
| Faulty | 37.19 | 17.94 |

According to these simulation results, the fault tolerance properties of the considered controllers seem to be reached, and they are slightly better than the original baseline regulator.

On the other hand, the robustness and reliability properties of the designed fuzzy controllers have been tested and assessed in simulation. Therefore the parameters of the wind turbine benchmark model described in Chapter 2 have been modified by 20% with respect to their nominal values. The obtained results summarized in Table 5.21 seem to show that the performances of the considered fuzzy controller are almost unchanged with respect to the nominal situation.

The remainder of this section shows also the results that can be achieved by using the adaptive control strategy used in connection with the online estimation scheme recalled in Chapter 4.

It is clear that, because of the underlying physical mechanisms and due to the switching control logic described in Chapter 2, the wind turbine system has nonlinear characteristics. Therefore also in this case the $i$th output $y(t)$ of the wind turbine ($i = 1, \ldots, 3$) continuous-time nonlinear model is approximated by a time-varying MISO discrete-time prototype with 2 inputs.

Note that the model online estimation scheme presented in Chapter 4 is tested using two different data sets. The first set consists of the estimation data, namely the fault-free sequences, whilst the second one contains the validation data, i.e., the faulty sequences. Therefore, the time-varying model parameters are estimated in order to minimize the so-called model–reality mismatch. In this way, the online estimated time-varying linear prototype should be able to provide the optimal fitting of both the fault-free and the faulty conditions.

Using this identified prototype, the model-based approach for determining the adaptive controller shown in Chapter 4 is exploited and applied to the actual wind turbine benchmark. Thus, according to Chapter 4, the parameters of the adaptive PI controllers have been computed. In particular, for each output, the identified time-varying prototypes consist of three MISO second order models; thus, the adaptive regulator parameters are computed analytically at each time step $k$.

It is worth highlighting the strategy applied for achieving the required active fault tolerance characteristics. With reference to the adaptive regulator models, the PI adaptive controller parameters are tuned via the Ziegler–Nichols rules, applied to the time-varying linear models, and by considering the *validation* data set, i.e., the faulty sequences. Therefore, the optimal controller performances with respect to set-point variations are enhanced also for the *faulty* working conditions. In this way, if both the model online parametric identification and the regulator tuning procedure are properly performed, the parameter adaptation mechanisms will lead to acceptable *active* fault tolerance properties. Moreover, by means of the considered adaptation mechanisms, the considered active FTC scheme is able to maintain good control performances, even when fault conditions different for those summarized in Chapter 2 are simulated.

In the following the considered adaptive FTC scheme, summarized in Fig. 5.24, together with the baseline wind turbine controller, has been implemented and compared in Matlab® and Simulink® environments, taking into account the benchmark uncertainty and the fault conditions. The experimental setup employs 3 adaptive PI regulators used for the control of the blade pitch angles and the generator control torque, in the partial and full load working conditions. The complete block scheme is shown in Fig. 5.24, whilst the initial conditions for the online estimation algorithm are reported in Table 5.22.

With reference to the switching logic depicted in Fig. 5.24, it consists of the strategy already recalled in Chapter 4. On the basis of the monitoring of the wind turbine variables $P_g(t)$ and $\omega_g(t)$, the control zone is correctly determined. Moreover, when the wind turbine works in the *partial load region*, the designed adaptive controller scheme is reported in Fig. 5.25.

Fig. 5.25 shows that, in the partial load working conditions, the baseline wind turbine torque controller is compensated via the adaptive regulator. In this way the adaptive controller is able to manage any possible uncertainty affecting the control

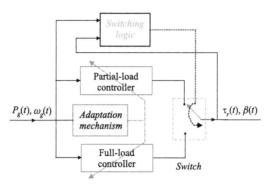

**FIGURE 5.24** Complete scheme of the adaptive FTC strategy.

**TABLE 5.22** Adaptive algorithm parameters.

| Variable | Value |
|---|---|
| $\Theta_0$ | $[0.1,\ 0.2,\ 0.3,\ 0.4]^T$ |
| $C_0$ | $10^9\ I_4$ |
| $\varphi_0$ | 1 |
| $\lambda_0$ | 0.001 |
| $\rho$ | 0.99 |
| $\nu_0$ | $10^{-6}$ |

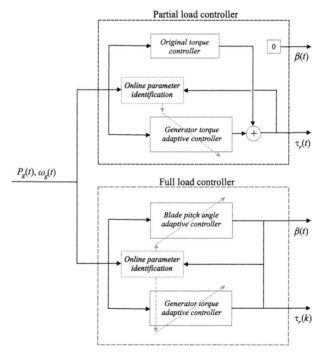

**FIGURE 5.25** Adaptive controller for the partial and full load with adaptation scheme.

parameters, or any faulty situations. In this case the adaptive PI controller exploits the output variable $y(k)$, namely $\omega_g(k)$, whilst the actuated input $u(k)$ is only $\tau_g(k)$, since $\beta_r(k) = 0$. The reference signal $r(k) = \omega_r(k)$ exploited by the adaptive controller allows improving the performance of the original control strategy.

On the other hand, when the wind turbine is working in the full load region, the considered control strategies are also represented in Fig. 5.25. Regarding again Fig. 5.25, one adaptive PI controller is exploited for the regulation of the blade pitch angle, $u(k) = \beta_r(k)$. This controller is fed by the monitored output $y(k) = \omega_g(k)$, and uses the constant reference

**TABLE 5.23** Controller $NSSE\%$ in partial and full load operations.

| Working Condition | Baseline Controller % | Adaptive PI Regulators % |
|---|---|---|
| Partial load fault-free | 39.34 | 24.62 |
| Partial load faulty | 42.19 | 28.73 |
| Full load fault-free | 19.53 | 22.95 |
| Full load faulty | 21.01 | 25.67 |

**TABLE 5.24** Fuzzy and adaptive FTC performances.

| Controller Type | Working Condition | Partial Load $NSSE\%$ | Full Load $NSSE\%$ |
|---|---|---|---|
| Fuzzy | Fault-free | 36.37 | 16.57 |
| Fuzzy | Faulty | 37.19 | 17.94 |
| Adaptive | Fault-free | 24.71 | 24.05 |
| Adaptive | Faulty | 27.52 | 25.72 |

signal $r(k) = \omega_{ref} = 162$ rad/s. The second PI adaptive controller generates the signal $u(k) = \tau_g(k)$, with $y(k) = P_g(k)$ and $r(k) = P_{ref} = 4.8 \times 10^6$ W.

Note that the control input $u(k) = \beta_r(k)$ is limited between $-2$ and 90 degrees, with a rate of change between $-9$ and 9 degrees/s. On the other hand, the second control input $\tau_g(k)$ must be lower that $3.6 \times 10^4$ N, with a maximal change rate between $-1.25 \times 10^4$ and $1.25 \times 10^4$ N/s.

In order to compare the advantages of the considered adaptive PI strategy, the obtained results are also compared with those achieved by using the original wind turbine benchmark regulators described in Chapter 2. The controller capabilities have been assessed again in simulation by considering different fault-free and faulty data sequences. In Table 5.23 the $NSSE\%$ values are computed for both the adaptive controllers and for different data sequences.

It is worth noting that in *partial load operation*, the performance is represented by the comparison between the power produced by the generator $P_g$ with respect to the theoretical maximum power output $P_{ref}$, given the instant wind speed. On the other hand, in *full load operation* the performance depends on the generator speed $\omega_g$ with respect to the nominal one, $\omega_{ref}$.

According to these simulation results, the fault tolerance properties of the considered adaptive PI controller seems to be reached, and they are slightly better than the benchmark regulators.

Finally, the robustness and reliability properties of the considered adaptive controllers have been tested and assessed in simulation. Therefore the parameters of the wind turbine benchmark model have been modified of 20% with respect to their nominal values. The obtained results summarized in Table 5.24 seem to show again the enhanced capabilities of the considered adaptive PI controllers, in both the fault-free and faulty conditions, when compared with the fuzzy regulators.

These simulation results can be used also for testing the design robustness, stability, and reliability of the considered adaptive methods with respect to modeling uncertainty and anomalous working conditions.

Finally, this section considered data-driven approaches oriented to the design of FTC solutions for regulating both the pitch angle and the generator torque of a wind turbine benchmark. This strategy was considered for enhancing the regulator design that could represent an alternative to standard model-based controllers, already implemented for wind turbine systems. The control design requires the knowledge of a dynamic model of the wind turbine system, which is achieved by means of data-driven schemes. On the other hand, the considered controller structures seem quite straightforward and easy to implement with respect to different model-based strategies considered in Chapters 2 and 4. Moreover, as remarked above, the considered design approach belongs to the so-called passive FTC schemes. The results obtained with the considered controllers are compared to those of the switching controller already implemented for the considered wind turbine benchmark. Further simulations will concern the application of model-based FTC methods to same wind turbine simulator, as summarized in Section 5.3.3.

### 5.3.3 Model-Based Fault Tolerant Control Examples

This section shows how the model-based nonlinear filter scheme recalled in Chapter 3 is used for fault estimation, even if in general could be exploited for fault detection and isolation. In particular, if the *fault isolation* task is required, a bank of

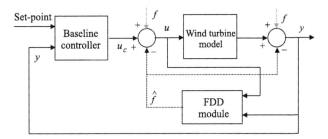

**FIGURE 5.26** Diagram of the adaptive FTC scheme.

fault estimation filters, for example, in the form of Eqs. (4.45), (4.46), and (4.47), can be exploited as residual generators, allowing to estimate the magnitude of different faults acting on the considered actuators.

The decoupling of the effect of both the disturbance $d$ and the fault $f$ (important for fault isolation) from the nonlinear filters is provided by the selection of the proper state component from the $\bar{x}_1$-subsystem. However, as the main point of the work is the design of an active FTC scheme with application to the considered benchmark, the problem of the fault isolation will not be addressed here. Moreover, the simulated benchmark considers only the occurrence of single faults, as described in Chapter 2.

In order to compute the simulation results described in the following, the FTC scheme has been completed by means of the benchmark controller described in Chapter 2. The logic scheme of the integrated adaptive fault tolerant approach is recalled in Fig. 5.26.

With reference to Fig. 5.26, the term $u$ represents the actuated inputs, $u_c$, the controlled inputs from the benchmark controller. On the other hand, $y$ is the measured outputs, $f$ the equivalent actuator or the sensor fault, whilst $\hat{f}$ the corresponding estimated signal.

Therefore, the logic scheme depicted in Fig. 5.26 shows that the FTC strategy is implemented by integrating the FDD module with the existing baseline controller. From the controlled input and output signals, the fault estimation module provides the correct estimate $\hat{f}$ of the $f$ fault, which is injected into the control loops, in order to compensate the effect of the actuator or sensor fault itself. After this correction, the baseline controller provides the nominal tracking of the reference signal.

Regarding the scheme shown in Fig. 5.26, as already remarked above, the considered strategy has been applied for the accommodation of additive and equivalent faults. However, the NLGA-based methodology could be extended to the case of sensor faults. In fact, sensor faults can be modeled as additive inputs in the measurements equation (i.e., the output equation) by means of the so-called sensor fault signature (often a column with only one element equal to "1", and the remaining entries being null). Moreover, sensor fault signatures could also be modeled as an input to the system in the same way as for actuator faults. This further task could be performed following the method addressed in Mattone and De Luca (2006).

Regarding the stability analysis of the overall FTC system, the simulation results addressed in this chapter highlight that the model variables remain bounded in a set, which assures control performance, even in the presence of faults. Moreover, the assumed fault conditions do not modify the system structure, thus guaranteeing the global stability.

However, a few more issues can be considered here. It should be clear that in steady-state conditions, when the fault effect is completely eliminated, the performances of the active FTC schemes are the same of the fault-free situation. Therefore, the performance of the complete system are the same of the fault-free nominal controller. The stability properties of the active FTC solution should be considered only in transient conditions, when the fault is not compensated. In fact, in these conditions, possible fault estimation errors correspond to signals injected into the feedback loop of Fig. 5.26. It is possible to show that the fault estimation error is limited and convergent to zero, thus the stability of the complete system is maintained. Moreover, a proper selection of the main design parameters for the FDD strategies described in Chapter 3 lead to accurate fault estimation, thus allowing to achieve robust FTC methodologies, as remarked in Chapter 4.

The remainder of this section describes the design and the simulations of the FTC applied to the wind turbine simulator. In more detail, Sections 5.2.1 and 5.2.2 summarized the results achieved from the FDD schemes. Once the disturbance decoupling has been achieved, the capabilities of the FTC strategies method are reported. Moreover, the performance evaluation of the developed FTC schemes with respect to modeling errors and measurement uncertainty is also considered. On the other hand, comparison of the considered FTC methodologies with respect to different FTC methods are analyzed.

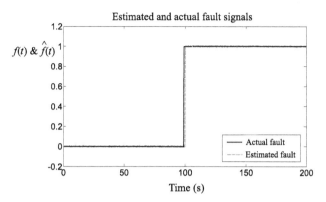

**FIGURE 5.27**  Estimate $\hat{f}$ of the actual fault $f$ (case 1).

As already remarked in Chapter 4, the $C_p$-map entering into the aerodynamic model of the wind turbine system has been approximated by using a two-dimensional polynomial of the form

$$\hat{C}_p(\lambda, \beta) = 0.010\,\lambda^2 + 0.0003\,\lambda^3\,\beta - 0.0013\,\lambda^3. \tag{5.8}$$

It is worth noting that the degrees and the coefficients of the two-dimensional polynomial described by Eq. (5.8) have been optimized comparing the approximation capabilities achieved with different sets of data directly acquired from the benchmark model. Moreover, the uncertainty structure modeled by Eq. (5.8) must lead to a wind turbine model in the affine form required by the model-based NLGA methodology recalled in Chapter 3.

It is worth noting that the arguments above provide an analytical description of the look-up table $C_p$ that takes into account all uncertainties, and not only the errors due to $C_p$ entry changes. However, since the $C_p$-map is used for the fault estimation filter design, any kind of uncertainty must be modeled. Moreover, it is followed the same procedure already proposed in Chen and Patton (2012), but developed only for linear state-space models. Under this consideration, the uncertainty distribution description $p_d(x)$ for the nonlinear model of Eq. (3.66) is identified using the input–output data from the wind turbine. The general assumption holding for this case is that the model–reality mismatch is varying more slowly that the disturbance signals, such as $d$. Moreover, the estimation approach does not exploit directly the wind $v(t)$ and the $c(t)$ control signals, but sequences that are filtered by the dynamic simulation process. Another important point regards the fact that the $C_p$-map estimation aims at describing the *structure* of the uncertainty, which should not depend on the wind size uncertainty. Only the so-called "directions" of the disturbances represent the important effect for disturbance decoupling, i.e., the $p_d(x)$ term, and not the "amplitude" of the uncertainty, i.e., the size of the disturbance $v(t)$ $(d)$.

In order to show the capabilities of the considered active FTC strategies, the benchmark has been tested with the reference signals required by the wind turbine benchmark, as described in Chapter 2. The designed NLGA adaptive filters recalled in Chapters 3 and 4 in the form of Eqs. (4.45), (4.46), and (4.47) allow estimating the magnitude of the different faults acting on the wind turbine benchmark.

As an example, for the actuator fault case 1 reported in Chapter 2, the nonlinear filter for $f$, which is decoupled from the effect of the disturbance $d$ representing both the wind speed $v(t)$ and the $C_p$-map uncertainty, has the form of Eq. (4.45) and is based on Eq. (5.26).

In order to compute the simulation results described below, the FTC scheme has been completed by means of the standard benchmark controller described in Chapter 2. The integrated FDD and FTC strategy corresponds to the logic scheme shown in Fig. 5.26. The following results refer to the simulation of the actuator fault case 1 modeled as a step signal with a size from 0 to 1. In particular, Fig. 5.27 shows the estimate of the actuator fault $f$ (solid line), when compared with the simulated actuator fault (dashed line).

As it will be shown in the following, after a suitable choice of the parameters of the FDD nonlinear filter, it provides an accurate estimate of the fault size, with minimal detection delay.

Fig. 5.28 shows the generator speed $\omega_g$ compared with its desired value $\omega_{nom}$. When the fault case 1 is compensated by the AFTCS scheme, the tracking error is small.

As highlighted in Fig. 5.28, during the time interval $100 < t < 200$ s only the baseline regulator without FTC is working. On the other hand, when the considered FTC scheme is acting in the control loop, during the interval $0 < t < 100$ s the tracking error due to the fault case 1 is much lower.

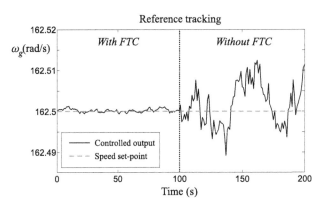

**FIGURE 5.28** Generator speed $\omega_g$ and its reference $\omega_{nom}$ with and without FTC.

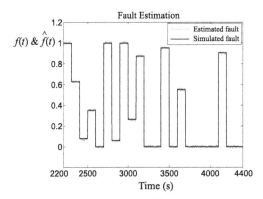

**FIGURE 5.29** Real-time estimate $\hat{f}$ of the intermittent actuator fault $f$ under full load.

The achieved simulation results summarized in Figs. 5.27 and 5.28 show the effectiveness of the presented integrated FDD and FTC strategy, which is able to improve the control objective recovery, and the reference tracking in the presence of actuator fault.

It is worth observing that the NLGA adaptive filter provides not only the fault detection, but also fault estimate, as highlighted in Chapter 4. Moreover, a fault modeled as additive step function has been considered, since it represents the realistic fault condition in connection with the benchmark model. However, the fault estimation module can be easily generalized to estimate, for example, polynomial functions of time, or generic fault signals belonging to a given class of faults, if the NLGA adaptive filters contain the internal model of the fault itself. The generalization to more general fault functions is beyond the scope of this monograph. On the other hand, a different fault scenario for the considered benchmark has been considered in the following. In particular, the case of *intermittent fault* is considered in the following.

Thus the results refer to the simulation of a fault $f$ modeled as a sequence of rectangular pulses with variable amplitude and length. Fig. 5.29 shows the estimate of the intermittent actuator fault $\hat{f}$ (dotted gray line), when compared with the simulated actuator fault $f$ (dashed black line). Also in this case, after a suitable choice of the parameters of the NLGA adaptive filter, the fault estimation module provides a quite good reconstruction of the fault signal.

Under this condition, Fig. 5.30 shows the reference $\omega_g$ compared with its desired value $\omega_{nom}$. The estimate feedback used by the considered active FTC is applied at $t = 3300$ s without any delay.

Also for the situation of an intermittent fault, Fig. 5.30 shows how the active FTC scheme is able to compensate and accommodate it.

In order to summarize the advantages of the considered FTC strategy, the performance of the wind turbine benchmark regulator recalled in Chapter 2 with and without the fault accommodation scheme has been evaluated in terms of $NSSE$ values already defined as

$$NSSE\% = 100 \sqrt{\frac{\sum_{k=1}^{N}(r(k) - y(k))^2}{\sum_{k=1}^{N} r^2(k)}}, \tag{5.9}$$

and considering different data sequences.

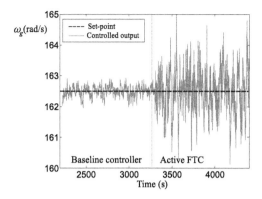

**FIGURE 5.30**  Reference $\omega_g(t)$ of the benchmark for the case of an intermittent fault under full load.

**TABLE 5.25** $NSSE\%$ fault accommodation performances.

| Fault Case # | Baseline Controller % | Active FTC % |
|---|---|---|
| 1 | 23.02 | 11.23 |
| 2 | 22.98 | 12.23 |
| 3 | 22.84 | 10.85 |
| 4 | 20.68 | 13.27 |
| 5 | 19.54 | 12.97 |
| 6 | 21.01 | 9.17 |
| 7 | 22.68 | 12.01 |
| 8 | 20.67 | 11.44 |
| 9 | 22.35 | 10.36 |

It is clear that in full load operation, the performance depends on the generator speed $\omega_g$ with respect to the nominal one, $\omega_{nom} = 162.5$ rad/s.

The simulation results of Table 5.25 show that in faulty conditions the control performances are improved when the considered model-based FTC strategy is exploited.

## 5.3.4 Performance Evaluation and Robustness Analysis

In this section, further experimental results have been reported regarding the robustness evaluation of the developed scheme with respect to modeling errors and measurement uncertainty. In particular, the simulation of different data sequences has performed by exploiting the wind turbine benchmark simulator, followed by a Matlab® Monte Carlo analysis.

Again, as already remarked above, the wind turbine simulator implemented in Matlab® and Simulink® environments is able to vary the statistical properties of the signals used for modeling process parameter uncertainty, and measurement errors. Therefore, also for the case of the fault tolerant controller validation and verification, the Monte Carlo analysis represents a viable tool for analyzing some properties of the considered FTC schemes. Under this assumption, Table 5.16 summarizes the nominal values of the considered wind turbine model parameters with respect to their simulated uncertainty that are simulated for this analysis.

Therefore, for performance evaluation of the control schemes, the average values of the $NSSE\%$ index were computed, and experimentally evaluated with 1000 Monte Carlo runs. The value of $NSSE\%$ is computed for several possible combinations of the parameter uncertainties of the wind turbine simulator.

It is worth noting that this analysis considers the uncertain parameters that have been simulated in order to analyze the *robustness* of the considered scheme with respect to *parameter variations*. In fact, the disturbance decoupling approach was used for removing the effect of the uncertain wind term $v(t)$, and not for handling the parameter variations summarized in Table 5.16.

Table 5.26 summarizes the results obtained by considering the considered FTC scheme integrating the original benchmark controller for the different fault cases.

**TABLE 5.26** Monte Carlo analysis for the considered model-based active FTC scheme.

| Fault Case # | $NSSE\%$ Average Value |
|---|---|
| 1 | 11.05 |
| 2 | 12.23 |
| 3 | 10.35 |
| 4 | 13.02 |
| 5 | 9.97 |
| 6 | 8.98 |
| 7 | 9.89 |
| 8 | 10.07 |
| 9 | 8.76 |

In particular, Table 5.26 summarizes the values of the considered performance index according to the average cases, with reference to the possible combinations of the wind turbine uncertainty parameters. Thus Table 5.26 shows that the considered active FTC scheme allows maintaining good control performances even in the presence of the fault cases, varying errors, and uncertainty effects.

Finally, the results demonstrate also that Monte Carlo simulation is an effective tool for experimentally testing the design robustness of the considered FTC methods with respect to modeling uncertainty.

### 5.3.5 Comparative Results and Stability Analysis

This section provides comparative results with respect to different FTC schemes recalled in Chapter 4. In more detail, the properties of the methods considered in this chapter have been analyzed with respect to alternative FDD approaches addressed in Chapter 3, in particular relying on the Sliding Mode Observer (SMO) and the Neural Network (NN) estimators.

Regarding the SMO used for the estimation of the fault acting on the system, it represents an alternative fault estimator module forming a part of the AFTCS strategy. Edwards and his coworkers (Edwards and Spurgeon, 1994; Edwards et al., 2000) proposed an SMO structure and design procedures, which obviate the use of symbolic manipulation, thus offering an explicit design algorithm. The SMO concept involves the design of observer gain matrices to ensure that the sliding surface is reached and maintained. During sliding the error between the system and the observer states remains close to zero, ensuring robust estimation under conditions of matched uncertainty. For this application, the SMO has two functions: (a) robust state estimation, and (b) using the state estimation the actuator fault is estimated using the concept of equivalent output injection (Edwards et al., 2000). The fault acting on the nominal system can be described via a linearized state-space model for the wind turbine benchmark described in Chapter 2. It is assumed that the system states are unknown and only the signals $u(t)$ and $y(t)$ are available. The SMO generates a state estimate and output estimate such that a sliding mode is attained in which the output error is forced to zero in finite time. The SMO structure of Edwards et al. (2000) can be obtained quite easily by choosing suitable gain matrices and the discontinuous switched component which is necessary to induce the sliding motion. The SMO design has been obtained by exploiting the Matlab® routines described in Edwards and Spurgeon (1998), Edwards (2004).

In particular, the linearized state-space model used for the design of the SMO for fault estimation has the form

$$\begin{cases} \dot{x}_e(t) = A_e\, x_e(t) + B_e\, u_e(t) + R_e\, f(t), \\ y_e(t) = C_e\, x_e(t), \end{cases} \tag{5.10}$$

where $u_e(t) = [\tau_{aero},\ \tau_r]^T$ and $y_e(t) = x_e(t) = [\omega_r,\ \tau_g]^T$. The matrices are defined as follows:

$$A_e = \begin{bmatrix} 0 & -1/J \\ 0 & -p_{gen} \end{bmatrix},\ B_e = \begin{bmatrix} 1/J \\ p_{gen} \end{bmatrix},\ R_e = \begin{bmatrix} 0 \\ 1 \end{bmatrix},\ C_e = \begin{bmatrix} 1 & 0 \\ 0 & 1 \end{bmatrix}. \tag{5.11}$$

The second fault estimation scheme exploited for comparison purposes has been developed by using the strategy relying on the Neural Network (NN) tool (Korbicz et al., 2004). The NN exploited as a fault estimator consists of a 4-input 1-output time-delayed three-layer Multiple-Layer Perceptron (MLP) NN with 3 neurons in the input layer, 5 neurons in the hidden

**TABLE 5.27** Active and passive FTC scheme comparison.

| Fault # | Active FTC % | SMO % | NN % |
|---------|--------------|-------|------|
| 1 | 11.05 | 13.33 | 15.74 |
| 2 | 12.23 | 14.45 | 16.33 |
| 3 | 10.35 | 11.78 | 12.89 |
| 4 | 13.02 | 14.56 | 15.98 |
| 5 | 9.97 | 10.75 | 12.04 |
| 6 | 8.98 | 13.33 | 15.74 |
| 7 | 9.89 | 10.29 | 12.01 |
| 8 | 10.07 | 12.02 | 13.56 |
| 9 | 8.76 | 9.76 | 10.29 |

layer, and 1 neuron in the output layer. The NN has been trained in order to provide the estimate of the fault signal on the basis of the training patterns and target sequences (Korbicz et al., 2004).

In order to provide a brief but clear insight into the above mentioned techniques, a comparison has been performed under the same previous working conditions and based on the $NSSE\%$ index. It is worth recalling the main features of the alternative FDD schemes implemented via the SMO or the NN estimator. In particular, the SMO strategy is able to decouple the model–reality uncertainty via the sliding motion, whilst the NN estimator was designed to *passively tolerate* disturbances and modeling errors.

From Table 5.27 it can be seen how the different model-based and data-driven FTC schemes (active and passive) are able to tolerate uncertainty and errors, trying to achieve different FTC properties.

The comparison of Table 5.27 highlights that the scheme using the SMO allows achieving better performances in terms of tracking error. However, the SMO can increase the computational time and the actuator signal activity with respect to the other solution.

A few further comments can be drawn here. When the modeling of the dynamic system can be perfectly obtained, in general model-based (active) FTC strategies are preferred. On the other hand, when modeling errors and uncertainty are present, alternative estimation schemes relying on adaptation mechanisms, or data-driven (passive) FTC schemes, have shown interesting robustness properties in the presence of unmodeled disturbances, modeling mismatches, and measurement errors. With reference to the NN estimator, in the case of a system with modeling errors, offline learning can lead to fair results. Other explicit disturbance decoupling techniques can take advantage of their robustness capabilities, but with quite complicated and not straightforward design procedures. The NN-based scheme relies on the learning accumulated from offline simulations, but the training stage can be computationally heavy.

The stability properties of the addressed FTC schemes have been discussed in Chapter 4. They have been analyzed and verified here by means of a Monte Carlo campaign based on the wind turbine benchmark nonlinear simulator. Initial conditions have been changed randomly and disturbance affecting the system have been simulated during the transient related to the stability analysis. All simulations have been performed by considering noise signals modeled as band-limited white processes, according to the standard deviations reported in Table 5.16.

As an example, Fig. 5.23 depicts, for a single Monte Carlo run, the main wind turbine model variables in full load working conditions.

Fig. 5.31 depicts the wind signal $v(t)$, the generator speed $\omega_g$ and power $P_g$, and the control input $\beta$. Also in this situation the main wind turbine variables remain bounded around the reference values, thus exhibiting the overall system stability in simulation, even in the presence of modeling errors and noise signals.

Finally, it is worth noting that the AFTCS design and the analysis procedures shown in this section were implemented using Matlab® and Simulink® software tools, in order to automate the overall simulation process. These feasibility and reliability studies are of paramount importance for real application of control strategies once implemented to future wind turbine installations.

## 5.4 WIND FARM MODEL APPLICATION

As remarked in Chapter 1, in recent years, many contributions have been proposed related to the topics of fault diagnosis of wind turbines and wind farms, see, e.g., Chen et al. (2011), Gong and Qiao (2013b). Some of them highlight the difficulties to achieve the diagnosis of particular faults, e.g., those affecting the drive-train, at wind turbine level. However, these faults

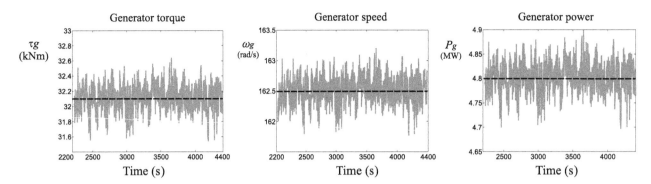

**FIGURE 5.31** Variables of the benchmark in full load operation.

are better dealt with at wind farm level, when the wind turbine is considered in comparison to other wind turbines of the wind farm (Odgaard and Stoustrup, 2013a). Moreover, FTC of wind turbines has been investigated, e.g., in Odgaard and Stoustrup (2015a), Parker et al. (2011), and international competitions on these issues arose (Odgaard and Stoustrup, 2012c; Odgaard and Shafiei, 2015b).

With reference to the benchmark model for the wind farm FDI that is considered in this monograph was previously proposed in Odgaard and Stoustrup (2013a). Based on this model, an international competition on wind farm FDI was also organized. Selected contributions were presented at the IFAC World Congress 2014, and the three top solutions for this competition were considered and evaluated in Odgaard and Shafiei (2015a). These three works were proposed in particular by Borcehrsen et al. (2014), by Blesa et al. (2014), and by Simani et al. (2014).

In particular, the FDI system designed in Borcehrsen et al. (2014) takes advantage of the fact that within a wind farm several turbines will be operating under similar conditions. To enable this, the turbines are grouped into several groups of similar turbines, then the turbines within each group are used to generate residuals for the turbines in the group. The generated residuals are then evaluated using dynamical CUmulative SUM (CUSUM). On the other hand, Blesa et al. (2014) addressed the FDI problem for the wind farm using Interval Parity Equations (IPE). Fault detection is based on the use of parity equations and unknown but bounded description of the noise and modeling errors. The fault detection test is based on checking the consistency between the measurements and the model by finding if the former are inside the interval prediction bounds. The fault isolation algorithm analyzes the observed fault signatures online, and matching them with the theoretical ones obtained using structural analysis. Finally, the same authors in Simani et al. (2014) proposed an FDI solution relying on fuzzy residual generators that are identified from the noisy measurements acquired from the simulated wind park. Note, however, that these fuzzy residual generators did not generate the reconstruction of the fault functions, as considered in this monograph. Therefore the fault diagnosis of wind turbine and wind farm systems has been proven to be a challenging task, and motivates the research activities carried out through this monograph.

By following the guidelines reported in Stamatis (2003b), the FMEA procedure has been performed on the wind farm simulator. The FMEA is a sensitivity analysis aimed at estimating the most sensitive measurements with respect to the simulated fault conditions. In practice, the monitored fault signals have been injected into the benchmark simulator, assuming that only a single fault may occur in the considered plant. Then, the $RMSE\%$ between the fault-free and faulty measured signals are computed, so that, for each fault $f_i(k)$, the most sensitive signals ($u_j(k)$ and $y_l(k)$) to that specific fault are selected. In particular, also in this case, the FMEA is conducted on the basis of the selection algorithm that is achieved by introducing the normalized sensitivity function $N_x$, defined by Eq. (5.1).

Also in this case its value represents the effect of the considered fault case with respect to a certain measured signal $x(k)$. The subscripts "f" and "n" indicate the faulty and the fault-free case, respectively. Therefore the measurements mostly affected by the considered fault imply a value of $N_x$ equal to "1." Otherwise, smaller values of $N_x$ mean that the signal $x$ is not practically affected by the fault. The signals characterized by the highest value of $N_x$ can be selected as the most sensitive measurements and they will be considered in the design of the fault diagnosis blocks.

The results of the FMEA sensitivity are reported in Table 5.28 for the wind farm benchmark. The selected signals for each fault included in the wind farm benchmark are divided as inputs and outputs.

As a result, the fault diagnosis blocks that are designed exploit the reduced subset of input and output signals $u_j(k)$ and $y_l(k)$ of Table 5.28, instead of using all the system measurements recalled in Chapter 2. Moreover, this leads to a noteworthy simplification of the residual generator structure, thus providing also a decrease in the computational effort.

**TABLE 5.28** FMEA and selected wind farm benchmark signals.

| Fault $f_i(k)$ | Most Sensitive Measurements $u_j(k)$ and $y_l(k)$ |
|---|---|
| 1 | $\beta_2, P_{g,2}, \beta_7, P_{g,7}, v_{w,m}$ |
| 2 | $\beta_1, \omega_{g,1}, \beta_5, \omega_{g,5}, v_{w,m}$ |
| 3 | $\beta_6, P_{g,6}, \beta_8, \omega_{g,8}, v_{w,m}$ |

Finally, the threshold test logic of the FDI/FDD modules described in Chapter 3 allows the achievement of the fault diagnosis task. The design of the FDI/FDD systems will be described in the following.

### 5.4.1   Control Design for Wind Farm

As described in Chapter 2, the wind and wake model provided the wind speed for each of the 9 turbines, contained in the vector $\mathbf{v}_w$, as well as for a measuring mast $v_{w,m}$. They are determined starting from a certain wind sequence (two different wind sequences are included in the simulator) and their elaboration takes into account the delay and the interaction among the turbines depending on wind direction. In particular, the wake effect is described as reported in Jensen (1983) by means of a static deficit coefficient of 0.9.

The overall model of the wind park represents the 9 wind turbines with the same submodel for each of them. It receives as input the $\mathbf{v}_w$ vector and the $\mathbf{P}_r$ vector containing the 9 reference signals from the controller. The outputs are the vectors $\mathbf{P}_g$, $\boldsymbol{\beta}$, $\boldsymbol{\omega}_g$ that contain the generated powers, the pitch angles, and the generator speeds, respectively, for each of the 9 turbines.

Inside the turbine submodel, the current wind speed is elaborated by means of a look-up table in order to compute the available power $P_w(t)$. Then the generated power is computed according to the relation

$$P_g(t) = P_c(t) + \gamma_p \sin(2\pi\sigma_p t) \tag{5.12}$$

where the first term $P_c(t)$ is equal to the current lower value between the filtered available power $\hat{P}_w$ and the filtered reference power $\hat{P}_r$:

$$P_c(t) = \begin{cases} \hat{P}_w(t), & \text{if } \hat{P}_w(t) < \hat{P}_r(t), \\ \hat{P}_r(t), & \text{if } \hat{P}_w(t) > \hat{P}_r(t). \end{cases} \tag{5.13}$$

The second term of Eq. (5.12) represents the oscillations caused by the drive-train, whose amplitude is $\gamma_p$ and frequency $\sigma_p$.

The filtered signals $\hat{P}_w(t)$ and $\hat{P}_r(t)$ differ from the input variables by means of a first order transfer function in the form:

$$\hat{P}_w(s) = \frac{\tau_w(v_w)}{s + \alpha_w(v_w)} P_w(s), \tag{5.14}$$

$$\hat{P}_r(s) = \frac{\tau_p}{s + \tau_p} P_r(s), \tag{5.15}$$

where the parameter $\tau_p$ is a fixed value, while the functions $\tau_w(\cdot)$ and $\alpha_w(\cdot)$ depend on the wind speed and are computed by means of a look-up table. Note that descriptions using maps rather than analytical functions allow providing a computationally simple, but at the same time also realistic, simulator of wind farm system. On the other hand, the relations of Eqs. (5.14) and (5.15) describe the filtering effect of the wind farm on the available and reference powers, respectively, which cannot change instantaneously.

The relation between the reference pitch signal and the actual pitch angle of the wind turbine of the wind farm has been described by a first-order transfer function in the form

$$\beta(s) = \frac{\tau_\beta}{s + \tau_\beta} \beta_r(s). \tag{5.16}$$

With respect to the second-order system already considered in Chapter 2, this means that the wind turbine blades of wind farm simulator are actuated by means of electric motors rather than hydraulic systems.

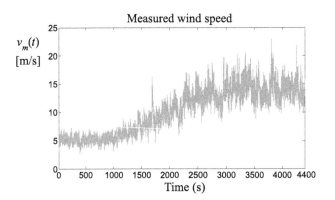

**FIGURE 5.32**    The mean wind speed sequence driving the simulations of the wind farm.

**TABLE 5.29** Faulty turbines and the fault activation times.

| Fault | # Affected Turbine: $(i, j)$ | Time (s) |
|---|---|---|
| 1 | # 7: (3, 1) | 1000–1100 |
|   | # 2: (1, 2) | 3000–3100 |
| 2 | # 1: (1, 1) | 1300–1400 |
|   | # 5: (2, 2) | 3300–3400 |
| 3 | # 6: (2, 3) | 1600–1700 |
|   | # 8: (3, 2) | 3600–3700 |

Then the generator speed of each turbine is modeled as described by

$$\omega_g(t) = f_\omega\big(P_c(t)\big) \left(1 + \frac{\gamma_p}{\omega_{g,max}} \sin(2\pi\,\sigma_p\,t)\right) \tag{5.17}$$

where $f_\omega(\cdot)$ is computed by means of a look-up table, whilst the oscillation term, due to the drive-train, has an amplitude equal to the ratio between the parameter $\gamma_p$ and the maximal generator speed $\omega_{g,max}$. In this way, the relation of Eq. (5.17) allows including both the generator controller and the effect of the drive-train oscillations, which depend on its time-constant $1/\sigma_p$ and the damping factor $\gamma_p$. This term is also reduced by the maximal generator speed $\omega_{g,max}$.

Finally, the wind farm controller forces each turbine to follow a reference power signal $P_r(k)$ that is 1/9 of the wind farm power reference. Moreover, in order to avoid fast variation of the control signal, the wind farm power reference is low-pass filtered to obtain $\hat{P}_{wf,r}$. The controller is modeled as discrete-time system and uses a sample frequency of 10 Hz:

$$P_r(k) = \frac{1}{9}\,\hat{P}_{wf,r}(k). \tag{5.18}$$

The following simulations refer to the wind farm benchmark model when the fault diagnosis schemes based on both the data-driven and model-based methodologies recalled in Chapter 3.

The mean wind sequence driving all the simulations covers the most common operative range from 5 up to 15 m/s, with a peak value of about 23 m/s. The wind and wake submodel described in Chapter 2 processes this sequence in order to generate the actual wind speed signal for all the turbines of the wind farm and for both the measurement musts associated with the two wind scenarios provided by the benchmark model, as shown in Chapter 2, also taking into account the disturbances and the interaction among turbines. Fig. 5.32 depicts the wind sequence leading the wind park system.

The available data consist of 440,000 samples of input–output measurements, acquired with a sampling rate of 100 Hz.

The three considered fault cases described in Chapter 2 affect different wind turbines at different times, by influencing the measured variables $\mathbf{u}(k)$ and $\mathbf{y}(k)$, and in particular the signals $\beta_i$, $\omega_{g,i}$, $P_{g,i}$ related to the $i$th wind turbine of the wind farm. These faults are difficult to detect at wind turbine level. However, they can be more easily detected at wind farm level by comparing the performance of different wind turbines. Table 5.29 shows the affected turbines per fault, together with the occurrence time.

These faults are simulated for the three cases of Table 2.4. Note that the turbines in Table 5.29 are denoted also by means of their row and column indices in the coordinate system, as reported in Chapter 2.

**TABLE 5.30** Fuzzy fault estimator performance.

| Data Set | Estimator RMSE (%) | | |
|---|---|---|---|
| | Fault 1 | Fault 2 | Fault 3 |
| Estimation | 0.90 | 0.87 | 0.92 |
| Validation | 1.03 | 1.01 | 1.05 |
| Test | 1.08 | 1.03 | 1.09 |

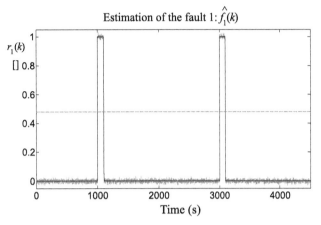

Estimation of the fault 1: $\hat{f_1}(k)$

**FIGURE 5.33** Fault 1 estimator residual $r_1(k)$ (continuous lines) and its threshold (dotted line).

## 5.4.2 Data-Driven Fault Diagnosis

The data-driven fault estimators recalled in Chapter 2 are designed by selecting $n_C = 5$ clusters and $o = 3$ as delay on the input and output regressors, related to the signals selected by the fault sensitivity analysis recalled above. On the other hand, the optimal $n_C$ and $o$ have been determined via extensive simulations that are exploited for minimizing the cost function of Eq. (3.94).

These data-driven fault estimators are based on the fuzzy prototypes relying on TS models that exploit Gaussian membership functions. Afterwards, the fault estimators are organized into a dedicated observer scheme of Chapter 3 that allows for the isolation of the three faults affecting the wind farm benchmark model described in Chapter 2.

Firstly, the capabilities of these three fuzzy fault estimators are evaluated in terms of RMSE, defined as the difference in percent between the measurements and the estimation, computed in fault free conditions. This index can be considered as the measure of the percentage of the data not correctly reconstructed by the estimator. As highlighted in Table 5.30, although this index increases when computed on data which are not used in the estimation phase (i.e., validation and test data sets), however, the percentage is always smaller than 5%, thus featuring a good modeling capability.

Then the detection of the three faults is achieved by means of the residuals $r_i(k) = \hat{f_i}(k)$ generated by the fault estimators, after the proper tuning of the threshold parameter $\delta$. In particular, Fig. 5.33 highlights the residual $r_1(k)$ relative to the fault 1 estimator, whose value is bounded by the thresholds when the fault is not active, while it is significantly over the threshold when the fault occurs on the two different turbines. Similar results are achieved by the fault 2 and 3 estimators, whose residuals $r_2(k)$ and $r_3(k)$ are depicted in Figs. 5.34 and 5.35, respectively.

In the same way as for the fuzzy estimator approach, 3 NN–NARX described in Chapter 3 have been designed to estimate the nonlinear behavior between the measurements selected by the fault sensitivity analysis recalled above and the considered fault cases. The selected architecture of the networks involves three layers, namely the input, hidden, and output layers. The number of neurons in the hidden layer has been fixed to $n_h = 16$. A number of $d_u = d_y = 5$ has been chosen for the input–output delays. Similarly to the fuzzy models, the neural networks modeling capabilities have been tested in terms of RMSE, and the results are reported in Table 5.31 in fault-free conditions for three different data sets (training, validation, and test). Also in this case a trial-and-error procedure is used to determine the optimal number of the delays $d_u$ and $d_y$, as well as the number of the neural network neurons, which leads to the minimization of the cost function of Eq. (3.107).

Note that in Table 5.31 three different data sets have been exploited. In fact, in order to verify and validate the achieved results, different data sets have to be used. The wind farm benchmark is able to generate different data sequences for each simulation, as described in Chapter 2.

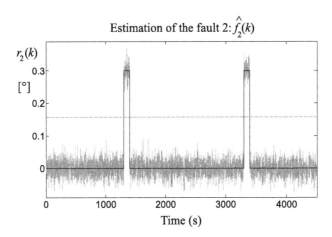

**FIGURE 5.34** Fault 2 estimator residual $r_2(k)$ (continuous lines) and its threshold (dotted line).

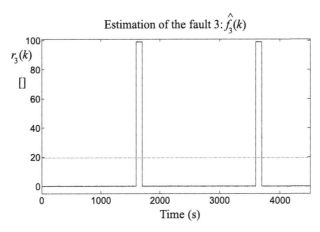

**FIGURE 5.35** Fault 3 estimator residual $r_3(k)$ (continuous lines) and its threshold (dotted line).

**TABLE 5.31** NN FDD performance.

| Fault Estimator # | 1 | 2 | 3 |
|---|---|---|---|
| $RMSE\%$ (training set) | 0.89 | 0.91 | 0.92 |
| $RMSE\%$ (validation set) | 0.91 | 0.93 | 0.95 |
| $RMSE\%$ (test set) | 1.06 | 1.04 | 1.23 |

**TABLE 5.32** The optimal values of the parameter $\delta$ used for fault diagnosis purposes by the designed residual generators.

| Residual $r_i(k) = \hat{f}_i(k)$ | $\delta$ for the Fuzzy Estimators | $\delta$ for the Neural Network Estimators |
|---|---|---|
| 1 | 2.15 | 2.34 |
| 2 | 2.23 | 2.45 |
| 3 | 2.34 | 2.89 |

The fault detection task is achieved by comparing the residual with a fixed optimized threshold, as for the case of the fuzzy estimators, reported in Table 5.32.

### 5.4.3 Model-Based Fault Diagnosis

This section describes the design and the simulations of the NLGA adaptive filters applied to the wind park benchmark. In particular, the results achieved from the estimation of the disturbance terms are presented first. Once the disturbance decoupling has been achieved, the performances of the NLGA fault estimators are reported.

In more detail, the $i$th $C_p$-map entering into the aerodynamic model of each of the 9 wind turbines of the wind farm described in Chapter 2 has been approximated by using a two-dimensional polynomial of the form

$$\hat{C}_p(\lambda_i, \beta_i) = 0.010\,\lambda_i^2 + 0.0003\,\lambda_i^3\,\beta_i - 0.0013\,\lambda_i^3 \tag{5.19}$$

for the $i$th turbine of the wind park. More details can be found in Chapter 2. By following the same procedure, the second disturbance term representing the $p_d(x)$ function in Eq. (3.66) is described by the polynomial $\hat{C}_{p_i}$ representing the wind wake from the $j$th turbine of the park affecting the $i$th turbine:

$$p_d(x) = 0.0027\,\lambda_j^2\,\beta_j - 0.0011\,\lambda_j^2, \tag{5.20}$$

with $i \neq j$. In general, note that the expression of both the $\hat{C}_{p_i}$-map and the disturbance term $p_d(x)$ in Eq. (5.20) depend on the $i$th wind turbine of the wind farm.

It is also worth observing that the considered scheme provides the analytical description of the disturbance effects due to all uncertainties, and not only the errors due to $C_p$ entry changes and the wind wake interferences among the wind turbines. However, since these terms are used for the fault estimation filter design, any kind of uncertainty must be modeled. A similar approach was proposed, e.g., in Chen and Patton (2012) but developed only for linear state-space models. Under these considerations, the uncertainty distribution description $p_d(x)$ for the nonlinear model of Eq. (3.66) is identified using the input–output data from the wind turbines of the wind farm. The general assumption holding for this case is that the model–reality mismatch is varying more slowly that the disturbance signals, such as $d$. Another important point regards the fact that the $p_d(x)$ estimation aims at describing the structure of the uncertainty, which should not depend on the wind size uncertainty. Only the so-called "directions" of the disturbances represent the important effect for disturbance decoupling, i.e., the $p_d(x)$ term, and not the "amplitude" of the uncertainty, i.e., the size of the disturbance $d$.

The designed nonlinear adaptive filters of Eqs. (4.45), (4.46), and (4.47) provide the estimate the magnitude of the different faults acting on the wind farm benchmark, as shown in Chapter 3. With reference to the overall input–affine model of the wind farm described in Chapter 2, the following terms can be determined:

$$n(x) = \begin{bmatrix} -\frac{\rho A}{2J}0.0010\,R^3\,x_1^2 - \frac{1}{J}x_2 \\ -p_{gen}\,x_2 \end{bmatrix}, \tag{5.21}$$

$$g(x) = \begin{bmatrix} 0 & \frac{\rho A}{2J}0.0003\,R^3\,x_2^2 \\ p_{gen} & 0 \end{bmatrix}, \tag{5.22}$$

and

$$\ell(x) = \begin{bmatrix} 0 & \frac{\rho A}{2J}0.0003\,R^3\,x_1^2 \\ 0 & 0.0001 \end{bmatrix}, \tag{5.23}$$

with reference to the $i$th turbine and the understanding that the subscript $i$ is dropped. Moreover, $p_d(x)$ is defined as

$$p_d(x) = \begin{bmatrix} \frac{\rho A}{2J}0.0010\,R^2\,x_1 & 0.0011 \\ 0.0002 & \frac{\rho A}{2J}0.0027\,x_2 \end{bmatrix}. \tag{5.24}$$

In the case of the model of Eq. (3.66), and recalling Eqs. (5.24), (5.23), and (5.22), it results in

$$S_0 = \bar{P} = \mathrm{cl}\big(p_d(x)\big) \equiv p_d(x). \tag{5.25}$$

If $\ker\{dh\} = \emptyset$, it follows that $\Sigma_*^P = \bar{P}$ as $\bar{S}_0 \cap \ker\{dh\} = \emptyset$. On the other hand, it is necessary to compute the expression $(\Sigma_*^P)^\perp = (\bar{P})^\perp$. However, it is worth noting that, for the case under investigation, the determination of the codistribution $(\Sigma_*^P)^\perp = (\bar{P})^\perp$ is enhanced due to the structure of Eq. (5.25).

**TABLE 5.33** RMSE of the NLGA fault estimator with disturbance decoupling.

| Fault Case | Fault 1 Estimator | Fault 2 Estimator | Fault 3 Estimator |
|---|---|---|---|
| RMSE (%) | 0.0007 | 0.0008 | 0.0009 |

**TABLE 5.34** Monte Carlo analysis parameter variations.

| Parameter | Nominal Value | Error Value % |
|---|---|---|
| $\rho$ | 1.225 kg/m$^3$ | $\pm 20$ |
| $J$ | $7.794 \times 10^6$ kg/m$^3$ | $\pm 30$ |
| $C_p$ | $C_{p0}$ | $\pm 50$ |

Finally, as an example, the design of the NLGA adaptive filter for the reconstruction of the fault case 2 is based on the expression

$$\dot{\hat{y}}_{1s} = M_1 \cdot f + M_2, \tag{5.26}$$

where

$$\begin{cases} M_1 = 0.8\,x_1^2 - 0.036\,x_1, \\ M_2 = 1.02\,x_2^2 + 15.7\,x_2 - 0.3\,x_1^3 + 0.77\,x_1^2. \end{cases} \tag{5.27}$$

The design of the NLGA adaptive filters for the reconstruction of the faults for the cases 1 and 3 is based on a different selection of the vector of Eq. (5.23), which leads to other expressions for the filter of Eq. (5.26). As an example, for the fault case 2 described in Chapter 2, the nonlinear filter for the reconstruction of $f_2$ decoupled from the disturbance $d$ representing the effect of both the wind $v_w(t)$ and the wake $v_{w,m}$ signals has the form of Eq. (4.45). After a suitable choice of the parameters in Eqs. (4.45), (4.46), and (4.47) the nonlinear filter provided an accurate estimate $\hat{f}_2(k)$ of the fault size, with minimal detection delay.

Finally, also the model capabilities of the three NLGA estimators with disturbance decoupling are evaluated in terms of $RMSE\%$, as summarized in Table 5.33, thus demonstrating the superiority of the model-based approach with respect to the data-driven methods.

The results achieved in Table 5.33 highlight the efficacy of the model-based NLGA methodology with disturbance decoupling with respect to the data-driven approaches.

### 5.4.4 Comparative and Robustness Analysis

The evaluation of the performances of the considered fault diagnosis strategies is based again on the computation of the FAR, MFR, TFR, and MFD indices recalled in Section 5.2.4. Also in this case, a proper Monte Carlo analysis has been performed in order to compute these indices and to test the robustness of the considered fault diagnosis schemes. Indeed, the Monte Carlo tool is useful at this stage, as the efficacy of the diagnosis depends on both the model approximation capabilities and the measurements errors.

In particular, extensive simulations based on a set of 1000 Monte Carlo runs have been executed, during which realistic wind turbine uncertainties have been considered. Some meaningful variables have been modeled as Gaussian white stochastic processes with nominal values and standard deviations corresponding to realistic error values summarized in Table 5.34.

The comparative analysis results are reported in Table 5.35. In particular, the different approaches to the fault diagnosis of the wind farm benchmark model, i.e., the Fuzzy Filters (FF), Neural Network Filters (NNF), and NLGA Adaptive Filters (NLGA–AF) for fault estimation are shown.

As remarked in Chapter 3, different FDI/FDD methodologies are considered here for comparison purposes and they are selected from the contributions submitted for the international competition on FDI and FTC of wind turbines and wind farm (Odgaard and Shafiei, 2015a). In particular, the CUSUM–Based Detection (CBD) scheme proposed in Borcehrsen et al. (2014) and the Interval Parity Equation (IPE) method presented in Blesa et al. (2014) are considered. Their performances are extrapolated by the results presented in Odgaard and Shafiei (2015a).

With reference to the solutions considered in this chapter, the results show the overall efficacy of the fault diagnosis solutions. In general, both the FF and NNF for fault estimation seem to achieve quite interesting results, and they have

**TABLE 5.35** FDI strategy comparison for the wind farm.

| Fault Case | Index | FF | NNF | NLGA–AF | CBO | IPE |
|---|---|---|---|---|---|---|
| 1 | FAR | 0.0010 | 0.0010 | 0.0006 | 0.0013 | 0.0024 |
|  | MFR | 0.0010 | 0.0010 | 0.0007 | 0.0000 | 0.0005 |
|  | TFR | 0.9990 | 0.9990 | 0.9994 | 0.9999 | 0.9995 |
|  | MFD (s) | 0.02 | 0.01 | 0.007 | 4.7 | 1.1 |
| 2 | FAR | 0.0010 | 0.2280 | 0.0004 | 0.0111 | 0.0015 |
|  | MFR | 0.0030 | 0.0010 | 0.0005 | 0.0015 | 0.0005 |
|  | TFR | 0.9970 | 0.9990 | 0.9996 | 0.9985 | 0.9995 |
|  | MFD (s) | 0.08 | 0.08 | 0.008 | ND | 12.4 |
| 3 | FAR | 0.0030 | 0.0010 | 0.0005 | 0.0055 | 0.0007 |
|  | MFR | 0.0080 | 0.0010 | 0.0006 | 0.0000 | 0.0010 |
|  | TFR | 0.9920 | 0.9990 | 0.9995 | 0.9999 | 0.9990 |
|  | MFD (s) | 0.02 | 0.01 | 0.006 | 13.1 | 15.2 |

a noteworthy performance level considering the MFD. Also the FAR and MFR are quite low, and in particularly neural networks present very low values of MFR for all the considered faults. However, for both the FF and NNF fault diagnosis design, optimization stages are required, for example, for the selection of the optimal thresholds. On the other hand, the NLGA–AF shows optimal results, with respect to the MFD index and the least TFR, compared to FF and NNF. However, the design complexity of NLGA–AF is much higher than for the other approaches, and in some cases the analytic solution cannot be determined.

The achieved results also lead to further considerations: the proper choice of the design parameters can lead to the least FAR and MFR, with very high TDR, and minimal MFD. Moreover, the Monte Carlo analysis validates the robustness and the estimation convergence properties of the considered FF and NNF fault diagnosis schemes, with respect to error, noise, and uncertainty. On the other hand, with respect to the two top FDI solutions selected in Odgaard and Shafiei (2015a), both the CBO and IPE present higher MFD and FAR than the FF and NNF methods, but in some cases, better TFR and MFR.

Note finally that all optimization procedures, i.e., the training of the neural networks and the estimation of the fuzzy models, are performed offline. On the other hand, the implementation of the achieved fuzzy or neural estimators used for the fault estimation tasks can be directly implemented as nonlinear difference (discrete-time) equations, as described, e.g., for the fuzzy systems in Rovatti et al. (2000), Simani (2005).

### 5.4.5 Sustainable Control for the Wind Farm Simulator

This section describes the design and the simulations of the main AFTCS solutions addressed in Chapter 3 applied to the wind park benchmark. In particular, the best results are achieved with the disturbance decoupling approach recalled in Chapter 3 that relies on the estimation of the disturbance terms appearing in Eq. (3.66). Once the disturbance decoupling has been achieved, the performances of the AFTCS method are summarized.

In more detail, the $C_p$-map entering into the aerodynamic model of the wind turbines of the wind park has been approximated by using a two-dimensional polynomial of the form

$$\hat{C}_p(\lambda, \beta) = 0.010\,\lambda_i^2 + 0.0003\,\lambda_i^3\,\beta_i - 0.0013\,\lambda_i^3 \tag{5.28}$$

for the $i$th turbine of the wind park. More details can be found in Chapters 2 and 3. By following the same procedure, the second disturbance term representing the $p_d(x)$ function in Eq. (3.66) and due to the wind wakes is described by the term $\hat{C}_{pi}$ in Eq. (5.20):

$$\hat{C}_{pi}(\lambda_j, \beta_j) = 0.0027\,\lambda_j^2\,\beta_j - 0.0011\,\lambda_j^2 \tag{5.29}$$

with reference to the wind wake from the $j$th turbine of the park affecting the $i$th turbine.

It is worth noting that the considered scheme provides the analytical description of the disturbance effects due to all uncertainties, and not only the errors due to $C_p$ entry changes and the wind wake interferences among the wind turbines. However, since these terms are used for the fault estimation filter design, any kind of uncertainty must be modeled. A similar approach was proposed, e.g., in Chen and Patton (2012) but developed only for linear state-space models. Under these considerations, the uncertainty distribution description $p_d(x)$ for the nonlinear model of Eq. (3.66) is identified using the input–output data from the wind turbine. The general assumption holding for this case is that the model–reality mismatch is varying more slowly that the disturbance signals, such as $d$. Another important point regards the fact that the $p_d(x)$ estimation aims at describing the structure of the uncertainty, which should not depend on the wind size uncertainty. Only the so-called directions of the disturbances represent the important effect for disturbance decoupling, i.e., the $p_d(x)$ term, and not the "amplitude" of the uncertainty, i.e., the size of the disturbance $d$.

The model-based approach to the design of the adaptive filters of Eqs. (4.45), (4.46), and (4.47) has provided an estimate of the magnitude of the different faults acting on the wind farm benchmark. The design of these filters for the compensation of the control fault is sketched in the following. More mathematical details of the general design procedure can be found in Chapter 3.

With reference to the input-affine model of Eq. (3.66), $x = [x_1 \, x_2]^T = [\omega_{i \, g} \, P_{i \, g}]^T$, $c = [P_{i \, r} \, \beta_i]^T$, and

$$n(x) = \begin{bmatrix} -\frac{\rho A}{2J} 0.0010 R^3 x_1^2 - \frac{1}{J} x_2 \\ -p_{gen} x_2 \end{bmatrix}, \tag{5.30}$$

$$g(x) = \begin{bmatrix} 0 & \frac{\rho A}{2J} 0.0003 R^3 x_2^2 \\ p_{gen} & 0 \end{bmatrix}, \tag{5.31}$$

and

$$\ell(x) = \begin{bmatrix} 0 & \frac{\rho A}{2J} 0.0003 R^3 x_1^2 \\ 0 & 0.0001 \end{bmatrix}, \tag{5.32}$$

with reference to the $i$th turbine and the understanding that the subscript $i$ is dropped. Moreover, $p_d(x)$ is defined as

$$p_d(x) = \begin{bmatrix} \frac{\rho A}{2J} 0.0010 R^2 x_1 & 0.0011 \\ 0.0002 & \frac{\rho A}{2J} 0.0027 x_2 \end{bmatrix}. \tag{5.33}$$

In the case of the wind farm overall model, with reference to Eq. (3.66), and recalling (5.24), (5.23), and (5.22), it results in

$$S_0 = \bar{P} = \mathrm{cl}(p_d(x)) \equiv p_d(x). \tag{5.34}$$

If $\ker\{dh\} = \emptyset$, it follows that $\Sigma_*^P = \bar{P}$ as $\bar{S}_0 \cap \ker\{dh\} = \emptyset$. On the other hand, it is necessary to compute the expression $(\Sigma_*^P)^\perp = (\bar{P})^\perp$. However, it is worth noting that, for the case under investigation, the determination of the codistribution $(\Sigma_*^P)^\perp = (\bar{P})^\perp$ is enhanced due to the structure of Eq. (5.25). Other mathematical details are similar to the derivation of the same filters, and they will not be reported here.

Finally, the design of the adaptive filter for the reconstruction of the fault $f$ affecting, for example, the actuator $\beta_i(t)$ (fault case 2) is based on the expression of Eq. (5.26):

$$\dot{y}_{1s} = M_1 \cdot f + M_2, \tag{5.35}$$

where

$$\begin{cases} M_1 = 0.8 x_1^2 - 0.036 x_1, \\ M_2 = 1.02 x_2^2 + 15.7 x_2 - 0.3 x_1^3 + 0.77 x_1^2. \end{cases} \tag{5.36}$$

The design of the NLGA adaptive filters for the reconstruction of the faults for the cases 1 and 3 is based on a different selection of the vector of Eq. (5.23), which will lead to other expressions for the filter Eq. (5.26).

As an example, for the fault case 2 of Table 5.29, the nonlinear filter for the reconstruction of $f$ decoupled from the disturbance $d$ representing the effect of both the wind $v_w(t)$ and the wake $v_{w,m}$ signals has the form of Eq. (4.45). After a suitable choice of the parameters in Eqs. (4.45), (4.46), and (4.47), the nonlinear filter provided an accurate estimate of the fault size, with minimal detection delay.

Thus the tests shown below refer to the simulation of the actuator fault $f$ modeled as a sequence of rectangular pulses with random amplitude and length. In particular, Fig. 5.36 shows the fault estimate $\hat{f}$ (dotted grey line), when compared with the simulated one (dashed black line).

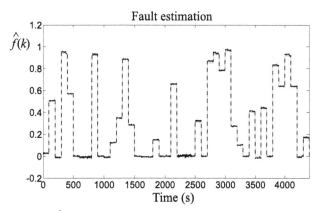

**FIGURE 5.36** (Dashed black line) fault estimate $\hat{f}$ of the (bold gray line) actual fault case 2.

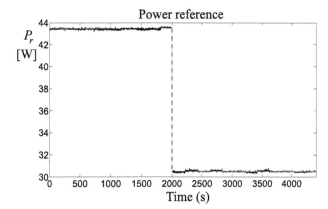

**FIGURE 5.37** $P_r$ reference signal for the fault case 2 with AFTCS.

Under this condition, Fig. 5.30 shows the power reference signal $P_r$ (continuous black line) compared with its desired value (grey dotted line), with fault accommodation.

In particular, Fig. 5.37 depicts the power reference to the wind farm, $P_r$, which is constant and equal to 43.6 MW until $t = 2000$ s, when it changes to about 30 MW.

It is worth observing that the model-based adaptive filter provided not only fault detection, but also the fault estimate. Moreover, a fault modeled as a sequence of pulses with variable amplitude and length has been considered, since it represents the realistic fault condition in connection with the wind turbine model. However, as already remarked, the fault estimation module can be easily generalized to estimate, for example, polynomial functions of time, or generic signals belonging to a given class of faults, if the NLGA adaptive filters contain the internal model of the fault itself, as recalled in Chapter 3.

In order to summarize the advantages of the FTC strategies available in the related literature, the performances of the AFTCS solutions applied to the wind farm benchmark were evaluated in terms of $NSSE\%$ and considering different data sequences.

The simulations were performed by exploiting the wind park simulator, followed by a Matlab® Monte Carlo analysis. Under this assumption, Table 5.16 reports the nominal values of the considered wind turbine model parameters with respect to their simulated uncertainty. The Monte Carlo analysis has been performed by modeling these variables as Gaussian stochastic processes, with zero-mean and standard deviations corresponding to realistic minimal and maximal error values of Table 5.36.

As remarked in Chapter 2, it was also assumed that the input–output signals $u$ and $y$ and the power coefficient map $C_p$ entries were affected by errors, expressed as standard deviations in percent of the corresponding nominal values $u_0$, $y_0$, and $C_{p0}$ also reported in Table 5.36.

Therefore, for performance evaluation of the control schemes, the best, average, and worst values of the $NSSE\%$ index were computed, and experimentally evaluated with 1000 Monte Carlo runs. The value of $NSSE\%$ is computed for several possible combinations of the parameter values reported in Table 5.36.

**TABLE 5.36** Wind farm variable uncertainty.

| Model Parameter | Actual Value | Min. Error % | Max. Error % |
|---|---|---|---|
| $\rho$ | 1.225 kg/m$^3$ | ± 0.1 | ± 20 |
| $J$ | 7.794 × 10$^6$ kg/m$^2$ | ± 0.1 | ± 30 |
| $C_p$ | $C_{p0}$ | ± 0.1 | ± 50 |
| $u$ | $u_0$ | ± 0.1 | ± 20 |
| $y$ | $y_0$ | ± 0.1 | ± 20 |

**TABLE 5.37** Comparison of the different FTC schemes.

| FTC Method/Fault | Case 1% | Case 2% | Case 3% |
|---|---|---|---|
| FTC with wind & wake decoupling | 10.33 | 11.56 | 10.47 |
| FTC with wind decoupling only | 14.07 | 15.06 | 15.34 |
| Data-driven FTC scheme | 13.74 | 14.37 | 15.01 |

It is worth noting that Table 5.36 describes the uncertain parameters that have been simulated in order to analyze the robustness and the reliability of the recalled AFTCS solutions with respect to parameter variations. In fact, the disturbance decoupling approach was used for removing the effect of the uncertain wind and wake effects, and not for handling the parameter variations in Table 5.16.

Table 5.37 summarizes the results obtained by considering the model-based FTC scheme integrating the original wind turbine farm for the different fault cases.

In particular, Table 5.37 summarizes the values of the considered $NSSE\%$ performance index, with reference to the possible combinations of the parameters described in Table 5.36. The results obtained with the model-based strategy relying on adaptive filters are also compared with those achievable by the approaches recalled in Chapter 3, which exploit the decoupling of the wind effect, as well as the passive FTC strategy using a data-driven scheme.

It is worth noting that, regarding the model-based AFTCS method with complete disturbance decoupling, Table 5.37 highlights how this solution seems to achieve better performance in terms of tracking error with respect to the two other methodologies.

## 5.5 SUMMARY

This chapter presented the simulations related to the considered benchmark systems, in which the considered solutions for FDI/FDD and FTC methodologies were tested and implemented. Firstly, the focus was placed on the single wind turbine benchmark, both the model-based and data-driven solutions for FDI/FDD and FTC were analyzed and validated by means of a Monte Carlo analysis. Then their performances were compared to those of other FDI/FDD and FTC methods, commonly adopted in the related literature. Furthermore, the tracking capability of the FTC schemes, based on both the fault accommodation and controller reconfiguration strategies, were evaluated.

Afterwards the wind farm benchmark system was considered. Similarly to the analysis carried out for the single turbine system, the addressed FDI/FDD and FTC methodologies were tested, validated, and evaluated with respect to the other model-based and data-driven solutions considered in this monograph.

Finally, in order to assess the considered systems in a more realistic framework, hardware-in-the-loop experiments were also analyzed, by means of an industrial computer interacting with the on-board electronics.

Chapter 6

# Matlab and Simulink Implementations

## 6.1 INTRODUCTION

The motivation for this chapter comes from the real need to have a description about the main challenges of modeling the wind turbine systems considered for fault diagnosis and sustainable control purposes in Matlab® and Simulink® environments. In 2009 these aspects began to stimulate research and development in the wide control community particularly for these installations, which require a high degree of sustainability. Note that this represents the key issue for offshore wind turbines and wind park installations, since they are characterized by expensive and/or safety critical maintenance work. In this case a clear conflict exists between ensuring a high degree of availability and reducing maintenance times, which affect the final energy cost.

On the other hand, as it will be highlighted in the following, wind turbines have highly nonlinear dynamics with a stochastic and uncontrollable driving force as input in the form of wind speed, thus representing an interesting challenge also from the modeling point of view, as highlighted in Chapter 2. In this chapter the modeling solutions will be described as implemented in Matlab® and Simulink® environments.

As described in Chapters 3 and 4, suitable fault diagnosis and fault tolerant control methods allow for sustainable solutions of the energy conversion efficiency over wider than normally expected working conditions. Moreover, suitable mathematical implementations of the wind turbine systems have highlighted the ability to capture the behavior of the process under diagnosis with arbitrary accuracy, thus providing an important point on the control design itself. In this way the fault diagnosis and control schemes developed in this chapter could guarantee prescribed performance, as demonstrated in Chapter 5. At the same time they give a degree of tolerance to possible deviation of characteristic properties or system parameters from standard conditions, if properly included in the wind turbine models, as considered here.

It is worth noting that due to the competitive nature of the wind turbine industry and possible confidentiality issues, the modeling available in the wind turbine literature is usually kept at a conceptual level. More detailed modeling of wind turbines can be found, e.g., in Burton et al. (2011).

Moreover, in the wind turbine area, there have been a number of IFAC and IEEE publications with sessions and special issues starting from 2009, based also on an international competition described, e.g., in Ostergaard et al. (2009), Odgaard and Stoustrup (2012d). These sessions and special issues have led to important results and publications that were recalled in this monograph, in order to give readers a basic research review. Moreover, the wind turbine benchmark simulators are described in the following sections, based on the description reported in Chapter 2.

## 6.2 WIND TURBINE SYSTEM BENCHMARK

Modern wind turbine systems have large, flexible structures operating in uncertain environments, thus representing interesting cases for modeling and advanced control solutions, as addressed in Chapters 2, 3, and 4. In particular, Chapter 4 demonstrated that advanced control solutions are able to achieve the desirable goal of decreasing the wind energy cost by increasing the efficiency, and thus the energy capture, or by reducing structural loading and increasing the lifetimes of the components and turbine structures.

This monograph aimed also at sketching the main challenges that exist in the wind industry and at stimulating new research topics in this area. Moreover, this book has focused only on HAWT, since they represent the most commonly produced large-scale installations today.

Another important topic discussed in Chapter 4 derives from the steadily increasing sizes and a growing complexity of wind turbines, thus giving rise to more severe requirements regarding the system safety, reliability, and availability (Odgaard et al., 2013). In this monograph, it was shown that the safety demand is achieved by introducing analytical (software) redundancy in the system architecture. The example considered here regards the pitch system for adjusting the angles of a rotor blade, as shown in Chapter 2. For each of the three blades, one totally independent pitch system is used, such that in the worst case of a malfunction in one or two pitch systems, the remaining one or two would still be able to bring the turbine to a standstill. This scheme improves the system safety, but it generates additional costs and possibly

**Fault Diagnosis and Sustainable Control of Wind Turbines. DOI: 10.1016/B978-0-12-812984-5.00006-7**

additional turbine down-times due to faults in the redundant system parts. The enhanced safety may lead to reduced system availability. This motivates the use of functional (analytical) redundancy, as remarked in Chapter 3.

It was remarked that even when reducing hardware redundancies, large wind turbines are prone to unexpected malfunctions or alterations of the nominal working conditions. Many of these anomalies, even if not critical, can lead to turbine shutdowns, again for safety reasons. Especially in offshore wind turbines, as remarked in Chapter 2, this results in a substantially reduced availability, because rough weather conditions may prevent prompt replacement of the damaged system parts. Therefore, the need for reliability and availability that guarantees continuous energy production motivated those sustainable control solutions considered in Section 4, whose implementations are outlined in this chapter.

The results summarized in Chapter 5 highlighted how these schemes were able to keep the turbine in operation in the presence of anomalous situations, perhaps with reduced performance, while managing the maintenance operations.

Apart from increasing availability and reducing turbine down-times, sustainable control schemes also obviate the need for more hardware redundancy, if virtual sensors could replace redundant hardware sensors, as suggested in Chapter 7. Moreover, the schemes considered in Chapter 3 and currently employed in wind turbines are typically on the level of the supervisory control, where commonly used strategies include sensor comparison, model comparison, and thresholding tests. These strategies enabled safe turbine operations, which involve shutdowns in case of critical situations, but they are not able to actively counteract anomalous working conditions. Therefore, the main aim of Chapter 5 was to revise these sustainable control strategies. They allowed obtaining a system behavior closer to its nominal situation despite of the presence of unpermitted deviations of any characteristic properties or system parameters from standard conditions (faults). Moreover, the schemes designed in Chapter 5 provide the reconstruction of the equivalent unknown input that represents the effect of a fault, thus achieving the fault diagnosis task as a by-product, as remarked in Chapter 3.

### 6.2.1 Wind Turbine Simulator Main Components

This section sketches the main components of the HAWT simulator, which are implemented in the wind turbine benchmark described in its analytical expressions in Chapter 2. It consists of a nacelle and a rotor, as depicted in Fig. 6.1 that implements the layout of Fig. 2.2. The nacelle comprises the generator, which is driven by a high-speed shaft. The high-speed shaft is in turn usually driven by a gear-box, which steps up the rotational speed from a low-speed shaft. The low-speed shaft is connected to the rotor, which includes airfoil-shaped blades. These blades capture the kinetic energy in the wind and transform it into the rotational kinetic energy of the wind turbine. These elements were represented in Fig. 2.2 and implemented in the simulator of Fig. 6.1.

The simulator was developed by Odgaard and his coworkers in Matlab® and Simulink® environments and described in detail in Odgaard and Stoustrup (2012c), Odgaard and Patton (2012), Odgaard et al. (2014), Odgaard and Stoustrup (2015a).

As shown in the Simulink® model of Fig. 6.1, the complete description of the wind turbine simulator consists of several submodels for the aerodynamics, as well as the dynamics of the pitch system and the generator/converter system, as sketched in Fig. 2.3. The generator/converter dynamics is described as a first order delay system, as highlighted in Fig. 6.2.

However, as remarked in Chapter 2, when the delay time constant is very small, an ideal converter can be assumed, such that the reference generator torque signal is equal to the actual generator torque. In this situation, the generator torque can be considered as a system input.

Finally, Fig. 6.1 includes also the main input and output signals that drive the wind turbine simulator, in particular, the wind speed and rotor torque. The simulator considers also the signals representing the rotor angular velocity, generator torque, generator angular velocity, and demanded generator torque. Finally, the model includes the pitch and its demanded value.

### 6.2.2 Aerodynamic Block

The aerodynamic block reported in Fig. 6.3 implements the expressions for the thrust force $F_T$ acting on the rotor and the aerodynamic rotor torque $T_a$. They are determined by the reference force $F_{st}$ and by the aerodynamic rotor thrust and torque coefficients $C_T$ and $C_Q$ described by the relations of Eq. (2.1), as implemented in the submodel of the Blade & Pitch System Simulink® block of Fig. 6.3 for each of the blades.

The reference force is defined from the impact pressure $\frac{1}{2} \rho v^2$ and the rotor swept area $\pi R^2$ (with rotor radius $R$), where $\rho$ denotes the air density, as described by (2.2). This expression is implemented as reported in Fig. 6.4

It is worth noting that for simulation purposes static wind speed is used. A more accurate model should exploit the effective wind speed, i.e., the static wind speed corrected by the tower and blade motion effects, as described in Section 2.3.3,

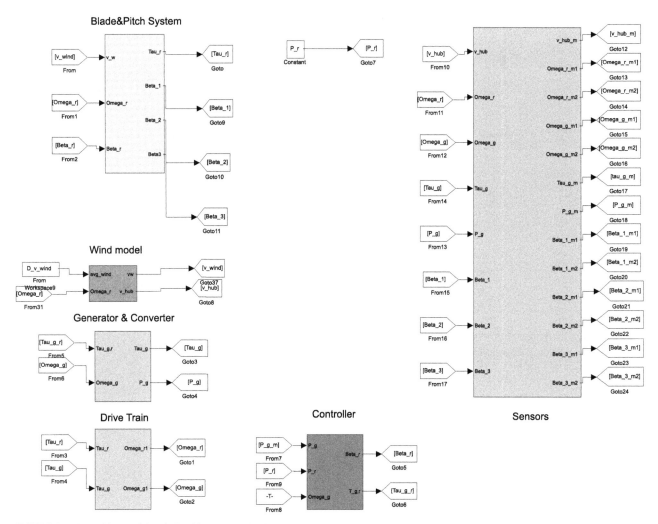

**FIGURE 6.1** Main blocks of the wind turbine system simulator.

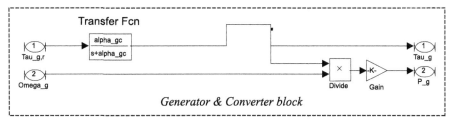

**FIGURE 6.2** Generator & Converter Simulink® block of the wind turbine simulator.

which were not included in the wind turbine simulator considered in this monograph. However, the aerodynamic maps used for the calculation of the rotor thrust and torque are represented as static two-dimensional tables, which already take into account the dynamic contributions of both the tower and blade motions, as sketched in the Simulink® submodel Fig. 6.4.

As highlighted in the expressions of Eq. (2.1), the rotor thrust and torque coefficients $(C_T, C_Q)$ depend on the tip-speed ratio $\lambda = \frac{\omega_r R}{v}$ and the pitch angle $\beta$. Therefore, the rotor thrust $F_T$ and torque $T_a$ assume the expressions of Eq. (2.3), implemented in the Simulink® submodel of Fig. 6.4.

These expressions implemented in Fig. 6.4 and representing Eq. (2.3) highlight that the rotor thrust and torque variables are nonlinear functions depending on the wind speed, rotor speed, and pitch angle. These functions are expressed as two-dimensional maps, which are also computed for the whole range of variation of both the pitch angles and tip-speed ratios. These maps represent a static approximation of more detailed aerodynamic computations that can be obtained using

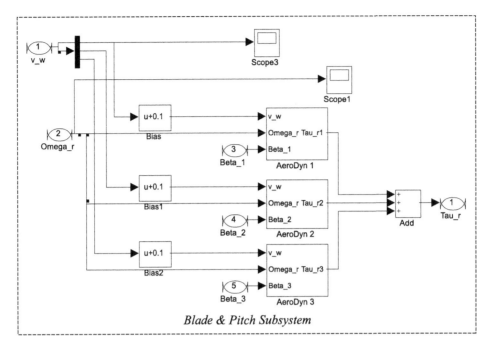

**FIGURE 6.3** Description of the aerodynamic model of the wind turbine simulator.

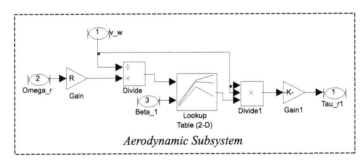

**FIGURE 6.4** Layout of the aerodynamic block for the $i$th blade of the wind turbine simulator.

more accurate wind turbine simulation codes, as, for example, the BEM method recalled in Chapter 2. This representation assumes that the aerodynamic lift and drag forces at each blade section are calculated and integrated in order to obtain the rotor thrust and torque (Gasch and Twele, 2012). More accurate maps can be obtained by exploiting the calculations implemented via the AeroDyn module of the FAST code, where the maps are extracted from several simulation runs (Laino and Hansen, 2002).

Finally, it is worth noting that for simulation purposes considered in this monograph, the tabulated versions of the aerodynamic maps are sufficient. On the other hand, for control design, the derivatives of the rotor torque (and thrust) are needed, thus requiring a description of the aerodynamic maps as analytical functions. Therefore, these maps can be approximated using combinations of polynomial and exponential functions, whose powers and coefficients can be estimated via data-driven methods, as recalled in Chapters 5 and 4.

### 6.2.3 Drive-Train Block

The drive-train system consists of a rotor, shaft, and generator, which is modeled as a two-mass inertia system, as described in Chapter 2. This system includes the shaft torsion, where the two inertias are connected with a torsional spring with spring constant $k_S$ and a torsional damper with damping constant $d_S$, as illustrated in Fig. 2.4 in Chapter 2. This system is implemented in the wind turbine simulator as sketched in Fig. 6.5.

With reference to Fig. 6.5, the angular velocities $\omega_r$ and $\omega_g$ are the time derivatives of the rotation angles $\theta_r$ and $\theta_g$. In this case, the rotor torque $T_r$ is generated by the lift forces on the individual blade elements, whilst $T_g$ represents the

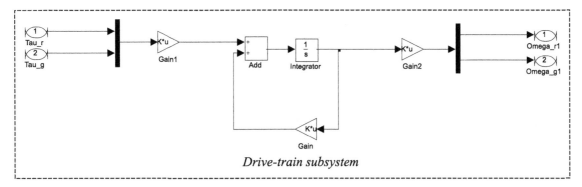

**FIGURE 6.5** Layout of the drive-train Simulink® subsystem.

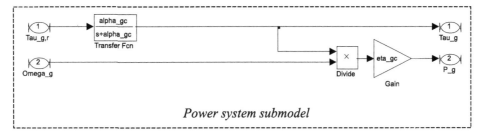

**FIGURE 6.6** Simulink® submodel of the power system.

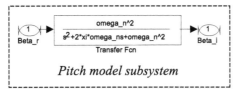

**FIGURE 6.7** Simulink® submodel of the pitch system.

generator torque. The ideal gear-box effect can be simply included in the generator model by multiplying the generator inertia $J_g$ by the square of the gear-box ratio $n_g$.

The motion relations were computed by means of Lagrangian dynamics, which first requires to define the generalized coordinates and external forces. In this way, the energy terms of the system are derived, as well as the motion descriptions, as illustrated in Chapter 2.

### 6.2.4 Power System Block

The model of the generator/converter dynamics is included in the wind turbine Simulink® benchmark. Note that for simulation purposes considered in this monograph, where the generator/converter dynamics are relatively fast, a simple first-order delay system is implemented here, as described by Eq. (2.13). This subsystem, which considers the first time derivative of the generator torque $T_g$ as a function of the demanded generator torque, $T_{g,d}$, and the delay time constant $\tau_g$, is represented in Fig. 6.6.

### 6.2.5 Pitch System Block

In pitch-regulated wind turbines, as those considered in this work, the pitch angle of the blades is controlled only in the full load working condition in order to reduce the aerodynamic rotor torque, while maintaining the turbine at the desired rotor speed.

On the other hand, the pitching of the blades to feather position (90 degrees) is used as main braking system to bring the turbine to standstill in critical situations. The wind turbine simulator exploited in this monograph implements the pitching technology that relies on a hydraulic pitch systems. For this implementation, the pitch system dynamics are described as a second-order delay model sketched in Fig. 6.7, which is also able to display oscillatory behavior.

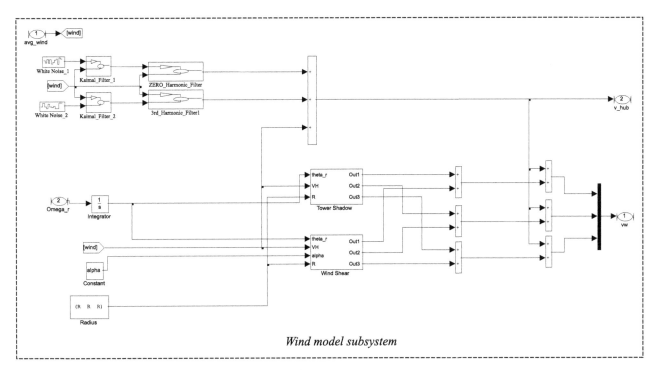

**FIGURE 6.8** Layout of the subsystem implementing the wind model process.

On the other hand, when electromechanical pitch systems are considered, first-order delay models are sufficient, as described by Eq. (2.14), that link the physical and the demanded pitch angle, $\beta$ and $\beta_d$, respectively, highlighted in Fig. 2.3. The parameter $\tau$ in Eq. (2.14) represents the delay time constant.

## 6.2.6 Wind Model Block

As already highlighted in Chapter 2, the differential heating of the earth's atmosphere is the driving mechanism for wind. Various atmospheric phenomena, such as the nocturnal low-level jet, sea breezes, frontal passages, and mountain and valley flows, affect the wind inflow across the wind turbine rotor plane.

The available wind resource are usually characterized by the spatial or temporal average of the wind speed; the frequency distribution of wind speeds; the temporal and spatial variation in wind speed; the most frequent wind direction, also known as the prevailing wind direction; and the frequency of the remaining wind directions, as described in Chapter 2.

To predict the capacity factor and maintenance requirements for a wind turbine, it is useful to understand wind characteristics over both long and short time scales, ranging from multiyear to subsecond. Therefore, it is very important for the wind industry to be able to model the variation of wind speeds. Turbine designers need the information to optimize the design of their turbines, so as to minimize generating costs. Turbine investors need the information to estimate their income from electricity generation.

Due to these reasons, the wind turbine simulator includes a subsystem able to provide an accurate description of the wind process behavior around the rotor plane, while taking into account also wind shear and tower shadow effects. The complete subsystem implementing the wind process is depicted in Fig. 6.8.

It is worth noting that detailed models of the wind are not usually exploited in the related literature, as remarked, e.g., in Odgaard et al. (2013). However, the wind turbine simulator considered in this work includes a typical wind description summarized in Chapter 2.

According to this representation, the wind process is modeled as the sum of a steady state mean wind and a perturbation wind, accounting for turbulence and/or gusts. The deterministic component of the wind field implements the transients specified by IEC 61400-1, the exponential and logarithmic wind shear models, and the tower shadow effects, which include the potential flow model for a conical tower, the downwind empirical model. On the other hand, the stochastic component of the wind field can be described according to the von Karman or Kaimal turbulence models, as highlighted in Fig. 6.8.

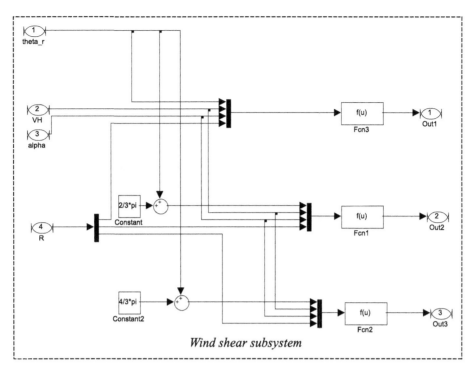

**FIGURE 6.9** Layout of the wind shear block.

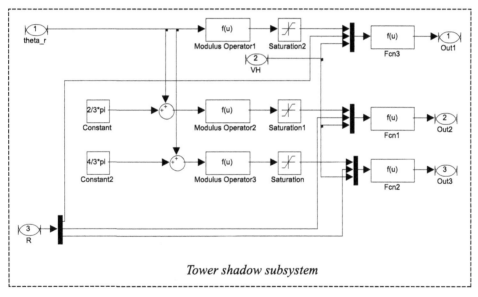

**FIGURE 6.10** Layout of the tower shadow block.

In this way the wind subsystem of Fig. 6.8 forms a scalar mean wind speed at hub height, a time-varying matrix that contains the wind speed for each point in the wind field, where the total wind speed field description includes the wind shear and the tower shadow components. The far wake component of one preceding wind turbine is relevant for the case of wind farms considered in the following.

Therefore, as describe by the relation of Eq. (2.28), the wind speed consists of four components, namely, a mean value, its stochastic component, wind shear, and tower shadow effects, reported in Figs. 6.9 and 6.10.

Actually, four wind speeds are involved in the benchmark model, one for the wind $v_{hub}$ acting on the hub, and one for each of the three blades $v_{w1}$, $v_{w2}$, $v_{w3}$. It is worth noting that the wind shear and tower shadow terms are considered only in the blade wind speeds.

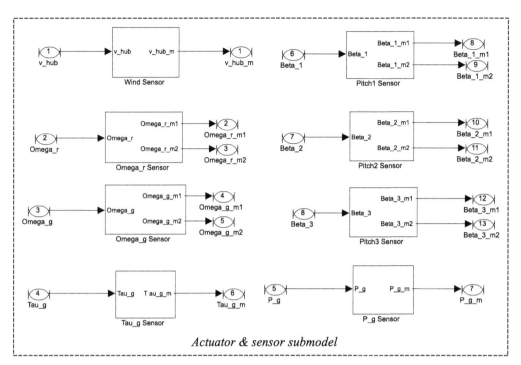

**FIGURE 6.11** Layout of the subsystem implementing the actuator and sensor models.

Finally, stochastic variables are added to the wind components except for tower shadow, giving a closer to reality parameterization of the wind speeds throughout the rotor plane. In this way, the wind field is converted to equivalent winds signals that act on two distinct parts of the blades, namely the tip and root sections.

### 6.2.7  Actuator and Sensor Model Block

The wind turbine simulators described here consider white noise added to all measurements acquired from the wind turbine actuator and sensor signals. The rationale relies on the assumption that noisy sensor signals should represent more realistic scenarios. In general, this assumption does not represent a realistic case, as a high-fidelity simulation requires a deeper insight into the measurement process and the knowledge of the sensor reliability. However, to the best of the authors' knowledge and from their experience with wind turbine systems, all measurements acquired from the wind turbine process are affected by weak noise processes with appropriate statistical descriptions.

Therefore, according to these considerations, the measurement process is described by the Simulink® system represented in Fig. 6.11.

### 6.2.8  Wind Turbine Controller Block

The main point of Chapters 2 and 4 was to introduce the technical challenges that exist in the energy conversion industry and to suggest alternative modeling and control strategies in this area, also for research purposes. In fact, wind turbines are complex structures operating in uncertain environments and lend themselves nicely to advanced control solutions. As shown in Chapter 4, advanced controller schemes help achieve the overall goal of decreasing the cost of wind energy by increasing the efficiency, thus the energy capture, and at the same time, increasing the lifetimes of the components and turbine structures.

In the case of a wind turbine, optimal blade pitch, $\beta$, and rotor velocity (via the tip-speed ratio, $\lambda$) are set based on the incident wind flow velocity, in order to maximize the power coefficient, $C_Q$. The manipulated variable for the pitch control is the power to the pitch actuators (voltage and/or current). For torque control, the generator excitation is used as a control actuator.

It is worth noting that the relationship between $\beta$, $\lambda$, and $C_Q$ is specific to each wind turbine, and must be determined for each particular case. However, this relationship is then fixed, though some slight variation may occur due, for example, to component wear or installation errors. Moreover, when a wind turbine reaches its rated power (i.e., above the rated wind

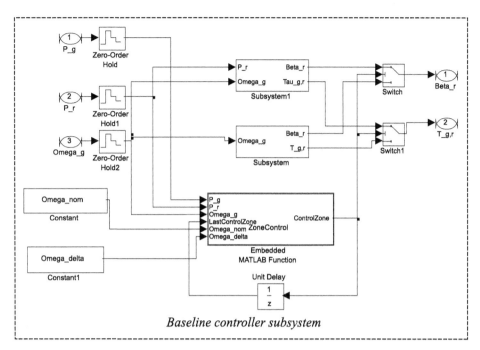

**FIGURE 6.12** The baseline controller implemented for the wind turbine simulator.

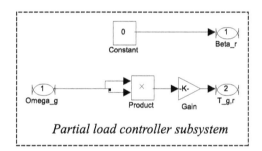

**FIGURE 6.13** The baseline controller for the partial-load conditions.

speed), the turbine needs to be "depowered" in order to avoid exceeding any rated specifications. In this situation, it is not required to maximize power conversion (i.e., the wind power that can be converted into electric energy) and, for variable pitch turbines, blade pitch can be adjusted in order to limit power converted.

After these remarks, control systems for wind turbines seem now well developed as described in Chapters 2 and 4, and the implementation of the baseline controller is sketched in the following, in order to help the reader understand the main features of the wind turbine simulator.

Chapters 2 and 4 highlighted that the primary Region 2 control objective for a variable-speed wind turbine is to maximize the power coefficient, and in particular the $C_Q$ look-up table of (2.1). The relationship between $C_Q$ and the tip-speed ratio $\lambda$ is a turbine-specific nonlinear function. $C_Q$ also depends on the blade pitch angle in a nonlinear way, and these relationships have the same basic shape for most modern wind turbines. Fig. 6.12 represents the baseline controller implemented for the wind turbine simulator.

The controller shown in Fig. 6.12 will operate the wind turbine at its highest aerodynamic efficiency point, at a certain pitch angle and tip-speed ratio. The pitch angle is easy to control, and can be reliably maintained at the optimal efficiency point. However, the tip-speed ratio depends on the incoming wind speed and therefore is constantly changing. Thus, the Region 2 controller is implemented in Fig. 6.13, which is concerned with varying the turbine speed to track the wind speed. When this strategy is used, the controller structure for partial load operation follows the sequential optimal calculation and regulation strategies remarked in Section 1.7, whose implementation is reported in Fig. 6.13.

On the other hand, Region 3 control is performed via a separate pitch control loop. In Region 3, the primary objective is to limit the turbine power so that safe electrical and mechanical loads are not exceeded. Power limitation is achieved by

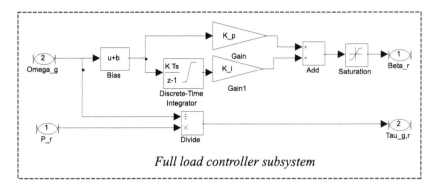

**FIGURE 6.14** The baseline controller for the full-load conditions.

pitching the blades or by yawing the turbine out of the wind, both of which can reduce the aerodynamic torque below what is theoretically available from an increase in wind speed. Therefore, in Region 3, the pitch control loop regulates the rotor speed (at the turbine "rated speed") so that the turbine operates at its rated power, as sketched in Fig. 6.14.

In this way the overall strategy of the wind turbine controller implemented in Fig. 6.12 uses two different controllers for the partial load region and the full load region. When the wind speed is below the rated value, the control system should maintain the pitch angle at its optimal value and control the generator torque in order to achieve the optimal tip-speed ratio (switch to Region 2).

At low wind speeds, i.e., in partial load operation, variable-speed control is implemented to track the optimum point on the $C_Q$-surface for maximizing the power output, which corresponds to the optimal value of the $\lambda$ parameter. The speed of the generator is controlled by regulating the demanded torque $T_{g,d}$ on the generator through the generator torque controller. In partial load operation it is chosen to operate the wind turbine at $\beta = 0$ degrees, since the maximum power coefficient is obtained at the pitch angle given by the expression of Eq. (2.24) and included in Fig. 6.13.

On the other hand, for higher wind speeds, i.e., in full load conditions, the desired operation of the wind turbine is to keep the rotor speed and the generated power at constant values. The main idea is to use the pitch system to control the efficiency of the aerodynamics, while applying the rated generator torque. However, in order to improve tracking of the power reference and cancel steady-state errors on the output power, a power controller is also introduced. Therefore, the speed controller is implemented as a PI controller that is able to track the speed reference and cancel possible steady-state errors on the generator speed, as shown in Fig. 6.14. Its transfer function is described by Eq. (2.25).

Note finally that new wind turbine control solutions can exploit further wind turbine state information from the sensing system, which has been suggested also within EU projects, as considered in Chapter 7. This improved state information is used to control the wind turbine blades and at the same time reduce the design bearing fatigue and extreme structural loads that are affecting the structure of the wind turbine. This control problem can be solved in a multivariate way, by optimizing the conflicting control objectives of power optimization while keeping the different loads below the design requirements. The control goal is to ensure that the controller is able to guarantee that extreme load requirements are not violated during eventually emergency stops of the wind turbine, as well as during severe wind gusts. An interesting challenge is to be able to use the rotor system to control the turbine, so that in effect the rotor performs like a "high level" sensor. In other words, the goal is to be able to use the rotor itself (along with the enhanced sensor set) to make the control system perform well. A part of this challenge is to ensure that real-time compensation of loading and gust disturbances is put into effect in a suitable time window, taking account of the close spectral content of the disturbance and control. This becomes a very significant challenge for very large rotor wind turbines (> 10 MW) as the required control and disturbance bandwidths become close, a problem similar to the structural filtering and control used in high performance combat aircraft (Shi and Patton, 2015).

It is worth noting that the control logic represented in Fig. 6.12 includes also the bumpless transfer mechanism described in Section 2.4.4 and illustrated in Fig. 2.10. The considered transition is that which brings the control system from partial load operation to full load operation, and vice versa. This task is performed by the Embedded MATLAB Function subsystem sketched in Fig. 6.12.

When the control system switches from partial load to full load operation, it is important that this transition does not affect the control signals, i.e., the generator torque and pitch angle. This procedure is known as a bumpless transfer, and is important because two controllers may not be consistent with the magnitude of the control signal at the time when the transition happens. If a switch between two controllers is performed without a bumpless transfer, a bump in the control signal may trigger oscillations between the two controllers, making the system unstable. The transition from partial to full

```
function ControlZone = ZoneControl(P_g, P_r, Omega_g,...
                        LastControlZone,Omega_nom,Omega_delta)
  if LastControlZone==0,
      if (P_g>=P_r) || (Omega_g>=Omega_nom)
          ControlZone=1;
      else
          ControlZone=0;
      end
  else
      if (Omega_g<(Omega_nom-Omega_delta))
          ControlZone=0;
      else
          ControlZone=1;
      end
  end
```

*Embedded MATLAB Function*

**FIGURE 6.15** Embedded MATLAB Function implementing the controller switching logic.

load operation must happen as the wind speed becomes sufficiently large. For stationary wind speeds this usually happens at about 12 m/s. However, it is not convenient to apply the wind speed as the switching condition, since the large inertia of the rotor causes the generator speed and output power to follow significantly later than a rise in the wind speed. Moreover, the wind speed is almost unknown. Therefore, it is more appropriate to exploit the generator speed as a switching condition. In particular, the switching from partial to full load condition is achieved when the generator speed $\omega_g(t)$ is greater than the nominal generator speed. On the other hand, the switching from full to partial load condition is applied if both the pitch angle $\beta(t)$ is lower than its optimal value and the generator speed $\omega_g(t)$ is significantly lower than its nominal value. Notice that a hysteresis is usually introduced to ensure a minimum time between each transition. This transition logic and the transition task is performed by the Matlab® function outlined in Fig. 6.15.

Note finally that the torque compensation is not important when operating above the rated wind speed, because the power controller has integral action. When operating below rated wind speed, the compensation torque is discharged to zero, as it would otherwise result in the optimal tip-speed ratio not being followed.

### 6.2.9 Wind Turbine Fault Blocks

The benchmark model considers three kinds of fault situation that can be simulated as described in Section 2.5.4, representing sensor, actuator, and system faults. They are modeled as additive or multiplicative faults with different degrees of severity, so that they can yield the turbine shutdown in case of a serious fault, or they can be accommodated by the controller, if the risk for the system safety is low.

As shown in Fig. 6.16, they affect the measurements of the pitch angles and the measurements of the rotor speed, in the form of a fixed value or a scaling error. They represent a common fault scenario of wind turbines, but their severity is low and they should be easy to identify and accommodate. In particular, electrical or mechanical faults in the pitch sensors, if not handled, result in the generation of a wrong pitch reference system by the controller with the consequence of a loss in the generated power. The speed of the rotor is measured by means of two redundant encoders, an offset faulty signal can affect these measurements when the encoder does not detect the updated marker, while a gain factor faulty signal represents the reading of excessive markers in each loop, due to dirt on the rotating part.

On the other hand, the considered actuator faults are modeled either as a fixed value or a changed dynamics of the transfer function. They affect the converter torque actuator, as well as the pitch actuator. In the former case, the fault is located in the electronics of the converter/generator module, as shown in Fig. 6.17.

In the latter case the fault is on the hydraulic system: it models the pressure drop in the hydraulic supply system (e.g., due to leakage in a hose or a blocked pump) or the excessive air content in the oil that causes variation of the compressibility factor. The actuator fault block regarding the pitch system is represented in Fig. 6.18.

Finally, as sketched in Fig. 6.19, the considered system fault concerns the drive-train, in the form of a slow variation of the friction coefficient in time due to wear and tear (months or year, but for benchmarking reasons in the model it has been accelerated up to some seconds). It results in a combined faulty signal affecting the rotor speed and the generator speed. It can be listed as a high severity fault, as it can yield the breakdown of the drive-train, but in a long time.

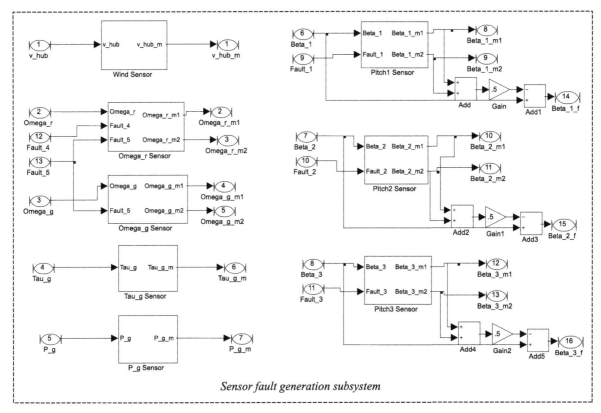

**FIGURE 6.16** The sensor fault simulation block included in the wind turbine benchmark.

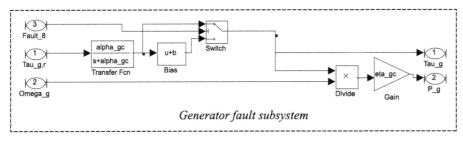

**FIGURE 6.17** The generator fault simulation block of the wind turbine benchmark.

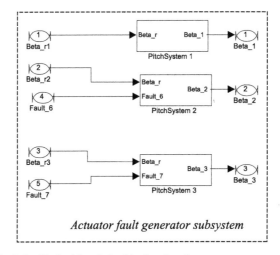

**FIGURE 6.18** The pitch actuator fault simulation block of the wind turbine benchmark.

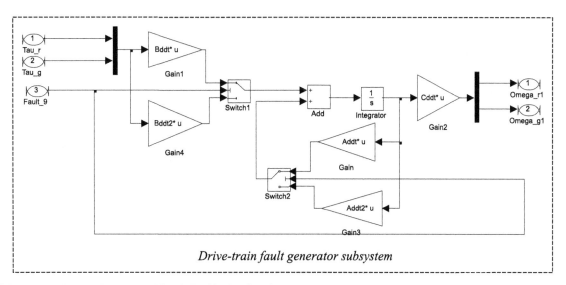

**FIGURE 6.19** The drive-train fault block of the wind turbine benchmark.

These fault conditions were summarized in Table 2.2 of Chapter 2, where a brief description of their typology and topology was also reported.

## 6.2.10 Wind Turbine Model Parameter Initialization

The parameters adopted in the benchmark model described in Section 2.5.1 were summarized in Table 2.3. Note that these nominal parameter values were modified in order to verify the robustness and reliability features of the developed FDI/FDD and FTC strategies addressed in Chapters 3 and 4.

The values of these parameters have been defined by means of the Matlab® script file reported in the following.

```
%%%%%%%%%%%%%%%%%%%%%%%%%%%%
% Script containing the parameters for the Benchmark Model
%
% Date 7.11.2008
%%%%%%%%%%%%%%%%%%%%%%%%%%%%

Ts=1/100;
Time=Ts:Ts:4400;


%%%%%%%%%%%%%%%%%%%%%%%%%%%%
% Wind data
%%%%%%%%%%%%%%%%%%%%%%%%%%%%

D_v_wind=[Time', 9+4*sin(0.01*Time)'];

load winddata
D_v_wind=[windtime windspeed];
%% Wind Model
seed1 = 256;
seed2 = 894;
turbulence_seeds=[seed1 seed2];
R=57.5;
H=87;
k =4.7;
a=2.2;
alpha = 0.1;
D_rotor=2*R;
r = D_rotor/2;
m = 1+alpha*(alpha-1)*R^2/(8*H^2);
Length_scale = 600; %[m]
```

```
L=Length_scale;
Turbulence_intensity = 12;
turb_int=Turbulence_intensity;
T_sample=0.05;

%%%%%%%%%%%%%%%%%%%%%%%%%
%% Pitch and Blade Model
%%%%%%%%%%%%%%%%%%%%%%%%%

omega_n=11.11;
xi=0.6;
rho=1.225;
R=57.5;
r0 = 1.5;

% Cq table
load AeroDynamics

%%Fault models%%

%fault 6
xi2=0.45;
omega_n2=5.73;

%fault 7
xi3=0.9;
omega_n3=3.42;

%transfers to ss models
[Apb,Bpb,Cpb,Dpb]=tf2ss([omega_n^2],...
[1 2*xi*omega_n omega_n^2]);
[Apb1,Bpb1,Cpb1,Dpb1]=tf2ss([omega_n2^2],...
[1 2*xi2*omega_n2 omega_n2^2]);
[Apb2,Bpb2,Cpb2,Dpb2]=tf2ss([omega_n3^2],...
[1 2*xi3*omega_n3 omega_n3^2]);

%%%%%%%%%%%%%%%%%%%%%%%%%
%% Drive Train Model
%%%%%%%%%%%%%%%%%%%%%%%%%

B_dt=775.49;
B_r=7.11;
B_g=45.6;
N_g=95;
K_dt=2.7e9
eta_dt=0.97;
J_g=390
J_r=55e6

Addt=[-(B_dt+B_r)/J_r B_dt/N_g/J_r -K_dt/J_r; ...
eta_dt*B_dt/N_g/J_g - (eta_dt*B_dt/N_g^2 + ...
B_g)/J_g eta_dt*K_dt/N_g/J_g; 1 -1/N_g 0];
Bddt=[1/J_r 0; 0 -1/J_g;0 0];
Cddt=[1 0 0;0 1 0];
Dddt=[0 0;0 0];

%fault models
%fault 9
eta_dt2=.92

Addt2=[-(B_dt+B_r)/J_r B_dt/N_g/J_r -K_dt/J_r;...
eta_dt2*B_dt/N_g/J_g - (eta_dt2*B_dt/N_g^2 + ...
B_g)/J_g eta_dt2*K_dt/N_g/J_g; 1 -1/N_g 0];
Bddt2=[1/J_r 0; 0 -1/J_g;0 0];
```

```
Cddt2=[1 0 0;0 1 0];
Dddt2=[0 0;0 0];

%%%%%%%%%%%%%%%%%%%%%%%%%
%% Generator & Converter
%%%%%%%%%%%%%%%%%%%%%%%%%

alpha_gc=1/20e-3;
eta_gc=0.98;

%Fault models
%fault 8
Constant_tau_gc=100;

%%%%%%%%%%%%%%%%%%%%%%%%%
%% Controller
%%%%%%%%%%%%%%%%%%%%%%%%%

K_opt=.5* rho* pi*R^2*eta_dt*R^3*0.4554/(N_g*8)^3

K_i=1;
K_p=4;
Omega_nom=162;
Omega_delta=15;
P_r=4.8e6;

%%%%%%%%%%%%%%%%%%%%%%%%%
%% Sensors
%%%%%%%%%%%%%%%%%%%%%%%%%

m_vw=1.5;
sigma_vm=0.5;
Np_vm=0.0071;
m_omega_r=0;
sigma_omega_r=0.004*2*pi;
Np_omega_r=2.3e-4;
m_omega_g=0;
sigma_omega_g=0.05;
Np_omega_g=5e-4;
m_tau_g=0;
sigma_tau_g=90;
Np_tau_g=.9;
m_P_g=0;
sigma_P_g=1e3;
Np_P_g=10;
m_Beta=0;
sigma_Beta=0.2;
Np_beta=1.5e-3;

% Fault models
Constant_Beta_1_m1=5
Gain_Beta_2_m2=1.2
Constant_Beta_3_m1=10
Constant_Omega_r_m1=1.4
Gain_Omega_r_m2=1.1
Gain_Omega_g_m1=0.9

%%%%%%%%%%%%%%%%%%%%%%%%%
%% Fault signals
%%%%%%%%%%%%%%%%%%%%%%%%%

D_Fault1=[Time' 1-[zeros(1,2000/Ts) ones(1,100/Ts)...
 zeros(1,2300/Ts)]'];
D_Fault2=[Time' 1-[zeros(1,2300/Ts) ones(1,100/Ts)...
```

```
 zeros(1,2000/Ts)]'];
D_Fault3=[Time' 1-[zeros(1,2600/Ts) ones(1,100/Ts)...
 zeros(1,1700/Ts)]'];
D_Fault4=[Time' 1-[zeros(1,1500/Ts) ones(1,100/Ts)...
 zeros(1,2800/Ts)]'];
D_Fault5=[Time' 1-[zeros(1,1000/Ts) ones(1,100/Ts)...
 zeros(1,3300/Ts)]'];
D_Fault6=[Time' 1-[zeros(1,2900/Ts) ones(1,100/Ts)...
 zeros(1,1400/Ts)]'];
D_Fault7=[Time' 1-[zeros(1,3400/Ts) (0:1/3000:2999/3000)...
 ones(1,40/Ts)...
 (2999/3000:-1/3000:0) zeros(1,900/Ts)]'];
D_Fault8=[Time' 1-[zeros(1,3800/Ts) ones(1,100/Ts)...
 zeros(1,500/Ts)]'];
D_Fault9=[Time' 1-[zeros(1,4000/Ts) ones(1,200/Ts)...
 zeros(1,200/Ts)]'];

%%%%%%%%%%%%%%%%%%%%%%%%%%
%% Noise Seeds
%%%%%%%%%%%%%%%%%%%%%%%%%%

min_seed = 0;
max_seed = 999;

% Generate random seeds for 13 'Random Number'
% Generators blocks

seed = min_seed + (max_seed-min_seed).*rand(13, 1);

% Round up to integer
seed = ceil(seed);
```

## 6.3 WIND FARM SYSTEM BENCHMARK

The wind farm simulator considered in this monograph and described in Chapter 2 was developed by the same authors of the wind turbine benchmark model. It consists of nine wind turbines arranged in a square grid of three rows and three columns. Two anemometers are placed in front of the first line of turbines for providing the measurements of the undisturbed wind speed. The considered wind turbines are 4.8 MW three-blade HAWTs, almost of the same type of the model previously described. Each of them is regulated by means of a local controller, but the wind farm benchmark includes a global control unit. Moreover, three fault scenarios are implemented.

The complete wind farm simulator consists of three subsystems already described in Chapter 2: the wind and wake model, the plant model, and the controller model, which are connected as depicted Fig. 6.20. The layout of the wind farm with nine turbines on the square grid and the masts along the wind direction was summarized in Fig. 2.14 of Chapter 2.

The wind farm benchmark considers two wind directions, namely, of 0 and 45 degrees, which can be selected by means of two switches sketched in Fig. 6.20. The wind turbines of the farm are defined by their row and column indices in the coordinate system, and they are illustrated in Fig. 6.21.

### 6.3.1 Wind and Wake Block

The wind and wake model generates the wind speed signals affecting each of the nine turbines and measured by the anemometers. These wind signals are computed starting from the considered wind sequence and their elaboration takes into account the delay and the interaction among the turbines depending on the wind direction. This model is included in the wind turbine subsystem for each of the nine wind turbines, as depicted in Fig. 6.22.

Note that the subsystem of Fig. 6.22 representing the $i$th wind turbine of the wind farm receives the wind sequence and the reference signal from the wind farm controller as inputs. The outputs are the generated power, pitch angle, and generator speed. As depicted in Fig. 6.22, the actual wind speed is elaborated by means of a look-up table in order to compute the available power. The generated power is computed according to the relation of Eq. (2.50). This expression takes into account also possible oscillations caused by the drive-train of the wind turbine.

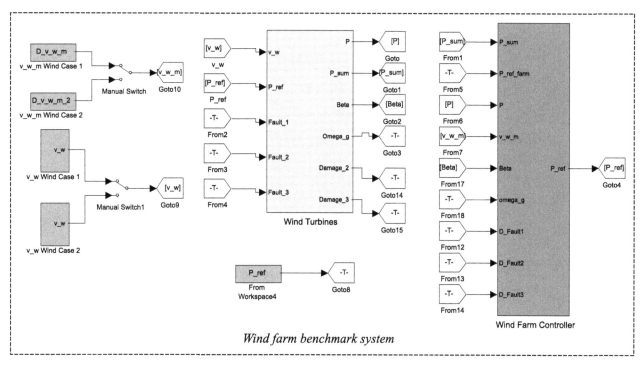

**FIGURE 6.20** Block layout of the wind farm benchmark.

The pitch system is similar to that of the turbine benchmark model of Eq. (2.36), but the transfer function between the reference pitch signal and the actual pitch angle has been reduced to the simpler first-order transfer function of Eq. (2.54). Moreover, the generator model of each turbine represented in Fig. 6.22 is described according to Eq. (2.55).

Finally, the wind farm controller forces each turbine to track the reference power signal, which is 1/9 of the wind farm power reference. Moreover, the wind farm controller is modeled as a discrete-time system in the form of Eq. (2.56) that uses a sample frequency of 10 Hz, as depicted by the Simulink® subsystem of Fig. 6.23.

### 6.3.2 Wind Farm Fault Block

As described in Chapter 2, three fault scenarios are considered for the wind farm benchmark. In general, as shown in Fig. 6.24, they affect the output variables of the plant system. Table 2.4 in Chapter 2 summarized the considered fault cases that are described below according to their topology and the wind farm subsystems.

In particular, the first fault case considered here represents the debris build-up on the blade surface. Its effect is a change of the aerodynamics of the affected turbine and the consequent decrease of the generated power that is modeled by a scaling factor of 0.97 applied to the generated power signal. The implementation of this fault is outlined in Fig. 6.25.

The second fault case is due to a misalignment of one blade caused by an imperfect installation. The effect is an offset between the actual and the measured pitch angle of the affected turbine. This fault can excite structural modes and creates undesired vibrations that can damage the system severely. The faulty signal involves an offset of 0.3 degrees on the pitch angle as outlined in Fig. 6.26.

Finally, the third fault represents the wear and tear in the drive-train. This fault affects the generated power that increases the amplitude of its oscillation by 26% with respect to the nominal value. Moreover, the generator speed increases the amplitude of its oscillation by 130%. The implementation of the fault case 3 is outlined in Fig. 6.27

### 6.3.3 Wind Farm Model Parameter Initialization

Finally, also in this case, Table 2.5 in Chapter 2 describes the parameters used in the wind farm benchmark model. Their values are defined by means in the Matlab® script summarized below.

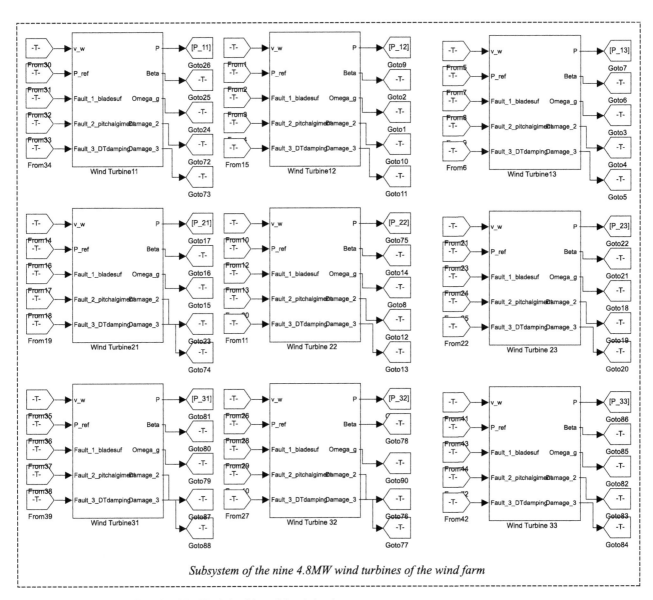

*Subsystem of the nine 4.8MW wind turbines of the wind farm*

**FIGURE 6.21** Subsystem of the nine 4.8 MW wind turbines of the wind park.

```
% Model parameters for simple wind turbine model for
% the wind park
%
% date 17.02.09

Ts=1/100; % Wind turbine sample time
Tsp=1/10; %Wind farm sample time
v_nom=12.5;

load windfarms % Wind sequence case 1 data

D_v_w_m=[time v_w_m];
D_v_w_11=[time v_w(:,1)];
D_v_w_21=[time v_w(:,2)];
D_v_w_31=[time v_w(:,3)];
D_v_w_12=[time v_w(:,4)];
D_v_w_22=[time v_w(:,5)];
D_v_w_32=[time v_w(:,6)];
```

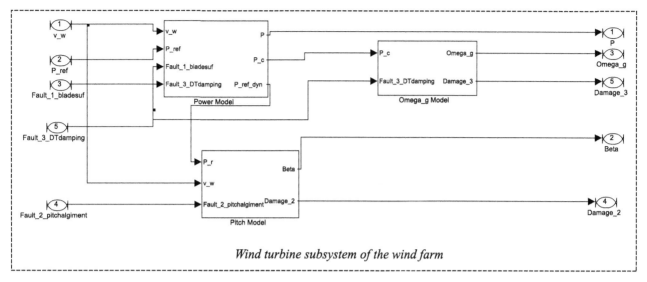

**FIGURE 6.22** Subsystem describing the $i$th wind turbine of the wind farm.

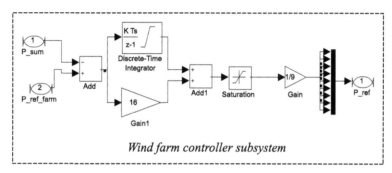

**FIGURE 6.23** Subsystem implementing the wind farm controller.

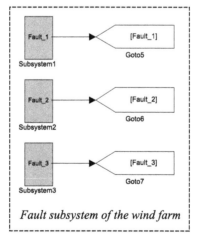

**FIGURE 6.24** Subsystem implementing the faults of the wind farm simulator.

```
D_v_w_13=[time v_w(:,7)];
D_v_w_23=[time v_w(:,8)];
D_v_w_33=[time v_w(:,9)];

load windfarms_case2 % Wind sequence case 2 data
```

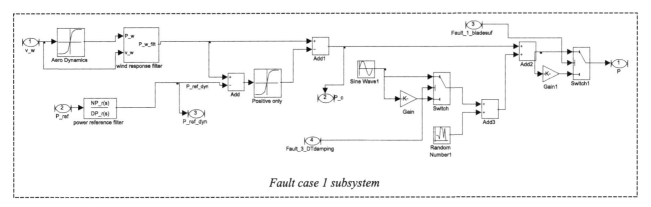

*Fault case 1 subsystem*

**FIGURE 6.25** Subsystem implementing the fault case 1 of the wind farm simulator.

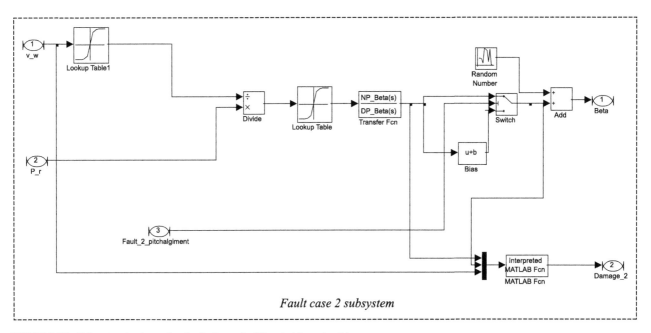

*Fault case 2 subsystem*

**FIGURE 6.26** Subsystem implementing the fault case 2 of the wind farm simulator.

```
D_v_w_m_2=[time v_w_m];
D_v_w_11_2=[time v_w(:,1)];
D_v_w_21_2=[time v_w(:,2)];
D_v_w_31_2=[time v_w(:,3)];
D_v_w_12_2=[time v_w(:,4)];
D_v_w_22_2=[time v_w(:,5)];
D_v_w_32_2=[time v_w(:,6)];
D_v_w_13_2=[time v_w(:,7)];
D_v_w_23_2=[time v_w(:,8)];
D_v_w_33_2=[time v_w(:,9)];

% Wind dependency filter with time-variant coefficient

p_w_in=v_nom-[8.5 8 7.5 7 6.5 6 5.5 5 4.5 4 3.5 3...
 2.5 2 1.5 1 0.5 0];
p_w_out=[0.0274 0.0345 0.0405 0.0460 0.0512...
 0.0575 0.064 0.0697 0.077...
0.0825 0.092 0.1 0.105 0.105 0.11 0.115 0.115...
 0.115];
```

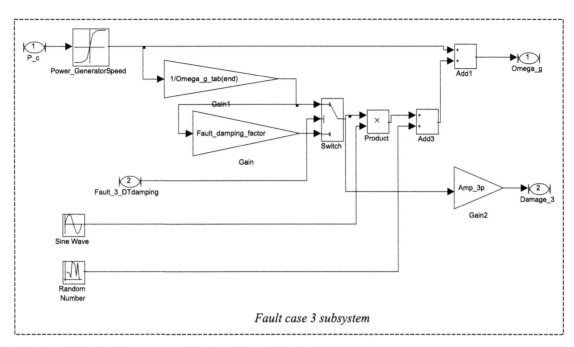

**FIGURE 6.27**  Subsystem implementing the fault case 3 of the wind farm simulator.

```
% Power reference dependency filter

NP_r=[1.2]
DP_r=[1 1.2]

% Max power

Pmax=4.8e6;

%Power noise models

Amp_3p_pow=1e3;
P_noise=2.5e5;

% Aerodynamic mapping

P_tab=Pmax*[0.0044 0.0096 0.0181 0.0306 0.0481...
 0.0713 0.1006 0.1375 0.1833...
 0.2359 0.3000 0.3702 0.4562 0.5521 0.6604 0.7854...
 0.9167 1];
v_w_tab=v_nom-[8.5 8 7.5 7 6.5 6 5.5 5 4.5 4 3.5 3...
 2.5 2 1.5 1 0.5 0];

% Pitch mapping
RC_p_tab=[7.54e-4 0.0429 0.0483 0.0815 0.0943...
 0.11 0.129 0.153 0.1839 0.223...
 0.274 0.3453 0.44 0.4961 0.5669 0.5989 0.6335...
 0.6518 0.6707 0.6905 0.7945 1];
BBeta_tab=[90 90 22 18 17 16.3 15.3 14.1 12.9...
 11.7 10.5 9.3 7.5 6.7 6 5.6 5.2...
 4 4.7 4.5 0 0 ];

% Pitch filter

NP_Beta=[1.6];
DP_Beta=[1 1.6];
```

```
% Pitch noise model
Beta_noise=.3;

% wind power mapping
rho=1.225;
A=pi*57.5^2;
vv_w_tab=0:.5:25;
P_w_tab=.5*rho*A*vv_w_tab.^3;

% omega_g

Omega_Power_tab=[1.37e5 2.17e5 3.24e5 4.64e5...
 6.32e5 8.45e5 1.09e6 1.39e6 1.74e6...
 2.15e6 2.61e6 3.13e6 3.71e6 4.37e6 4.51e6 4.65e6...
 4.8e6]
Omega_g_tab=[48.3 56.2 64.3 72.4 80.5 88.5 96.6...
 104.7 112.8 120.8 128.9 137...
 145.1 153.1 154.7 156.4 158]
Amp_3p=0.4
Freq_3p=4.5 % rad/s
Omega_noise=0.1

% WT Farm control

WF_cont_num=[0.09516]
WF_cont_den=[1 -0.9048]

%%% Faults

% Fault 1
Fault_Blade_debris_factor=.97;

% Fault 2
Fault_PitchMisalignment_offset=.3;

% Fault 3
Fault_damping_factor=2.3

Time=Ts:Ts:4400;
D_Fault1_11=[Time' 1-[zeros(1,2000/Ts)...
 0*ones(1,100/Ts) zeros(1,2300/Ts)]'];
D_Fault1_12=[Time' 1-[zeros(1,3000/Ts)...
 ones(1,100/Ts) zeros(1,1300/Ts)]'];
D_Fault1_13=[Time' 1-[zeros(1,2000/Ts)...
 0*ones(1,100/Ts) zeros(1,2300/Ts)]'];
D_Fault1_21=[Time' 1-[zeros(1,2000/Ts)...
 0*ones(1,100/Ts) zeros(1,2300/Ts)]'];
D_Fault1_22=[Time' 1-[zeros(1,2000/Ts)...
 0*ones(1,100/Ts) zeros(1,2300/Ts)]'];
D_Fault1_23=[Time' 1-[zeros(1,2000/Ts)...
 0*ones(1,100/Ts) zeros(1,2300/Ts)]'];
D_Fault1_31=[Time' 1-[zeros(1,1000/Ts)...
 ones(1,100/Ts) zeros(1,3300/Ts)]'];
D_Fault1_32=[Time' 1-[zeros(1,2000/Ts)...
 0*ones(1,100/Ts) zeros(1,2300/Ts)]'];
D_Fault1_33=[Time' 1-[zeros(1,2000/Ts)...
 0*ones(1,100/Ts) zeros(1,2300/Ts)]'];

D_Fault2_11=[Time' 1-[zeros(1,1300/Ts)...
 ones(1,100/Ts) zeros(1,3000/Ts)]'];
D_Fault2_12=[Time' 1-[zeros(1,2300/Ts)...
 0*ones(1,100/Ts) zeros(1,2000/Ts)]'];
D_Fault2_13=[Time' 1-[zeros(1,2300/Ts)...
 0*ones(1,100/Ts) zeros(1,2000/Ts)]'];
```

```
D_Fault2_21=[Time' 1-[zeros(1,2300/Ts)...
 0*ones(1,100/Ts) zeros(1,2000/Ts)]'];
D_Fault2_22=[Time' 1-[zeros(1,3300/Ts)...
 ones(1,100/Ts) zeros(1,1000/Ts)]'];
D_Fault2_23=[Time' 1-[zeros(1,2300/Ts)...
 0*ones(1,100/Ts) zeros(1,2000/Ts)]'];
D_Fault2_31=[Time' 1-[zeros(1,2300/Ts)...
 0*ones(1,100/Ts) zeros(1,2000/Ts)]'];
D_Fault2_32=[Time' 1-[zeros(1,2300/Ts)...
 0*ones(1,100/Ts) zeros(1,2000/Ts)]'];
D_Fault2_33=[Time' 1-[zeros(1,2300/Ts)...
 0*ones(1,100/Ts) zeros(1,2000/Ts)]'];

D_Fault3_11=[Time' 1-[zeros(1,2600/Ts)...
 0*ones(1,100/Ts) zeros(1,1700/Ts)]'];
D_Fault3_12=[Time' 1-[zeros(1,2600/Ts)...
 0*ones(1,100/Ts) zeros(1,1700/Ts)]'];
D_Fault3_13=[Time' 1-[zeros(1,2600/Ts)...
 0*ones(1,100/Ts) zeros(1,1700/Ts)]'];
D_Fault3_21=[Time' 1-[zeros(1,2600/Ts)...
 0*ones(1,100/Ts) zeros(1,1700/Ts)]'];
D_Fault3_22=[Time' 1-[zeros(1,2600/Ts)...
 0*ones(1,100/Ts) zeros(1,1700/Ts)]'];
D_Fault3_23=[Time' 1-[zeros(1,1600/Ts)...
 ones(1,100/Ts) zeros(1,2700/Ts)]'];
D_Fault3_31=[Time' 1-[zeros(1,2600/Ts)...
 0*ones(1,100/Ts) zeros(1,1700/Ts)]'];
D_Fault3_32=[Time' 1-[zeros(1,3600/Ts)...
 ones(1,100/Ts) zeros(1,700/Ts)]'];
D_Fault3_33=[Time' 1-[zeros(1,2600/Ts)...
 0*ones(1,100/Ts) zeros(1,1700/Ts)]'];

P_ref=[Time' [43.2e6*ones(1,2000/Ts)...
 30.3e6*ones(1,2400/Ts)]'];

% Noise Seeds

min_seed = 0;
max_seed = 999;

% Generate random seeds for 27 'Random Number'
% Generators blocks

seed = min_seed + (max_seed-min_seed).*...
rand(27, 1);

% Round up to integer
seed = ceil(seed);
```

### 6.3.4 Fault Diagnosis Module Implementation

The problem of the residual generator design for the FDI of the wind turbine systems is implemented as described in this section.

The wind turbine system is assumed to be modeled by the description provided in Section 2. The main signals used in the simulation blocks described by the variables $u(k)$ and $y(k)$ represent the controlled inputs and the system outputs, respectively. As remarked in Chapter 3, the model–reality mismatch in fault-free conditions can be represented by the difference $y(k) - \hat{y}(k)$. In fact, it takes into account measurement errors, parameter variations, and disturbances. The reconstruction of the measurement $y(k)$, i.e., $\hat{y}(k)$, is obtained from an identified model, as described in Chapter 3. According to the description provided in Chapters 2 and 3, in practice, the signals $u^*(k)$ and $y^*(k)$ are acquired by measurement sensors, which are inevitably affected by errors.

On the other hand, if the sensor dynamics are neglected, also faults affect the measurement process, which are modeled as highlighted in Fig. 6.28, where the faults $f_u(k)$ and $f_y(k)$ are additive signals affecting the measurements $u(k)$ and $y(k)$.

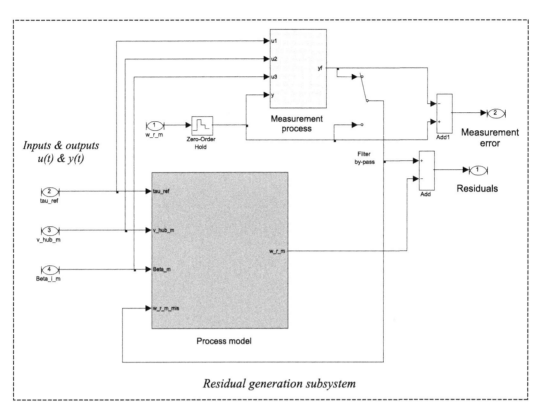

**FIGURE 6.28** The residual generation block for FDI.

Regarding the FDI task, this monograph recalled both the data-driven and model-based approaches described in Chapter 3 that are exploited for residual generators from the redundant input and output signals $u(k)$ and $y(k)$. In this way, Fig. 6.28 shows that proper residual signals are computed as the different between the actual $y(k)$ and its reconstruction $\hat{y}(k)$, as represented in Fig. 6.29.

After the residual generation task, its evaluation is performed for detecting any fault occurrence, and for isolating the faulty actuator or sensor signals.

A direct geometric threshold comparison is recalled in Chapter 3 to perform the fault detection stage. However, a detection delay can be present due to the fault modes. The fault detection logic is performed according to the test described in Chapter 3 and implemented by the Simulink® block outlined in Fig. 6.30.

Actually, the residual $r(k)$ is modeled as a stochastic variable, whose mean and variance values are estimated by means of the constant blocks shown in Fig. 6.30. These functions evaluate the mean and variance values of the fault-free residual signals. Note that these parameters could be exactly computed from the $r(k)$ statistics, usually unknown.

A robustness and reliability degree is introduced for distinguishing the normal and the faulty behaviors, which is represented by the tolerance parameter $\delta$ (normally $\delta \geq 2$) remarked in Chapter 3. As shown by the results achieved in Chapter 5, the technique relying on the Monte Carlo tool is used here in order not to obtain conservative results. In particular, extensive simulations lead to the optimal value of $\delta$ that minimizes the false alarm probability and maximizes the true detection rate. This topic was addressed in Chapter 5.

Another important issue concerns the fault isolation task, and it is achieved using a bank of residual generators properly designed, which is based on the Generalized Observer Scheme (GOS). This task can be easily solved as recalled in Chapter 3, since different faults can affect different input or output measurements. In this way, when the outputs are fault-free, the input fault $f_u(k)$, possibly affecting one of the inputs $u(k)$, is diagnosed with the bank of estimators depicted in Fig. 6.31.

In general, the number of residual generators coincides with the number of faults to be diagnosed. Fig. 6.31 shows that the $i$th residual generator is fed by all but the $i$th input measurement (or even more input signals, if necessary) and all output measurements. The generated residual signal is thus sensitive to all but the $i$th fault $f_u(k)$. These residual generators were described in Chapter 3. In particular, the $i$th estimator that does not depend on the $i$th input measurement is obtained using $y(k)$ and all but the $i$th input measurement $u_i(k)$ ($i = 1, \ldots, r$).

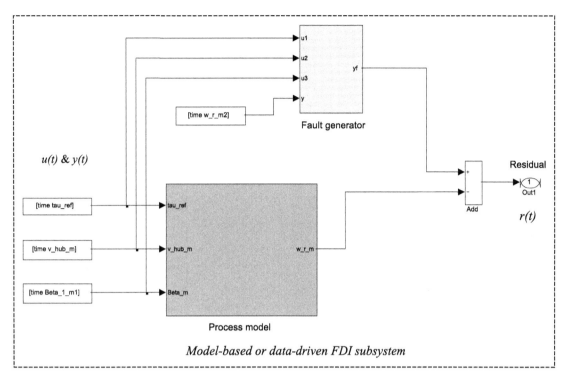

**FIGURE 6.29** The residual generation module for FDI.

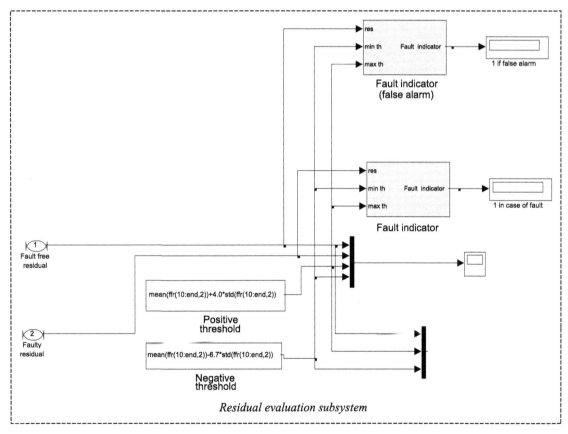

**FIGURE 6.30** The residual evaluation block for FDI.

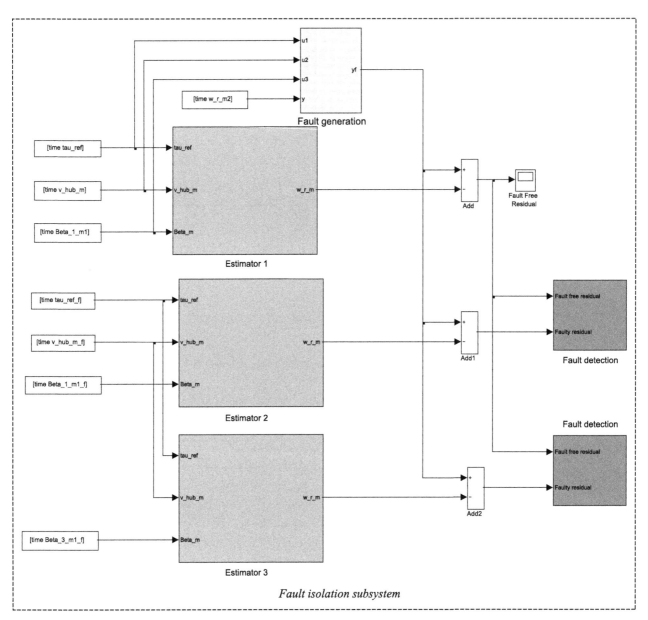

**FIGURE 6.31** Simulink scheme for the isolation of the system faults.

In the same way, when the input variables are fault-free, a fault $f_y(t)$ affecting the output measurement is diagnosed with another output estimator bank. The fault isolation scheme is based on a residual evaluation logic whose implementation is depicted in Fig. 6.32.

With reference to the scheme of Fig. 6.31, when a counter block has the value of "1", it means that the residual is affected by the fault, whilst "0" indicates that the corresponding residual does not depend on the particular fault.

Finally, as remarked in Chapter 3, according to the scheme of Fig. 6.31, note that multiple faults are isolable, as only the $i$th output signal feeding the residual generator is affected by the fault on $y_i$. On the other hand, multiple faults on the inputs $u_i$ are not isolable as the residuals depend on the faults affecting different inputs.

### 6.3.5 Fault Tolerant Control Module Implementation

This section consists of three parts. The first part describes the implementation of estimation procedures used for achieving the robustness of the FTC solutions recalled in Chapter 4; they are required for the design of the active and passive FTC

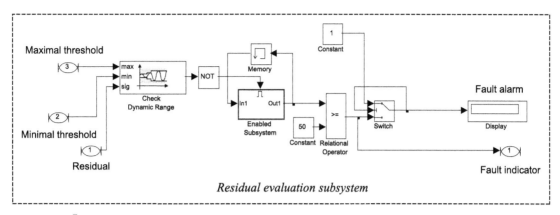

**FIGURE 6.32**  Simulink® scheme for the evaluation of the residuals generated by the estimator banks.

schemes that are considered in this monograph; finally, the complete structure of the FTC strategy is summarized at the end of this section.

Regarding the disturbance estimation issue, as described in Chapters 3 and 4, in order to achieve a robust FTC solution, the disturbance acting on the system has to be decoupled. As remarked in Chapter 2, the main sources of uncertainty and disturbance are represented by the wind signal. It is often assumed that the wind speed is uniform across the rotor plane. However, when instantaneous wind fields are analyzed near the rotor plane, the wind input may vary in space and time over the rotor plane itself. The deviations of the wind speed from the nominal wind speed across the rotor plane can be considered disturbances for control design. Therefore, since the wind speed varies across the rotor plane, wind speed point measurements convey only a small part of the information about the wind inflow.

Chapter 2 highlighted that the disturbance signal $v(t)$ affects the model through the power coefficient $C_p$. Initially, the $C_p$-surface values are most often not actually measured but computed, and if measured, only a few blades in an entire series are considered, but without performing any measurement on the actual turbine. Secondly, these values can be assumed to change slowly with time, though only a few percent per year. These changes are due to wear and tear of the blades, as well as debris build-up on them.

Solutions dealing with this problem were based on the estimation of the $C_p$ values and the wind speed $v(t)$, as recalled in Chapters 3 and 4. It was assumed that their effect could be separated, due, for example, to the slow change in the $C_p$ values, while the wind speed variations are relatively faster. Fault diagnosis schemes based on unknown input observers were also suggested in Chapter 3.

On the other hand, different disturbance decoupling strategies were based on nonlinear schemes, as proposed in Chapter 4. The approach follows the same concept of the estimation of the disturbance distribution matrix proposed for the linear case in Chen and Patton (2012). In particular, as described in Chapter 3, this approach required analytical knowledge of the nonlinear disturbance distribution relation of the unknown input, i.e., the wind measurement $v(t)$.

In more detail, the $C_p(\beta, \lambda)$-map appearing in Eq. (2.1) is estimated by means of a two-dimensional polynomial representation, which is a function of the tip-speed ratio $\lambda$ and the blade pitch angle $\beta$. This polynomial structure (coefficients and degrees) was determined in order to minimize the errors between the actual and simulated measurements, which were acquired from the wind turbine simulator described in Chapter 2.

The Parameter Estimation™ toolbox of the Simulink® environment was used for performing the suggested estimation. In particular, the $C_p$-map entries were estimated in order to minimize the model–reality mismatch, as represented in Fig. 6.33, i.e., in order to minimize the difference between the monitored outputs $y(t)$ from the wind turbine simulator and the outputs generated by the wind turbine model containing the $C_p$-map described by Eq. (2.1).

Therefore, the $C_p$-map entering into Eq. (2.1) was approximated by using a two dimensional polynomial included in the subsystem of Fig. 6.33, where the unknown coefficients of the polynomial function, as well as its degrees, were estimated as described in Chapter 4. However, in order to obtain a model that can be used for the design of the residual generator or fault estimator filters, as described in Chapter 3, some coefficients were forced to be equal to zero.

On the other hand, concerning the robust fault estimation strategy, the disturbance decoupling schemes recalled in Chapter 4 can belong to both linear and nonlinear frameworks. In principle, the design methodologies exploited suitable coordinate transformations that highlighted a subsystem affected by the fault, but decoupled by the disturbances. These models were the starting point to design passive and adaptive FTC filters for fault detection or their estimation. These filters

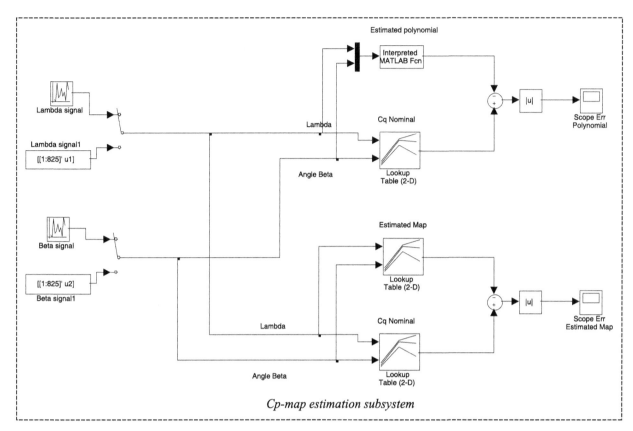

**FIGURE 6.33** Subsystem used for the approximation of the $C_p$-map.

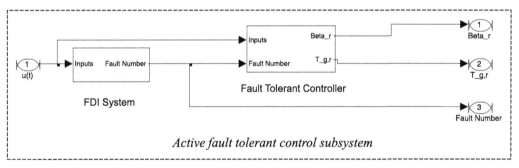

**FIGURE 6.34** Subsystem implementing the robust FTC methodologies.

were able to both detect additive fault acting on a single actuator and estimate its magnitude. It is worth observing that, by means of these robust design approaches, the fault estimates were decoupled from the disturbance $d$.

The proposed approaches were applied to the general nonlinear models of the wind turbine systems described in Chapter 2. The design of the robust fault estimators for FTC purposes was based on the scheme outlined in Fig. 6.34.

As remarked in Chapters 2 and 4, this methodology can be applied to the system under diagnosis if and only if suitable fault detectability conditions are satisfied. In this case, the subsystem represented in Fig. 6.34 in the new reference frame was decomposed into 3 submodels, where the first one was always decoupled from the disturbance $d$ and affected by the fault $f$.

On the other hand, with reference to the adaptive strategy recalled in Chapter 4, adaptive filters can be designed if suitable conditions and constraints were verified. Under these conditions, the design of the adaptive filters was achieved, with reference to the wind turbine model, in order to provide a fault estimate $\hat{f}(t)$, which asymptotically converges to the magnitude of the fault $f$. This scheme is outlined in Fig. 6.35.

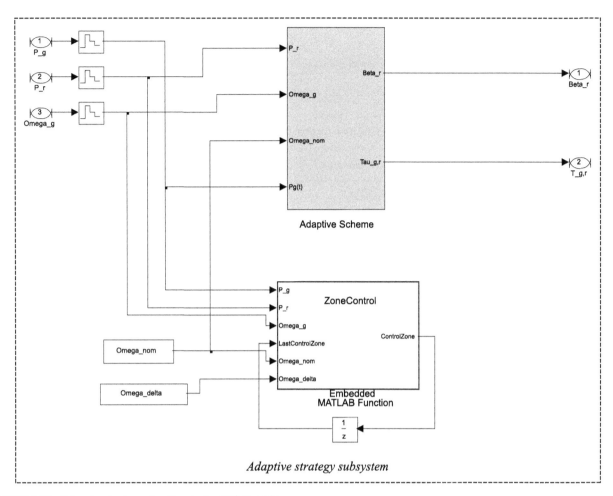

**FIGURE 6.35**    Subsystem implementing the adaptive FTC strategies.

The proposed filters were based, for example, on the adaptive algorithm with forgetting factor (Ioannou and Sun, 2012), and they were described by the adaptation laws recalled in Chapter 4, which are functions of the output estimate and the corresponding normalized estimation error. Proper design parameters are represented by $\lambda > 0$, i.e., the bandwidth of the filters, the forgetting factor $\beta \geq 0$, and the normalization factor $N^2 = 1 + \check{M}_1^2$.

Chapter 4 highlighted interesting properties between the asymptotic relation of the normalized output estimation error and the fault estimation error. Moreover, Chapter 5 showed the results regarding the adaptive filters described in Figs. 6.34 and 6.35. In particular, these fault estimation filters provided estimates $\hat{f}(t)$ that asymptotically converged to the magnitude of the actual faults $f$. This point is important since, as remarked in Chapter 4, the fault estimation error $f - \hat{f}(t)$ has to be bounded and asymptotically converge to zero, in order to guarantee the stability of the complete FTC system.

Finally, a different strategy relying on passive FTC schemes is shown in Fig. 6.36.

On the other hand, concerning the controller accommodation strategy, the remainder of this section describes the implementation of the complete structure of the FTC strategies, which required the information provided by the fault estimation module of Fig. 6.34.

It is worth noting that the filter schemes proposed in Chapter 4 were implemented here for fault estimation, even if they in general could be exploited for fault detection and isolation. In particular, if the fault isolation task needs to be solved, the bank of filters outlined in Fig. 6.31 can be exploited as residual generators, allowing to estimate the magnitude of different faults acting on the considered system. The decoupling of the effect of both the disturbance $d$ and the fault $f$ (important for fault isolation) from the fault estimator filters is provided by the selection of the proper state component from the subsystem model. However, as the main point of the monograph is the design of robust FTC schemes with application to the considered benchmarks, the problem of the fault isolation is not fundamental. Moreover, the simulated benchmarks considered only the occurrence of single faults, as described in Chapter 2.

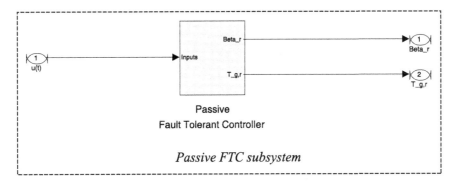

**FIGURE 6.36** Subsystem implementing the passive FTC methods.

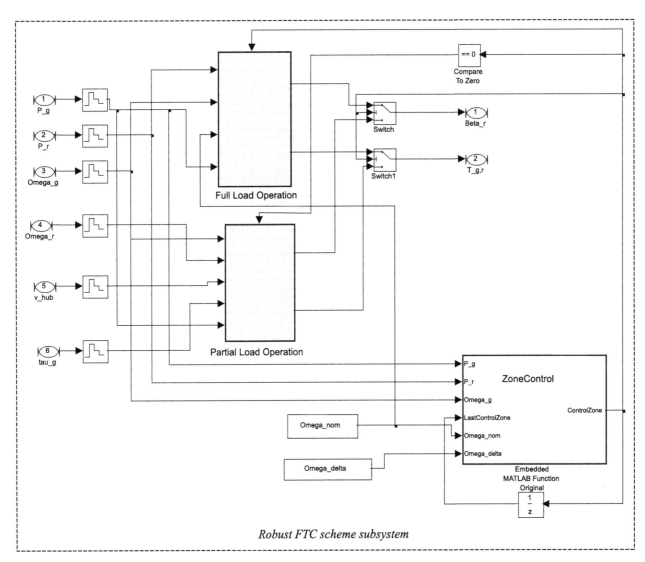

**FIGURE 6.37** Subsystem implementing the complete FTC strategy.

In order to compute the simulation results described in Chapter 5, the FTC schemes were completed by means of the benchmark controllers recalled in Chapter 2. The subsystem implementing the integrated FTC approach is shown in Fig. 6.37.

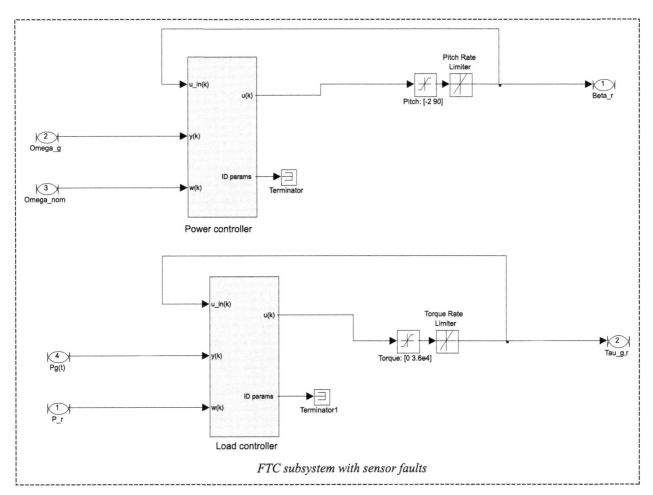

*FTC subsystem with sensor faults*

**FIGURE 6.38** Subsystem implementing the FTC strategy for sensor faults.

With reference to Fig. 6.37, the actuated inputs can be different from the controlled inputs due to the presence of faults. Moreover, also the actuated inputs can be different from the output of the benchmark controllers, which include a switching logic. On the other hand, also the measured outputs can be affected by the faults $f$, but they are corrected by means of the corresponding estimated signals, $\hat{f}$, provided by the fault diagnosis module of Fig. 6.28.

Therefore the scheme depicted in Fig. 6.37 shows that the FTC strategy was implemented by integrating the fault estimator module with the existing benchmark control system. From the controlled input and output signals, the fault estimation module provides the correct estimate $\hat{f}$ of the faults $f$, which was injected into the control loop, in order to compensate the effect of the faults themselves. After this correction, the benchmark controllers provided the nominal tracking of the reference signals.

Regarding the scheme shown in Fig. 6.37, as remarked in Chapter 4, the proposed strategy was applied for the accommodation of additive input faults. However, the proposed FTC methodology could be extended to the case of sensor faults. In fact, sensor faults can be modeled as additive inputs in the measurements equation (i.e., the output equation) by means of the so-called sensor fault signature (often a column with only one element equal to "1", and the remaining entries being null). Moreover, sensor fault signatures could also be modeled as an input to the system, in the same way of actuators faults. This further task was performed following the method (a further coordinate change) described in Chapter 4. Its implementation is shown in Fig. 6.38, with reference to the full-load controllers of the wind turbine system.

With reference to the stability analysis of the proposed FTC solutions, the simulation results shown in Chapter 5 highlighted that the model variables remained bounded in a set, which assured the control performances, even in the presence of faults. Moreover, the assumed fault conditions did not modify the system structure, thus guaranteeing the global stability.

However, a few more issues can be considered here. It should be clear that in steady-state conditions, when the fault effect was completely eliminated, the performances of the implemented FTC were the same as those of the fault-free situation. Therefore, the performances of the complete systems were the same as those of the fault-free nominal controllers. The stability properties for most of the considered FTC solutions were considered only in transient conditions, when the faults were not compensated. In fact, in these conditions, the fault estimation errors corresponded to a signal injected into the feedback control loop of Fig. 6.34. As highlighted in Chapter 4, it was possible to show that the fault estimation errors were limited and converging to zero, thus the stability of the complete systems was maintained. Moreover, the main design parameters of the proposed solutions were addressed in Chapter 5.

### 6.3.6 Monte Carlo Simulation Tool

The stability properties of the overall FTC strategies presented in Chapter 5 were verified and validated by means of a Monte Carlo analysis based on the wind turbine benchmark simulators when controlled by means of the considered baseline regulators described in Chapter 2. In fact, as pointed out in Chapter 2, the wind turbine systems contain the power coefficients map $C_p$ that cannot be described by any analytical model obtained via first principles. Thus, the Monte Carlo analysis represented the only method for estimating the stability of the developed solutions when applied to the monitored processes.

Initial conditions were changed randomly and disturbances affecting the system were simulated during the transient related to the stability analysis.

All simulations were performed by considering noise signals modeled as band-limited white processes, according to the standard deviations reported in Chapter 2.

As an example of the Monte Carlo simulation tool, Fig. 6.39 highlighted that the main wind turbine model variables, such as the generator torque $\tau_g$, the tip-speed ratio $\lambda$, and the generator power $P_g$, remain bounded around the reference values, proving the overall system stability in simulation, even in the presence of disturbances and uncertainty. This subsystem refers to the case of the partial load working conditions.

The results achieved by means of the subsystem of Fig. 6.39 shown in Chapter 5 highlighted that in the first part of the simulation the output power $P_g$ becomes larger than the theoretical one $P_{g,max}$, as the kinetic energy from the rotor shaft is converted into electrical energy produced by the generator. On the other hand, $P_{g,max}$ can be above the generated power, since the inertia of the rotor is accelerated before $P_{g,max}$ can be matched.

As a further example of the Monte Carlo simulation tool, the subsystem working above rated wind speed (full load conditions) is depicted in Fig. 6.40.

Fig. 6.40 depicts the subsystem using the generator speed $\omega_g$ and the control input $\beta_r$. Also in this situation the main wind turbine variables remained bounded around the reference values, thus establishing the overall system stability in simulation, even in the presence of modeling errors and noise signals.

It is worth noting that the design schemes followed by the analysis tools summarized in this chapter were developed using Matlab® and Simulink® software tools, in order to automate the overall simulation processes. As remarked in Chapter 7, these feasibility and reliability studies are of paramount importance for real application of control strategies once implemented to future wind turbine installations.

To this aim, Section 6.3.7 finally illustrates how the designed control algorithms are assessed through the Hardware-In-the-Loop (HIL) test-bed that was used to evaluate the capabilities of the solutions reported in Chapter 5 in more realistic experimental situations.

### 6.3.7 Hardware-In-The-Loop Tests

In order to evaluate the potential of utilizing the proposed solutions in real applications and investigate their capability to on-board implementation, this section briefly summarizes the Hardware-In-the-Loop (HIL) test-bed.

This tool served to validate definitely the desired requirements attributed to the designed algorithms considering almost real conditions that the wind turbine systems may experience during their working situations.

For this purpose, the HIL tool was developed according to Fig. 6.41 as described in Odgaard and Stoustrup (2015a), which provided capabilities to validate the developed control algorithms under almost real-time conditions.

This laboratory facility proposed by Odgaard and his coworkers, e.g., in Odgaard and Stoustrup (2015a) consists of the following three modules:

1. **Computer simulator**, which was created in Labview® software from Matlab® and Simulink® environments, provided the modeling of the wind turbine dynamics considering the factors such as uncertainty, disturbance, measurement errors, and wind turbine component models, as described in Chapter 2. This software tool allowed also to monitor the parameters related to the FDI and FTC methodologies, as well as analyze their performance.

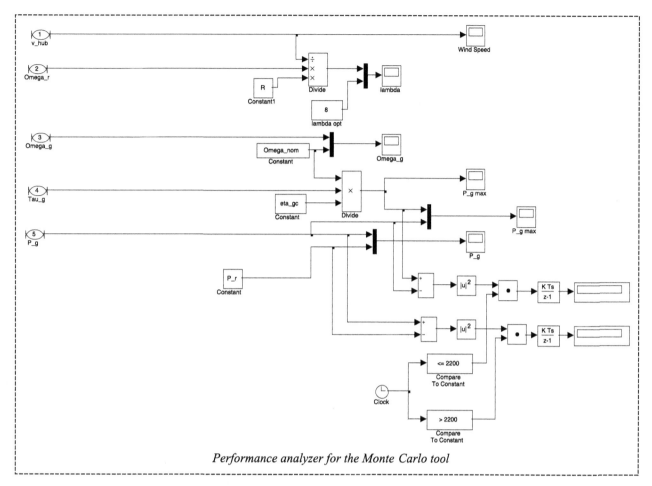

*Performance analyzer for the Monte Carlo tool*

**FIGURE 6.39** Subsystem of the wind turbine benchmark working at partial load.

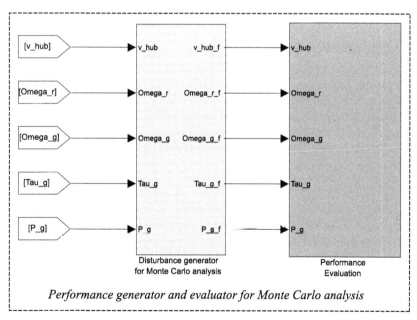

*Performance generator and evaluator for Monte Carlo analysis*

**FIGURE 6.40** Subsystem of the wind turbine benchmark operating at full load.

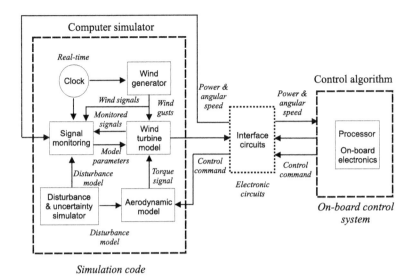

**FIGURE 6.41** Main components of the HIL tool.

2. **On-board electronics**, which implemented the FDI and FTC solutions. The electronic device utilized in this module is the AWC 500 system, which in addition satisfies standard wind turbine technical specifications. This element also provided the flexibility to implement and evaluate the different solutions considered in this monograph. As it can be seen from Fig. 6.41, the on-board electronics receive the wind turbine power and generator angular rate as inputs, and after digital signal processing, generate the generator torque and pitch command output signals transmitted to the computer simulator. The generator torque and the pitch commands were generated by the proposed FDI and FTC strategies and were applied to the wind turbine benchmark to guarantee its stability and the required specifications.

3. **Interface circuits**, consisting of appropriately selected input–output cards, which receive the output signals from the computer simulator and transmit the output signals generated by the digital signal processing algorithms.

The results achieved from this test-bed were reported in Chapter 5. It was shown that some deviations were found between the simulated and experimental results, but consistent with the ones from the Monte Carlo analysis recalled in Section 6.3.6. In fact, the performances in the simulation case were somehow better than the HIL experimental cases, which was reasonable due to the issues detailed below:

- The accuracy of the float calculations in the on-board electronics processor was more restrictive than of the CPU of the computer simulator.
- The major deviation between the achieved results originated from the analog-to-digital and digital-to-analog conversions.

Moreover, since the data must be transferred between the on-board electronics and the computer simulator, a 16 bit conversion was inevitable, so this conversion error may lead to the deterioration of the experimental results. Note also that, since the real situations do not need to transfer data between the on-board electronics and the computer, this error is not a problem and is consistent with the results already achieved via the Monte Carlo tool of Section 6.3.6. Moreover, although there were some deviations between the simulation and the experimental results, due to the reasons stated above, these deviations were not critical, and the results obtained were accurate enough for motivating real wind turbine applications.

## 6.4 SUMMARY

This chapter described the main implementation and computational aspects of the most important modeling and control blocks used for the development of the wind turbine and wind farm benchmarks in Matlab® and Simulink® environments. An insight into the wind turbine and farm modeling, as well as their control loops, was provided with the goal of providing overviews of the typical actuation, sensing and control implementations available for wind turbine simulators. The chapter intended to provide also an updated and broader perspective by covering not only the modeling and control aspects in Matlab® and Simulink® environments of individual wind turbines, but also outlining a number of solutions of advanced control approaches. In summary, wind energy is a fast growing industry, and this growth has led to a large demand for better modeling and control of wind turbines by means of suitable simulation codes. Uncertainty, disturbances, and other deviations from normal working conditions of the wind turbines make the control challenging, thus motivating the need for advanced implementations and a number of sustainable control realizations that should be exploited to reduce the cost of wind energy. By enabling this clean renewable energy source to provide and reliably meet the world's electricity needs, the tremendous challenge of solving the world's energy requirements in the future will be enhanced by using advances simulation codes and their practical implementations. The wind resource available worldwide is large, and much of the world's future electrical energy needs can be provided by wind energy alone if the technological barriers are overcome. The development of sustainable control solutions for wind energy systems was demonstrated in this chapter by means of proper simulation codes, whilst there are many fundamental and applied issues that were addressed by the systems and control community to significantly improve the efficiency, operation, and lifetimes of wind turbines. Finally, this chapter considered the implementation of the fault situations in practical ways with reference to the severity of their end-effects, in order to perform a motivated selection of the faults that were diagnosed and accommodated.

Chapter 7

# Conclusions

## 7.1 INTRODUCTION

With increasing concerns about climate change and the depletion of fossil fuels, this monograph highlighted that renewable energy became a topical area for research. Significant penetration of, for example, wind and solar electricity generation systems took place in an effort to supplement thermal power and other traditional modes of electricity generation. However, the costs associated with these developing technologies are high and favorable feed-in tariffs are usually employed to encourage development. Moreover, the variability of the production from such intermittent renewables incurs extra grid integration costs. Nevertheless, renewable technologies such as wave energy are required to satisfy increasing world energy demand in a climate neutral and cost effective way. Therefore the key issue is to reduce the cost of renewable energy, in order to compete with, and eventually displace, significant number of conventional electricity generation plants.

With this economic goal in mind, it is important that renewable energy converters incorporate sufficient intelligence to allow them to convert the available energy as efficiently as possible for a given capital investment, while prolonging the life of the conversion systems and their components. It is clear that control systems technology has a strong role to play in optimizing the operation of these conversions systems. In particular, this monograph considered a selection of modeling, control system, supervision approaches, and methodologies in the wind application area as a sample of the spectrum of possibilities. In addition, since this application area and the application of control technology in the application area have grown up relatively independently, we have examined some key aspects of both the modeling and control challenges of this application area and the nature of the control systems being developed. In addition, the monograph took the opportunity for cross-pollination between domains, especially in view of the fact that wind energy is now relatively mature from the modeling point of view, while presenting challenging aspects in the fields of control, supervision, fault diagnosis, and fault tolerance.

The monograph covered different aspects of the wind conversion systems, spanning also the application area of wind energy. First, we provided an introduction to wind conversion systems, presenting an overview of the conversion systems and principles, the mathematical models that are used to describe them, and the wind and wave resources that drive them, together with some control possibilities. An important feature of this monograph is the comparisons and contrasts made between the different techniques and technologies in this application area. Second, the work was focused on both model-based and data-driven approaches to the modeling and control of wind turbines. Different data-driven schemes for monitoring and fault detection in wind turbines were considered, which use the measurements already available in wind turbines, thus avoiding the introduction of expensive auxiliary condition monitoring modules. The main advantage with respect to the model-based approaches is that only the data of the system behavior are exploited. On the other hand, model-based strategies that may require accurate analytical descriptions of the systems under investigation were able to provide more efficient regulation and quicker reaction and recovery against disturbance and uncertainty affecting the system, with respect to data-driven schemes available in the literature. On the other hand, this monograph analyzed modeling, supervision, and control approaches ranging from the more traditional techniques, typically found in feedback regulatory systems, to more advanced methods, also appearing in the solution of optimal and nonlinear control. Optimal and suboptimal methodologies were considered as viable approaches to maximize the energy capture for wind energy devices, but also focused on the important issue of the system availability. Finally, the application examples considered in the monograph adopted standard statistics approaches, where the wind excitation and the uncertainty affecting the system were modeled as a Gaussian stochastic disturbance. On the other hand, the experimental examples were also obtained by considering high-fidelity simulators or hardware-in-the-loop tools which exhibit realistic complex and nonlinear behaviors of wind turbine systems. Key considerations concerned the calculation of optimal solutions, which take into account of nonideal effects and model–reality features of the energy conversion systems. In summary, the topics addressed in this monograph, including several overview studies, enhanced the view of the relative extent to which control technology has penetrated wind energy. It could reasonably be argued that the greater advancement of control technology reflected the greater commercial development of wind energy. Aside from maximizing converted power in view of the physical constraints of the system and a need to maximize the lifetime of device components, there was clear value in advanced modeling, monitoring, and control strategies.

The contents of the chapters of the monograph are briefly summarized in the following, in order to recall the most important achievements of the book.

- Chapter 2 introduced the wind turbine systems considered in this work, describing their characteristics, categories, and components, together with some global statistics that highlight the importance of the discussed topics. Then two realistic test cases were presented in detail, which represented high-fidelity simulators of a single wind turbine and of a wind farm, respectively. They were developed using the analytical description of the system behaviors, and provided also the models of typical fault cases. This chapter described also the faults affecting the wind turbine systems and determined their effect on the system behavior. Redundancies in these systems were also highlighted in order to define the detectable faults, and the remedial actions that must be conducted to stop the propagation of the fault were established.

- Chapter 3 recalled the most important fault diagnosis strategies applied to wind turbine systems. These techniques can be used for both condition monitoring purposes and fault diagnosis, in order to provide the information required by the fault tolerant control module. Model-based and data-driven strategies were also recalled, which rely on measured, as well as estimated, variables from the systems under investigation.

- Chapter 4 summarized the fault tolerant control algorithms that were considered and applied to wind turbine systems. They were based on the signal correction principle, which means that the control system is not modified since the inputs and outputs of the baseline controller are compensated according to the estimated faults. The fault tolerant control algorithms recalled here exploited the fault diagnosis designs of Chapter 3. Passive and active fault tolerant control systems were also discussed and compared, in order to highlight the achievable performances and the complexity of their design procedures. Controller reconfiguration mechanisms were finally analyzed, which were able to guarantee the system stability and satisfactory performance.

- Chapter 5 summarized the performances achieved by developed fault diagnosis and fault tolerant control systems, analyzed and applied to both the single wind turbine and wind farm benchmark models. The chapter recalled also the results obtained in comparison with the different fault diagnosis and fault tolerant control schemes considered in Chapters 3 and 4. The robustness and reliability features of these solutions were also analyzed by means of the Monte Carlo tool, which was able to take into account possible disturbance and uncertainties affecting the systems under diagnosis.

- Chapter 6 reported the Matlab and Simulink implementations of the proposed data-driven and model-based approaches to the fault diagnosis and fault tolerant control solutions of wind turbine and wind farm benchmark models. The main Simulink modules implementing the suggested strategies were briefly discussed and analyzed. These simulation tools are available from the link of the authors' home-page. Finally, Hardware-In-the-Loop tests were carried out in order to highlight the controller performances in more realistic and effective frameworks.

- Chapter 7 summarized and discussed the main achievements and the most important results of the monograph. It suggested and analyzed some future investigations, research directions, and open problems of the topics addressed in this monograph.

## 7.2 CLOSING REMARKS

It was remarked that the increased level of wind-generated energy in power grids worldwide has raised the levels of reliability and sustainability required of wind turbines. Wind farms should be able to generate the desired value of electrical power continuously, depending on the current wind speed level and on the grid demand. Therefore, possible faults affecting the system must be properly identified and treated, before they endanger the correct functioning of the turbines or become critical faults. In fact, large rotor wind turbines are extremely expensive systems, therefore their availability and reliability must be high, in order to assure the maximization of the generated power while minimizing (O&M) services. Alongside the fixed costs of the produced energy, mainly due to the installation and the foundation of the wind turbine, the O&M costs could increase the total energy cost by about 30%, particularly considering the offshore installation.

These considerations motivated the use of proper fault diagnosis solutions coupled with fault tolerant controllers. It seems that most of the wind turbine installations featured simply conservative strategies against faults, consisting of the shutdown of the system to wait for maintenance service. Hence effective strategies coping with faults were considered in this monograph for improving the turbine performance, particularly in faulty working conditions. As highlighted in this work, the main benefits concern the prevention of critical failures that jeopardize wind turbine components, thus avoiding unplanned replacement of functional parts, as well as the reduction of the O&M costs and the increment of the energy production. On the other hand, the advent of computerized control, communication networks, and information techniques brings interesting challenges concerning the development of novel real-time monitoring and fault tolerant control design strategies for industrial processes. Hence the modeling, fault diagnosis, and sustainable control issues for wind turbine systems proved to be a challenging task, thus motivating the research activities carried out through this monograph.

It was also highlighted that modern wind turbines have large, flexible structures operating in uncertain environments, thus representing interesting cases for advanced control solutions. Advanced controllers can help achieve the desirable goal of decreasing the wind energy cost by increasing the efficiency, and thus the energy capture, or by reducing structural loading and increasing the lifetimes of the components and turbine structures. The monograph remarked that HAWTs have the advantage that the rotor is placed atop a tall tower, where it can take advantage of larger wind speeds higher above the ground. Moreover, HAWTs used for utility-scale installations include pitchable blades, improved power capture, and structural performance, as well as no need for tensioned cables used to add structural stability. On the other hand, VAWT solutions are more common for smaller turbines, where these disadvantages become less important and the benefits of reduced noise and omnidirectionality become more pronounced. It is worth noting that the generating capacity of modern and commercial turbines ranges from less than 1 kW to several MW. The proper wind turbine system modeling oriented to the design of a suitable control strategy is more cost-effective for large wind turbines.

Another important point considered in this monograph derived from the steadily increasing sizes and growing complexity of wind turbines, thus giving rise to more severe requirements regarding the system safety, reliability, and availability. The safety demand can be commonly achieved by introducing redundancy in the system architecture, like additional sensors, which become vital for a safer operation of wind turbines. Classic examples refer to the pitch systems for adjusting the angles of a rotor blade. For each of the three blades, one totally independent pitch system is used, such that in the worst case of a malfunction in one or two pitch systems, the remaining one or two would still be able to bring the turbine to a standstill. It was shown that this solution improved the system safety, but it generates additional costs and possibly additional turbine down-times due to faults in the redundant system parts. The enhanced safety may lead to a reduction of the system availability.

The monograph briefly described the main components of a wind turbine. In particular, the blades interact with the wind producing movement of the rotor shaft. Their profile is designed in order to obtain a good value of the aerodynamic lift with respect to the aerodynamic resistance and, at the same time, to oppose the proper stiffness to the applied variable mechanical loads that determine the wear and tear effect in time. Construction materials should be light, such as plastic materials reinforced with glass, aluminium or carbon, depending on the blade size.

The wind turbine hub constrains the blades to the rotor shaft, transmitting the extracted wind power. It can contain the pitch actuator that forces the blade to have a certain orientation relative to the wind main direction, for control purposes. It is worth noting that the use of brake systems (mechanical, electrical, hydraulic) can avoid the rotor movement when the turbine has to be kept in nonoperating conditions.

The wind turbine nacelle includes one (or more) stage gear-box that is used to adapt the mechanical power of the rotor shaft to the generator shaft, by increasing the rotational speed and by decreasing the torque, in order to permit an efficient conversion of energy. The design of the gear box involves epicycloidal or parallel axis gears. We have also discussed a generator that converts the mechanical energy of the connected shaft to electrical energy. It can be an asynchronous or synchronous machine, the former comprising an induction three-phase motor that works as generator providing energy to the grid, as its speed is higher than the synchronous speed and the applied torque is motive. The usual structure of an asynchronous generator involves a squirrel cage rotor, implying a low difference between the synchronous speed and the actual speed in the nominal working region, so that it can be considered a constant speed machine. The configuration called doubly-fed, with a power converter located between the rotor and the grid, allows the functioning of the system at a variable rotor speed. Otherwise, the synchronous generator provides a voltage with frequency proportional to its rotational speed. In the configuration of a full-converter, similarly to the doubly-fed generators, a power converter has to be interposed between the generator and the grid, in order to permit variable speed functioning. The output power of a modern turbine can be up to five megawatts.

The wind turbine nacelle contains the shafts, brakes, gear box, generator, and control equipment. It is located at the top of the tower and also contains the anemometer, which is the sensor providing the current wind speed. Its measurements are exploited by the control system. The yaw mechanism includes several electrical motors that orientate the heading orientation of the turbine parallel to the main wind direction. Finally, the wind turbine tower represents the load carrying structure that supports the nacelle, hub, and blades. The height of the tower determines the height of the hub, which is a prominent value in the generation of power, since the wind speed increases with the distance from the earth's surface.

The monograph highlighted that wind turbine control goals and strategies depend on the turbine configuration. HAWTs may be "upwind", with the rotor on the upwind side of the tower, or "downwind." The choice of upwind versus downwind configuration affects the choice of yaw controller and the turbine dynamics, and thus the structural design. Wind turbines are also variable pitch or fixed pitch, meaning that the blades may or may not be able to rotate along their longitudinal axes. Although it was remarked that fixed-pitch machines are less expensive initially, the reduced ability to control loads and change the aerodynamic torque means that they are becoming less common within the realm of large wind turbines. Variable-pitch turbines may allow all or part of their blades to rotate along the pitch axis.

It was shown that wind turbine working conditions affect their performances. In fact, variable-speed turbines tend to operate closer to their maximum aerodynamic efficiency for a higher percentage of the time, but require electrical power processing so that the generated electricity can be fed into the electrical grid at the proper frequency. As generator and power electronics technologies improve and costs decrease, variable-speed turbines are becoming more popular than constant-speed turbines at the utility scale. The main difference between fixed-speed and variable speed wind turbines appears for mid-range wind speeds in Region 2, which normally encompasses wind speeds between 6 and 11.7 m/s. Except for one design operating point (10 m/s), a variable-speed turbine captures more power than a fixed-speed turbine. The reason for the discrepancy is that variable-speed turbines can operate at maximum aerodynamic efficiency over a wider range of wind speed than fixed-speed turbines.

It is worth noting that even a perfect wind turbine cannot fully capture the power available in the wind. In fact, actuator disc theory shows that the theoretical maximum aerodynamic efficiency, which is called the Betz Limit, is approximately 60% of the wind power. The reason that an efficiency of 100% cannot be achieved is that the wind must have some kinetic energy remaining after passing through the rotor disc. If it did not, the wind would by definition be stopped, and no more wind would be able to pass through the rotor to provide energy to the turbine.

With reference to the fault diagnosis methods considered in this monograph, it was remarked that there exist several ways in which faults can affect the system. A closed-loop system can be viewed as the union of interconnected elements, namely the main process, actuators, controller, input sensors, and output sensors. Each of these components can be associated to a fault, so that process faults, actuator faults, controller faults, and sensor faults can be considered. The simplest approach to fault diagnosis involves hardware redundancy. It consists in the usage of multiple sensors, actuators, or components to measure, or control, a particular signal. The related diagnosis is based on the comparison among the different redundant hardware information, hence a voting technique is adopted to decide if and where a fault occurs. Although the hardware redundancy can be very effective, it involves extra cost concerning the equipment and the maintenance operations, which could be in conflict with the sustainability requirements.

On the other hand, analytical redundancy solutions do not need additional hardware components, rather they exploit a model of the system under investigation for estimating the value of a particular variable on the basis of the preexisting sensors, so that a residual signal can be generated and a diagnosing logic can be inferred. In particular, the residual signal should be close to zero in normal operating conditions and significantly different from zero when a fault occurs. In principle, the analytical redundancy consists of a model-based approach, because of its dependence on the system model. A model-based module is thus implemented via software on a process control computer, without additional costs on the equipment, as already remarked. Some drawbacks of the model-based approach concern the accuracy of the generated estimates, as well as the disturbances and the noise affecting the process, that can lead to a false alarm or missed fault.

It was highlighted that the detection of a fault by means of the model-based approach exploits the generation and the evaluation of the residual. This signal is firstly generated as the error between the measured and the estimated variable. The latter comes from the model elaboration of the input–output measurements. The residual, at this stage, should be independent of system input and output, as it is ideally zero for every input–output condition in the fault-free case. Then the residual has to be evaluated, in terms of fault likelihood, and a decision rule has to be applied to determine the fault occurrence. At this stage the residual can be simply compared to a fixed threshold (such procedures are also called geometrical methods), or preprocessed through a proper function, e.g., a moving average or more complex statistical method, ahead of the threshold test.

The monograph considered a particular class of the model-based methods, i.e., the signal-based technique that occurs when only output measurements are available. It includes the case of vibration analysis (e.g., related to rotating machinery) that is performed by means of band-pass filters or spectral analysis.

Among the most important model-based FDI techniques recalled in this monograph, the output-observer, parity equation, and parameter estimation were considered. They require adequate knowledge of the state-space or input–output behavior of the system under investigation, which is expressed in terms of analytical relationships. However, most of the real industrial processes have a strongly nonlinear behavior that cannot be modeled using a single model for all the operating conditions. Indeed, the parameters of the system may vary with time and the unavoidable disturbance and noise effects have unknown characteristics. The result is a discrepancy between the plant behavior and its mathematical description, even in fault-free conditions, that can lead to the impossibility of generating a residual, as the fault may be hidden by the modeling errors. The Unknown Input Observer (UIO), the eigenstructure assignment, and the parity relation methods are FDI strategies that take into account these key issues.

Therefore it was argued that the proper mathematical description of the system under investigation, required by classical model-based FDI strategies, is very difficult to derive in practice, sometimes even impossible. Therefore, because of these considerations, data-driven modeling approaches offer a natural tool to handle poor knowledge of the system, together with disturbances and noises. Indeed, their implementations exploit input–output data directly acquired from the system, for deriving the relations among those variables. For example, fuzzy logic theory allowed the representation of the process under investigation by means of a collection of local affine fuzzy models, among which the transitions are handled by identified fuzzy parameters. Furthermore, also neural networks can handle complex nonlinear behaviors, as they have the capability of learning the system functioning on the basis of the information provided by the training data. These considerations motivate the studies of data-driven strategies applied to the wind turbine context, because their behavior is characterized by very complex and nonlinear dynamics, as described in this monograph.

The monograph also summarized the most important control solutions, which were able to cope with possible malfunctions and fault situations (indicated as FTC system), as they possess the capability to automatically manage the faults affecting the components. FTC methods include two main approach categories, namely Passive FTC and Active FTC. They can require knowledge of the fault functions. It was remarked that Passive FTCs require neither fault diagnosis nor controller reconfiguration, as they are designed to be robust against a set of possible faults. However, although they do not involve the problem related to fault diagnosis, such as false alarms or missed faults, they have limited fault tolerance capabilities. They make use of robust control techniques to ensure that the considered closed-loop system remains insensitive to faults, using fixed controller parameters, without the information regarding the fault occurrence. This is accomplished by designing a controller that is optimized for fault-free situations, while satisfying some graceful degradation requirements in the faulty cases. On the other hand, Active FTCs react to the faults actively by reconfiguring the control actions so that the performance of the fault-free system can be maintained, even in the presence of faults. In particular, an AFTC is mainly based on a fault diagnosis block that provides real-time information about the faulty, or fault-free, status of the system monitored. The controller exploits a further control loop aimed at the compensation of the faulty signals, the fault accommodation. The main advantage of this approach is that the controller can be designed considering only the nominal operating conditions. Therefore, as remarked in this monograph, the need of advanced control solutions for these very demanding systems motivated also the requirement of reliability, availability, maintainability, and safety over power conversion efficiency. These issues have begun to stimulate research and development of FTC systems.

In particular, wind turbines in the megawatt size range are expensive, and hence their availability and reliability must be high in order to maximize the energy production. This issue could be particularly important for offshore installations, where O&M services have to be minimized, since they represent one of the main factors of the energy cost. The capital cost, as well as the wind turbine foundation and installation, determines the basic term in the cost of the produced energy, which constitutes the energy "fixed cost." The O&M represents a "variable cost" that can increase the energy cost by about 30%. At the same time, industrial systems have become more complex and expensive, with less tolerance for performance degradation, productivity decrease, and safety hazards. This leads also to an ever increasing requirement on reliability and safety of control systems subjected to process abnormalities and component faults. As a result, it is extremely important to consider the fault diagnosis tasks, as well as the achievement of fault-tolerant features for minimizing possible performance degradation and avoiding dangerous situations. With the advent of computerized control, communication networks, and information techniques, it became possible to develop novel real-time monitoring and fault-tolerant design techniques for industrial processes, but this also brings challenges.

Therefore the purpose of this monograph is to revise the basic solutions to sustainable control design, which are capable of handling faults affecting the controlled wind turbine. For example, changing dynamics of the pitch system due to a fault cannot be accommodated by signal correction. Therefore it should be considered in the controller design, to guarantee stability and a satisfactory performance. Among the possible causes for changed dynamics of the pitch system, one can mention a change in the air content of the hydraulic system oil. This fault is considered since it is the most likely to occur, and since the reference controller becomes unstable when the hydraulic oil has a high air content. Another important aspect raised is when the generator speed measurement is unavailable and when the controller should rely on the measurement of the rotor speed, which is contaminated with much more noise than the generator speed measurement. This makes it necessary to reconfigure the controller to obtain a reasonable performance of the control system.

In order to outline and compare the controllers developed using AFTC and PFTC design approaches, this monograph showed how to derive them using the same procedures in the fault-free case. In this way any differences in their performance or design complexity would be caused only by the fault tolerance approach, rather than the underlying controller solutions. Furthermore, the controllers should manage the parameter-varying nature of the wind turbine along its nominal operating trajectory caused by the aerodynamic nonlinearities. Usually, in order to comply with these requirements, the controllers are designed, for example, using Linear Parameter-Varying (LPV) modeling or fuzzy descriptions.

This monograph highlighted that the key issue between AFTC and PFTC schemes is that an active fault-tolerant controller relies on a fault diagnosis system, which provides information about the faults to the controller. In the considered case the fault diagnosis system FDD contains the estimate of the unknown input (fault) affecting the system under control. The knowledge of the fault $f$ allows the AFTC to reconfigure the current state of the system. On the other hand, the FDD is able to improve the controller performance in fault-free conditions, since it can compensate for, e.g., the modeling errors, uncertainty, and disturbances. On the other hand, the PFTC scheme does not rely on a fault diagnosis algorithm, rather it is designed to be robust towards any possible faults. This is accomplished by designing a controller that is optimized for the fault-free situation, while satisfying some graceful degradation requirements in the faulty cases. However, with respect to the robust control design, the PFTC strategy provides reliable controllers that guarantee the same performance with no risk of false FDI or reconfigurations.

In general, it was argued that the methods used in the fault-tolerant controller designs should rely on output feedback, since only part of the state vector is measured. Additionally, they should take the measurement noise into account. Moreover, the design methods should be suited for nonlinear systems or linear systems with varying parameters. The latest proposed solutions for the derivation of both active and the passive fault-tolerant controllers rely on LPV and fuzzy descriptions, to which the fault-tolerance properties are added, since these framework methods are able to provide stability and guaranteed performance with respect to parameter variations, uncertainty, and disturbance. Additionally, LPV and fuzzy controller design methods are well-established in multiple applications, including wind turbines. Moreover, to add fault-tolerance to the common LPV and fuzzy controller formulation, different approaches can be exploited. For example, the AFTC scheme can use the parameters of both the LPV and fuzzy structures estimated by the FDD module for scheduling the controllers. On the other hand, different approaches can be used to obtain fault-tolerance in the PFTC methods. For this purpose, the design methods considered in this monograph were also modified to cope with parametric uncertainties. Alternatively, other methods could have been used, which preserve the nominal performance. Generally, these approaches rely on solving some optimization problems where a controller is calculated subject to maximizing the disturbance attenuation.

It is worth noting that, prior to designing and applying any new control strategies on a real wind turbine, the effectiveness of the solutions has to be tested in detailed simulation models. Several simulation packages exist that are commonly used in academia and industry for wind turbine load simulation. As an example, the monograph recalled the use of one of the most used simulation package, the FAST code (Jonkman and Buhl, 2005), provided by NREL. It represents a reference simulation environment for the development of high-fidelity wind turbine prototypes that are taken as a reference test-cases for many practical studies. Moreover, FAST provides a high-fidelity wind turbine model with 24 degrees of freedom, which is appropriate for testing the developed control algorithms but not for control design. On the other hand, for other purposes, a reduced-order dynamic wind turbine model, which captures only the dynamic effects directly influenced by the control, can be exploited and used for the preliminary analysis of model-based control designs.

The monograph highlighted that control systems for wind turbines are now well established, and many of these schemes were successfully exploited in wave energy devices. In the case of wind turbine devices, the general problem is to maximize energy capture, subject to grid and environmental constraints. However, the objective of energy capture maximization might be modified to that of maximization of economic return, which requires a balance to be achieved between maximizing energy capture and minimizing wear on components. For the current analysis, in order to retain a focus on the fundamental control issues, the general problem of energy capture maximization was considered. There are two broad approaches which may be taken to solve the energy maximization problem, i.e., the overall extremum seeking control and the determination of an optimal set-point for the system, which gives maximum energy capture, followed by a regulation to make sure this set-point is achieved.

It was remarked that extremum seeking control approach is attractive from the point of view of the lack of requirement for a detailed model, but may have dynamic performance limitations in convergence rates, and may have difficulty finding a global maximum over a nonconvex performance surface. On the other hand, the determination of the optimum can require an accurate mathematical description of the system under diagnosis.

It is clear that the control scheme attempts to devise algorithms that force a system to follow a desired path, objective, or behavior modality. The control issue is also regarded as a tracking problem, where the objective is for the system output to follow the reference input. While problems of this type do occur in energy conversion applications, for example, in speed control of wind turbines, it is more useful to broaden the set of problem descriptions and potential solutions a little, in order to assess the potential of control engineering in the general energy conversion context. A generic problem framework, consisting of an upper (optimal) set-point generation stage and a lower control loop to ensure tracking of the set-point. Both sets of control calculations must be mindful of physical constraints in the system. In the wind energy case, for variable speed turbines, an optimal rotational speed is first calculated, and torque and/or blade pitch control are used to achieve the required rotational speed.

In general, the control problem definition required maximization or minimization if prescribed performance objective was subject to proper system constraints, thus leading to a constrained optimization problem. The definition considered here is not inconsistent with the purpose of a classic controlled system with a feedback loop, where the objective function is usually a quadratic measure of the difference between the controller output and its desired value, i.e., the tracking error, with respect to the reference or the set-point. In this way, the desired performance of the tracking system in closed-loop can be specified in a variety of ways.

As an example, the minimization of the sensitivity of the closed-loop system to variations in both the system description and external disturbance are related to the system robustness, and specific control methodologies to address these objectives have been developed since the late 1970s. In most cases, control design methods provide an explicit solution for the feedback controllers, while some methods solve the more general optimization problem defined at each time step. In this monograph, specific or general solutions useful in the FDI/FDD and FTC of wind turbines were recalled and analyzed.

Finally, it was observed that some control methods required a mathematical model of the system, in order to determine the control algorithm, and such methods were termed model-based. The requirement for an accurate mathematical system model often involved considerably more work than the calculation of the controller itself, though system identification techniques were also proposed to determine a black-box model. This model has no structural relationship to the physical system. The combination of data-driven techniques with a mathematical procedure for controller determination was thus considered to develop adaptive controllers, with the capability to autonomously adapt to unknown or time-varying systems. Adaptive control schemes based on linear system models demonstrated the capability to track variations in a linear model due to the presence of nonlinearity, though nonlinear systems are best controlled with a dedicated fixed-parameter nonlinear controller. Significant care and attention was thus paid to adaptive schemes to ensure stability and convergence over all operating regimes.

## 7.3 FURTHER WORK AND OPEN PROBLEMS

The main goal in this section consists in proposing some new ideas regarding the development of advanced control and innovative sensing strategies for wind turbine plants that allow both maximizing the energy conversion and reducing their installation (i.e., the load-carrying structure), operation, and maintenance costs (Bossanyi et al., 2009). Therefore the primary focus should be oriented to the design of suitable control schemes that enable slender load-carrying structures with larger rotors and improved working conditions. This is possible by mitigating the wind-induced vibrations and fatigue damage. In this way, if the load-carrying structures are reduced, this directly contributes to the cost-effectiveness of wind energy, whilst also improving system reliability.

The control methods that should be considered for this purpose have to maximize the wind energy capture, which requires a better understanding of the wind resources and the nacelle issues, as well as better wind forecasting methods (Bianchi et al., 2007; Burton et al., 2011; Garcia-Sanz and Houpis, 2012; Rivkin and Silk, 2012). Control systems play an important part in this. In fact, variations in load due to blade "flapping", unsteady blade aerodynamics, wind shear effects, vibration due to pitch errors, and fatigue stress should be identified by numerical and experimental activities, since a suitable design is strongly related to nacelle issues. In general, it is clear that control strategies should be chosen and designed so that the blade rotational frequency and its harmonics do not match the natural frequencies of the substructure and wind spectra, in order to reduce fatigue stress and peak loads in load-carrying structures (Bianchi et al., 2007; Fischer et al., 2012).

These advanced control strategies require more reliable information about the state of the wind turbine system, the local wind field around the rotor, aerodynamic loads on the blades, and the drive-train loading. For this to be possible, sensor measurements must be combined to produce the relevant information about the state of the system. In order to make this possible, more sensor information than that normally available is required, for example, more accurate estimation of effective wind speed and structural modes, etc. Measurement reliability thus represents a key issue. Moreover, the load-carrying structure is also integrated with appropriate sensing devices. However, sensors and measurements are prone to faults, and if the control system relies on these for information, and no action is taken, this could lead to higher, rather than lower, loads and/or lower availability (lost production time or derating).

These techniques, which are based on the development of proper measurement devices and sensing methods for improving the system monitoring, require the use of disturbance decoupling schemes, in order to cope with uncertain external conditions. Note that these advanced control and innovative sensing concepts should be first designed and tested on a single wind turbine, as described in Chapter 5. A further challenge that should be addressed is the extension of these novel concepts to wind farms, where the whole design is considered as a global optimization problem. Therefore the overall challenge is to design an integrated approach to advanced operation of a wind turbine and farm, to enhance performance

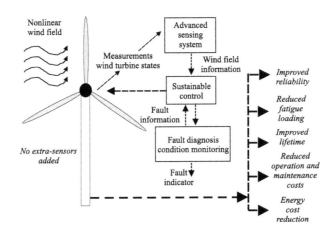

**FIGURE 7.1** Key points of the sustainable control design scheme.

for new and operating wind power plants and improve control of the wind turbine and farm, while reducing the fault rate, and therefore resulting in smaller operation and maintenance cost.

A good understanding of typical faults in wind turbine components (sensors, actuators, etc.) is exploited to build in tolerance to faults in the control system. This is an application of fault-tolerant control, as proposed in Chapter 4 (van Engelen et al., 2011; Odgaard et al., 2013; Shi and Patton, 2015) (i.e., the sustainable control concept), which is fundamental for successfully improving the control performance of wind turbines in the presence of degradation or faults, and for enabling the potential cost of energy reductions to be avoided. In fact, the synergistic use of vibration condition monitoring, predictive and proactive maintenance, and load identification can potentially lead to increase the wind turbine installation lifetime, by reducing operation and maintenance costs. Also the coupling between wind turbine control for power optimization and fatigue/extreme loads with respect to lifetime control has to be considered. High-fidelity simulation codes and the test benches considered in Chapter 5 serve to assess the effectiveness and the feasibility of the solutions when applied to wind turbines and wind farms. In this way the final challenge is to design an integrated approach to advanced operation of wind turbines and wind farms, to improve the performance for new and operating wind power plants and to improve their control, by reducing the fault rate, therefore resulting in smaller operation and maintenance costs.

### 7.3.1 Sustainable Control Design Objectives

The main idea should be based on the development of innovative sensing and control methodologies for wind turbines for substantially reducing the costs of the installations and making maintenance schedules more proactive and predictive. Therefore, when considering a wind farm, the main focus is on the control optimization of the entire wind plant using the integrated sensing design, which is very effective for both maximizing the wind energy efficiency and reducing the load-carrying structure, operation, and maintenance costs.

In fact, the European offshore wind business is currently facing challenges in relation to costs of installed offshore wind farms, mainly costs directly (fabrication) and indirectly (installation) related to support structures (Fischer et al., 2012; Willecke and Fischer, 2013). However, although the cost of fabrication and installation can be substantially reduced using this approach, new control solutions for wind turbines must be developed that provide improved mitigation of structural loads both in terms of fatigue and extreme loads (severe gusting, etc.). On the one hand, this enables lighter load-carrying structures with larger rotors, thus reducing the energy cost. On the other hand, the proposed control solutions reduce operation and maintenance costs, while optimizing the wind energy conversion. These key topics should be considered in the proposed solutions, as highlighted in Fig. 7.1.

Therefore the key objectives of sustainable control solutions driven from ambitious industry set goals related to the current industrial practice should aim at: (i) reducing lost production factor, (ii) decreasing the cost of energy, and (iii) reducing maintenance costs of the main turbine components (Bossanyi et al., 2009; Odgaard, 2012). The lost production factor is defined as the actual production with respect to the possible production, and covers lost production caused by the wind turbine and nonproduction loses due external events like grid error or derating by the operator. The cost of energy includes all the expenses incurred in putting up the turbine, keeping it operating, and taking it down again divided by how much the turbine produces during its lifetime.

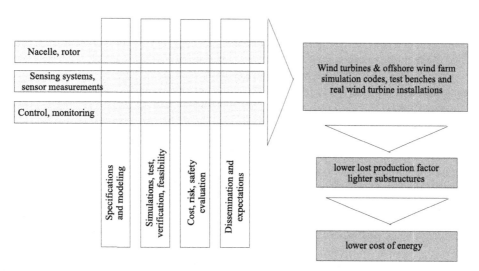

**FIGURE 7.2** Links between objectives, overall goals, and impacts.

These targets can be achieved by a sustainable control strategy relying on wind turbine state estimation schemes that combine models of sensor signals available on state-of-the-art industry turbines, with control actions and models of relevant dynamical component parts of the system. Appropriate sensing devices must also be included in the wind turbine installation (nacelle and load-carrying structure). The mathematical models must include simple fatigue data to enable fatigue loads to be reduced by the wind turbine control system. Confidence and reliability of the estimated information is to be addressed by applying condition monitoring, fault diagnosis, and fault tolerant control methodologies. This enables the controller to cope with damage and failure in components such as blades and their sensors, while providing the desired level of stable performance. A by-product of this scheme is fault diagnosis and condition/vibration monitoring of all the modeled system parts, as described in Chapter 3. Fault information can be used for wind turbine operators, and in a suitable format for scheduled, predictive, and proactive (with root cause analysis) maintenance operations. Note that another aspect requiring attention regards the system safety. In fact, if no special actions are undertaken, the wind turbine would simply have to be shutdown in case of sensor faults.

Therefore the main tasks and objectives of the sustainable control methodology should include:

- The specifications and models for rotor system, tower, and structures related to load cases;
- The models of available sensors and actuators and faults occurring in the sensors and actuators;
- Sensing systems, which combine sensor systems and condition, vibration monitoring, fault diagnosis, and fault tolerance of the main turbine units;
- Load reducing control system design, which incorporates new sensing system outputs into sensor and actuator fault tolerant control designs;
- High-fidelity and benchmark wind turbine test and verification assessment of solutions in terms of fatigue and extreme loads (evaluated in accordance with IEC 61400) and turbine lifetime;
- Lost production factors and contribution to the cost decrease.

The links between tasks and objectives, the contribution to overall goals, and the impact on the costs are sketched in Fig. 7.2.

As already described in Chapter 2, a wind turbine system is a complex energy conversion fluid flow machine, which entails coupled hydro-aero-electro-mechanical and control issues. The control design should include specific competences and know-how to realize tailored solutions to the highlighted goals and objectives, focusing on increased reliability and risk of wind turbine designs with lower operating cost and improved energy efficiency.

Different open problems and future research issues regarding energy conversion systems have been proposed recently by the European Community. In particular, the interest should be focused on wind energy. This issue requires the development of advanced control of large scale wind turbines and farms, which are needed to improve efficiency and to further reduce the cost of wind energy, as well as to increase the value of wind energy, by improving the response to power system disturbances or electricity market conditions. On the one hand, one challenge to be achieved by the sustainable control strategy is the development of new controls systems that treat the entire wind plant as a control optimization problem. On the other hand, the wind energy capture maximization for individual assets with the wind-turbine-centric controls and the development of a better understanding of the wind resources and better wind forecasting methods have to be carefully taken into account. Therefore the overall challenge is to design an integrated approach to advanced operation of a wind turbine and farm, to improve performance for new and operating wind power plants and improve their control, while reducing the failure rate and therefore resulting in smaller operation and maintenance costs.

The proposed sustainable control design suggests the development of advanced control, innovative sensing, vibration, and condition monitoring systems for wind turbines using all available sensor information to build a better understanding of the local conditions around both the rotor system and load-carrying structure. This information is then exploited by the controller to actively reduce the structural loads associated with lighter installations. Therefore, the methodology used to lower the support structure investment cost is achieved through mitigation of aerodynamic and hydrodynamic induced fatigue loads through an integrated design approach, as well as site and design specific wind turbine control concepts. Relying on control rather than stronger, heavier, and more expensive structures to be possible, confidence, reliability, and effectiveness of the solutions must be demonstrated. The current idea confronts this by developing a fault tolerant control system approach. It is not based on future, yet untested sensors, but uses existing sensor technology, where faults and failures are studied in detail using past experience, and serve for applications of control that have considerable impact even in the short term.

Results from applying new methodologies/technologies from estimation, condition and vibration monitoring, fault diagnosis, and control significantly increase the viability of the proposed solutions in areas that have been identified as specifically important by the wind industry. These solutions should bring the exploited technologies to a level where it has to be tested and validated on high fidelity wind turbine simulators and test benches as described in Chapter 5. The extension of the proposed solutions to offshore wind turbines and wind farms have to be achieved through the use of simulation codes and real operating installations. With reference to offshore wind farms, the effect of marine environment, wind, current, and wave effects has to be properly taken into account in simulation. Business cases, normative, health, and safety issues have to be addressed from a life-cycle perspective. These activities align with the industry procedures already in place, and relevant business units are involved to support professional assessments of risks of the technological solutions from this design strategy, but also the business processes. These solutions have set performance and cost targets that align with the expected impacts of the design requirements, and key performance indicators should be developed to access the impact of the technological results on these targets early in the design solutions.

## 7.3.2 Sustainable Control Concepts and Approaches

The sustainable control design concepts rely on the enhancement and improvement of the knowledge level of the states around the wind turbine system (nacelle, tower, rotor system, and load-carrying structures), which are exploited for the design of suitable condition and vibration monitoring, as well as advanced control solutions. In fact, this information should be used to increase the controllability and maintainability of the turbine, thus enabling the same wind turbine nacelle on a lighter load-carrying structure and substantial reduction of maintenance costs.

During the wind turbine service lifetime – this is fundamental for offshore installations – load-carrying supports are subjected to continuously changing sea currents and extreme load peaks induced by the nacelle, the wind and sea wave actions (Burton et al., 2011; Fischer et al., 2012). A sound structural design must face these different types of changing loads, partially fulfilling requirements that can sometimes have opposite effects. The load-carrying supports must be designed to withstand vibrations of very significant intensity; moreover, small and continuous vibrations must be taken into account due to the concept of fatigue. In addition, the dynamic analysis of lighter structures with larger rotors under combined actions such as wind and sea waves is very complex and needs to be handled only making use of reliability analysis tools commonly used in the fault-tolerant control framework (Blanke et al., 2006; Isermann, 2005; Odgaard et al., 2013; van Engelen et al., 2011). The structural shape and the material choice are important at the design stage of the installation; however, the current state-of-the-art proves that this design approach is limited and cannot easily lead to a definitely robust and cost-effective structural solution (Burton et al., 2011; Fischer et al., 2012). An important step forward can be made by equipping the structure with suitable sensing devices that provide further information to the so-called "semiactive" control system to minimize the overall dynamic response under extreme conditions (Bottasso et al., 2013). In this way, regarding the load-carrying supports, the use of advanced sensing strategies applied to the overall wind turbine system and exploited by the novel control schemes is beneficial for the load-carrying structure itself.

Therefore, on the one hand, mitigation of structural loads to a higher degree is done by the joint use of advanced sensing systems and novel control strategies. On the other hand, advanced control, condition and vibration monitoring, as well as fault diagnosis and fault tolerant control strategies, have to be used to increase the reliability and availability of the system, thus leading to a lower lost production factor. In addition, fault diagnosis, prognosis, and root-cause analysis can be employed to increase the efficiency level of the operation and maintenance service planning, as faults can be diagnosed, predicted, and eliminated in advance. In this case the faults can be considered also as the loads and structural stress that affect the load-carrying structure of the wind turbine.

The required wind information used to increase the controllability of the wind turbine consists of the distributed wind information in the sectors of the rotor plane, in terms of wind velocity and turbulence level. On the other hand, properly installed sensors on the load-carrying structure must be also considered for condition and vibration monitoring. Since standard industry sensors have to be used as inputs by suitably chosen estimation schemes, this information should be computed for the specified sectors in the rotor plane, around the blades and for the load-carrying structure. As information is required for the control of turbines with lighter load-carrying structures and larger rotors, it is crucial that the information from the nacelle, tower, and structure is reliable. The wind information can be obtained by extending work on effective wind speed estimators and wind flow field prediction to provide wind information in sufficient detail around the rotor, see, e.g., Friis et al. (2011), Ostergaard et al. (2009).

By combining the wind information with blade and tower loads, the turbine controller is thus able to optimize the power, while keeping fatigue and extreme loads of the lighter substructure under design specifications with original nacelle concepts. These loads are taken into account in the control design by including simple representations of them in the overall model design (Bianchi et al., 2007; Fischer et al., 2012).

As the control strategy increasingly relies on the sensors and state estimation from the wind turbine (Odgaard et al., 2013), any faults in the measurements must be detected, isolated, and properly handled by the control solution. The list of relevant sensors includes – but is not necessarily limited to – ultrasonic wind sensors, Light Detection And Raging (LiDAR) technology, mechanical wind sensors, pitch position sensors, blade load sensors, rotor position (azimuth) sensors, rotor speed sensors, as well as tower-top and load-carrying structure accelerometers (Lio et al., 2015; Towers and Jones, 2014). The scope for sensors is limited to state-of-the-art industry sensors, supporting realistic but high technology levels for the key parts of the overall system. Note that in the perspective of already industrial proven sensors for advanced control applications, LiDAR models can be considered for comparison purposes, in order to verify if advanced sensing techniques using industry sensors is able to provide the same information achieved by means of remote sensing technologies (Towers and Jones, 2014; Valencia-Palomo et al., 2014).

In order to increase availability, sensor faults and misreading have to be considered. It is equally important to diagnose and accommodate faults in the pitch actuator, as this is the main actuation used for controlling structural loads on the rotor, tower, and load-carrying structures. The designed controller should consequently be tolerant towards pitch actuator faults, and as well ensure that load levels during emergency shutdowns are within acceptable extreme load levels (Bianchi et al., 2007; Bottasso et al., 2013). Fault-tolerant control designs that are able to simultaneously accommodate both sensor and actuator faults in the presence of varying structural loads over a wide range of wind speed conditions must be developed.

In order to reduce sensor inaccuracy due to site and turbine specific conditions and installation induced offset, auto-calibration schemes are also part of the scope of the design approach. In particular, these problems are expected to be handled by the monitoring and the fault tolerant part of the control related issues (Odgaard et al., 2013; Shi and Patton, 2015).

Note finally that the sustainable control solutions must be validated on wind turbine high-fidelity simulation codes and test benches. Moreover, the verification and the validation on offshore wind farms should be achieved thorough the use of suitable simulators and real installations. In this way also the effect of marine environment, wind, current, and wave effects should be properly taken into account in simulation for offshore wind farms.

### 7.3.3 Sustainable Control Approaches and Working Methods

The challenging activities of a sustainable control strategy are aimed at developing innovative sensing advanced control strategies that allow the use of more cost-effective design solutions, thus leading to lighter structures with larger rotors on new turbines with an overall installation, maintenance, and energy cost reductions. The proposed solutions have to be verified and validated through the use of high-fidelity simulators of wind turbines and offshore wind farms, as well as test benches and possibly to real operating installations.

The sustainable control design follows an iterative approach, where existing and new sensors installed on both the nacelle and the load-carrying structure are combined with software sensors (observers) that estimate all the dynamic variables, i.e., the states of the turbine, as recalled in Chapter 2. The information given by proper state observers is used to monitor the whole system both by comparing the estimates with some additional sensors and by checking thresholds associated to not measured variables. These new sensors are a key aspect of the sustainable control strategy, as they are used to both validate

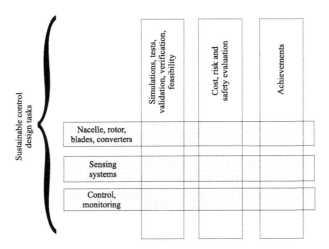

**FIGURE 7.3** A sketch of the sustainable control design tasks.

the state observers and give redundant information useful for congruity tests. This sensing system represents a powerful way to achieve wind turbine condition monitoring and fault diagnosis. The redundant information also enables the health of actuators to be diagnosed; the development of abnormal stresses can be also detected. All this increased information provides a powerful background not only for fault monitoring and diagnosis, but also to develop advanced rotor control to achieve high performance even in the presence of variable load conditions. The sustainable control scheme has fault tolerant control and robustness characteristics built in, so that the control performance can be maintained in the presence of certain faults, disturbances, and uncertainties, as remarked in Chapter 4. Specific fault and disturbance estimation techniques can be implemented to compensate faults and disturbances in the feedback control using available measurements and robust state estimates, as suggested in Chapter 3. As the level of measured and estimated information increases, the wind turbine control can be modified with confidence, reliability and overall effectiveness, leading to realistic objectives of designing lighter load-carrying structures, to reduce the lost production factor, maintenance, and an overall cost of energy in new wind turbine installations (Plumley et al., 2014).

The overall sustainable control design consists of a number of activities organized into several tasks where the iterations are indicated as cross-couplings in Fig. 7.3.

These design objectives are also organized into suitable tasks, as described below.

1. Description of the identified problems for lighter wind turbine load-carrying structures enhanced by advanced control and sensing systems; control requirement characterization;
2. Load-carrying structures and nacelle related issues: blades, rotor, converter, sensor, and actuator modeling;
3. Analysis and development of new sensing systems, which combine sensors, observers, condition and vibration monitoring, fault diagnosis, and fault tolerance;
4. Design of advanced control systems incorporating the new sensing solutions into an integrated approach for fatigue and extreme stress mitigation, so as to show tolerance features with respect to sensor and actuator faults;
5. Test and verification assessment of solutions in terms of fatigue, extreme load and lifetime, lost production factors, and contribution to a cost of energy decrease in both high-fidelity wind turbine and offshore wind farm simulators, and by testing on the wind turbine test benches;
6. Cost evaluation, turbine risk, and safety analysis;
7. Final result analysis.

The modeling of the nacelle, e.g., the rotor blades and nacelle structure, is very important to evaluate the efficacy of the control schemes on the induced stresses on the system. The modeling of the sensors and actuators of the wind turbine nacelle with and without faults was also considered in Chapter 2 using software codes, as demonstrated in Chapter 5. The availability of either reliable sensors or state estimates also allows for the estimation of the wind field in the vicinity of the rotor blades such that the efficiency of the wind turbine can be increased and/or severe damages prevented. The nacelle models and the load-carrying description, provided in Chapter 2 and modeled via the simulation codes provided in Chapter 6 are the basis for state estimation and wind turbine control issues addressed in Chapters 3 and 4, respectively. Furthermore, modeling the faults in sensors and actuators can be exploited for developing proper fault detection, isolation, and estimation methods via model-based, adaptive, geometric, and artificial intelligence principles, as shown in Chapter 5.

Using the achievements recalled in Chapter 2, Chapter 3 suggested the development of new sensing system. It provided the optimal estimation of the required wind turbine state information (wind speeds, wind turbulence, rotor speeds, pitch position and speeds, blade, load-carrying structure, and drive-train moments and loads) given the status of the sensors. In Chapter 3, the sensor faults and sensors that externally cause misreadings are diagnosed by using proper unknown input observers based on adaptive thresholds and robust control techniques. Appropriate fault accommodation methods were considered in Chapter 4, such that the sensing system provides the best state estimation. Fault accommodation in the sensing system can be achieved using estimator reconfiguration and/or fault estimation and compensation methods based on geometric and fuzzy approaches. Condition monitoring schemes can be also considered for providing further information on the load-carrying structure of the wind turbine via frequency methods. Moreover, the sustainable control design strategy could also address the optimization of the sensor placement in the wind turbine system in order to maximize the achievable performances.

The sustainable control strategy can exploit the latest research results in terms of integrated structural design and advanced turbine controls, which are very effective for load mitigation (and thus lighter load-carrying structures), as well as develop new specific controllers. These new controllers use the wind turbine state information, provided by the sensing system, and are based on robust control approaches for uncertain dynamic systems. The control goal is to ensure that extreme load requirements are not violated during eventually emergency stops of the wind turbine, as well as during severe wind gusts. Furthermore, during almost normal operations, the optimization problem arising from the conflicting control objectives of power optimization and maximum admissible loads/stresses is considered. In order to achieve such a goal, the control of the wind turbine blades, reducing the design bearing fatigue and extreme structural loads affecting the lighter load-carrying structure of the wind turbine is mandatory.

The new sustainable wind turbine controller can be thus designed to be robust with respect to sensors and actuator faults by using appropriate robust fault tolerant control techniques combined with the multivariate control design. The interesting challenge is to be able to use the rotor system to control the turbine, so that, in practice, the rotor performs like a "high level" sensor. In other words, the goal is to be able to use the rotor itself (along with the enhanced sensor set) to make the control system perform well. A part of this challenge is to ensure the real-time compensation of loading and gust disturbances exploiting fuzzy tools or using disturbance estimation methods based on geometric approach, as recalled in Chapters 3 and 4. This becomes a very significant challenge for wind turbines (> 10 MW) as the required control and disturbance bandwidths become close, similar to the structural filtering and control problem in high performance aircrafts (Castaldi et al., 2014). The developed sensing and control systems are to be tested and verified on wind turbine and wind farm high-fidelity simulators and technological demonstrators of wind turbines. These tests and evaluations should ensure that the fatigue and extreme loads are within the required limits and they are consistent with the load-carrying structure with larger rotors. In the high-fidelity wind turbine and wind farm simulators a number of load cases have to be tested to emulate lifetime fatigue loads, as well as extreme load cases subjected to faults, in accordance with the IEC 61400 standard.

Cost evaluation, turbine risk and safety analysis need to be performed according also to the specifications defined in Chapter 2.

These activities align with the industry procedures already in place, to support professional assessments of turbine risk and risks of the technological solutions from the sustainable control design, but also the business processes.

Fig. 7.4 presents the links between the sustainable control design, whose tasks and targets are performed according to the design phases, as addressed in Chapter 4.

Finally, the overall design includes also the problems of cost evaluation, risk and safety analysis. It evaluates how the current design methodology influences the final cost of energy of the wind turbine, due to the reduction of the wind turbine down-times, and the implementation of a lighter load-carrying structure. Business cases, normative issues and health and safety issues should be addressed from a life-cycle perspective. The link between the objectives, the contribution to overall goals of the integrated design, and impact on the cost of both the energy and the whole installation are sketched in Fig. 7.4. The synergistic integration among the design tasks enhances the achievement of the required targets and tasks of the overall sustainable control design. By addressing these objectives, the life cycle is explored from models, over design to verification and experimentation. The result is a number of specific contributions that lead to assess the design scheme at a suitably increased technology and manufacturing levels.

### 7.3.4 Sustainable Control Design Ambition

The main challenge of wind turbine manufactures is to achieve a lower cost of energy. A key way for obtaining this is to decrease the weight of the load-carrying structures, while preserving the capability of carrying the nacelle, blades, rotor, and converters. However, this is not a viable solution by itself, as lighter load-carrying structures may not be consistent

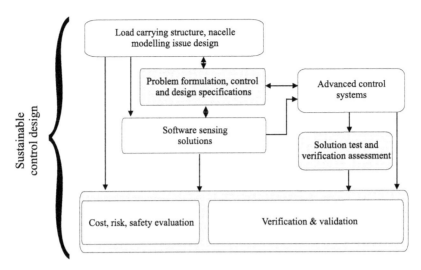

**FIGURE 7.4** Graphical presentation of tasks and targets of the overall sustainable control design strategy.

with the nominal structural stresses, both static and dynamic. Therefore, in order to allow for lighter structures, load and fatigue stresses must be mitigated (Bianchi et al., 2007; Fischer et al., 2012). This requires better control of the wind turbine dynamics without modifying the energetic performance. Improvements of state-of-the-art load mitigating control strategies are thus required, which may use, e.g., individual pitch (Bossanyi, 2003; Bottasso et al., 2013; Lio et al., 2015) and wind speed preview by Light Detection And Raging (LiDAR) technology (Lio et al., 2015). However, this setup should make the wind turbines more expensive and more sensitive to faults, as the load mitigating control system depends on these devices. It also potentially results in lower availability of the wind turbines, if not accommodated (Odgaard, 2012), especially at very large rotor diameters, operating offshore.

Taking into account the current state-of-the-art, it is apparent that a gap must be filled to reduce the cost of wind turbines. Such a goal can be achieved by developing suitable control strategies, able to limit structural loads both in terms of fatigue and extreme loads. However, the implementation of effective controls to limit these structural loads needs a many state measurements, i.e., wind speed in different rotor sectors or at different blades, turbulence levels, rotor position (azimuth angle), structure vibration monitoring, blade and tower moments, etc. (Bossanyi and Hassan, 2000; Bottasso et al., 2013).

The basis of the sustainable control design challenge consists in the innovation related to how sensors are used in the wind manufacturing industry (Bianchi et al., 2007). Additional wind turbine states are provided by a combination of suitable state observers, designed to tolerate both possible faults in different sensors and misreadings from sensors due to external causes (Odgaard and Stoustrup, 2013b; Johnson and Fleming, 2011). In case the sensor faults are not accommodated, the control system is not able to further guarantee the required load levels or the availability of the turbine such that undesired wind turbine shutdown could occur increasing the loss production factor. Wind turbine shutdown often leads to problematic extreme stresses, which should be avoided either by proper modulations of the load during emergency stops or by avoiding shutdowns using controllers designed to mitigate the effects of these stresses (Bossanyi and Hassan, 2000; Bottasso et al., 2013). It is apparent that this task relies on the availability of an increased number of wind turbine states (or measurements) that can be provided by the comprehensive use of the existing sensors integrated by suitable state estimation techniques or by additional industrial available sensors. Current works on wind turbine state estimation, see e.g., Bottasso et al. (2007), are properly extended in the current design scheme to account for structural stresses over a wide range of wind speed conditions. These tools should provide the state information required by the wind turbine compensator to control the blades of the rotor and the nacelle, and limit the mechanical stress on the lighter load-carrying structures of the wind turbine.

Advanced control methods have stimulated research in a wide range of industrial control communities and academia, particularly for those systems that demand a high degree of reliability and availability, and at the same time are characterized by expensive and/or safety critical maintenance work (van Engelen et al., 2011). In particular, with reference to wind turbines, these recently developed installations are examples that have complex nonlinear aerodynamics and with a stochastic and uncontrollable driving force (i.e., wind speed), see, e.g., (Bianchi et al., 2007). Moreover, the offshore wind turbine site accessibility and system availability is not always guaranteed during or soon after malfunctions, primarily due to changing weather conditions. In the literature, fault tolerant control, together with fault detection and diagnosis schemes, has been recognized as the proper solution to ensure these desired requirements, i.e., to guarantee the sustainability of wind turbines by increasing both the availability and the power generation.

Therefore, developing the new software sensing tools, condition monitoring, advanced control and fault diagnosis methodologies/technologies, applying all these tools to wind turbine sensors, actuators, and systems, and demonstrating their effectiveness represent one of the main ambitions of the proposed sustainable control design. The suggested solutions can be validated using high-fidelity simulators of wind turbines and wind farms, as described in Chapter 2.

The requirement to increase the knowledge level of the wind turbine state variables (or states) concerns both the area around the wind turbine rotor and its load-carrying structure (substructure and foundations). This improved information is used to enhance the controllability of the turbine thus enabling lighter load-carrying structure with the same nacelle, rotor, blades and converters, evaluated on selected industrial wind turbine, and developed in accordance with the IEC 61400 standard. At the same time, this enables the wind turbine performance to be improved by controlling the wind turbine closer to its optimal operation. Better fault diagnosis, condition and vibration monitoring, using the increased information level, can increase the reliability and availability features, thus leading to lower lost production factor. In addition, fault diagnosis and vibration monitoring can be used to increase efficiency level of maintenance of the service organization as faults are diagnosed, predicted and eliminated well in advance of service. Performing effective automatic fault prediction and fault/failure root cause analysis during wind turbine operation represents extremely highly ambitious aspect of the overall sustainable control design.

The new required information consists of the wind field distribution in the rotor plane sectors, in terms of wind velocity and turbulence level. Standard industrial sensors can feed suitable estimation modules computing these measurements for the specified sectors in the rotor plane and around the blades. Since these measurements and their reliability are extremely important for the control of turbines with lighter load-carrying structures, the effectiveness of the wind information estimate is crucial. In this way, the combination of the wind information with the nacelle and structural loads enhances the controller task to optimize the power generation while keeping fatigue and extreme loads of the blades and the load-carrying structure under design specifications. This point represents an ambitious aspect of the proposed sustainable control design strategy.

Since the control of the wind turbine relies on more information of the state around the rotor, the reliability of the sensors and the further state estimates are of higher importance. This means that any faults in these sensors must be detected, isolated, and accommodated, if possible, based on the commonly used sensors summarized in Chapter 2. It is worth observing that this set of sensors could be modified if a more accurate, reliable, and better quality state estimation is required. The optimal placement of these sensors can be also considered. On the other hand, in order to reduce sensor inaccuracy due to site, turbine specific conditions, and installation induced offset, self-calibration schemes can be considered as a scope of the present design. Note also that the required beneficial state measurements and fault information come directly as a by-product from the same fault diagnosis and control modules that are exploited for improving the reliability and availability of the measurement sensor system itself, as shown in Chapter 3. The fault information is also used for operation monitoring and maintenance, thus leading again to potential saving in energy cost. This point represents a further ambitious task for the proposed design strategy.

It is important to ensure that the wind turbine can be controlled as required, since the load mitigation cannot be ensured if the actuators in the wind turbine (here the pitch actuators are particularly relevant) are not operating as expected. Fault diagnosis and fault tolerant control with respect to pitch actuator faults are consequently an important goal in the sustainable control design approach, solved with the tools recalled in Chapters 3 and 4. Extreme loads during emergency shutdown are typically design baring load cases, it is consequently important to ensure that the wind turbine at all times is able to perform an emergency shutdown without violating the extreme structural load levels at the specific turbine, as demonstrated by exploiting the methods summarized in Chapter 4.

The availability, sustainability, and reliability of desired characteristics are especially important for the cost-efficiency of offshore wind farms. This means that the lighter load-carrying structure and the increased controller dependency on the increased state information level should not result in decreased availability of the turbine. The availability should instead be increased at least with respect to sensor faults or measurement sensor misreading to external phenomena. The relative inaccessibility of offshore sites, forces minimum maintenance-related financial impacts. Health and condition and vibration monitoring, fault diagnosis and fault tolerant control strategies are the technology tools that must be used to develop reliable, available, and efficient wind turbines with lighter load-carrying structures. Good progress can depend in turn on the design of suitable sensing and actuation schemes, hardware for acquisition, and with appropriate methodologies. For this very ambitious purpose, different technologies can be exploited within the tools recalled in Chapter 4. Therefore, an important ambitious point of this design scheme is that the fault tolerant control task is not only exploited against actuator and sensor faults, but also for coping with the unwanted dynamic behavior represented by the fatigue and loads of the lighter load-carrying structures. One of the most important problems of the fault tolerant control issues is that they have never been related to real problems.

Another ambitious task of the sustainable control design is to consider the best combination of sensors and actuators in order to guarantee a prescribed degree of fault tolerance, thus leading to an increase of the availability and reliability of the wind turbine installation. Proper optimization strategies can be exploited for the most suitable placement of the sensors and the selection of the design parameters.

The combination of the proposed novel techniques could lead to a reduction in the lost production factor. Practical verifications of the developed techniques should be also planned. The simulation of these methodologies is usually investigated on high fidelity industry standard wind turbine simulation platforms. The experiments deal with specified internal faults and external malfunctions caused by misreading in different key sensors and in the pitch actuator sensing.

Thus the proposed sustainable control design aims to achieve these goals by fulfilling the following objectives, which are comparable with the best industrial practice and have important for the research and innovation aspects:

- Facilitating lighter load-carrying structures, while maintaining nacelle (rotor, blades, drive-train, and converters) and lifetime. In fact, if in general the loads are reduced such that load-carrying structures can be lighter, the remaining structure is also affected by lower loading levels;
- Reducing the lost production factor, considered as an average over the wind turbine system; this point can be achieved by implementing both advanced sensing and fault tolerant control strategies;
- Reducing maintenance costs of the main turbine components;
- Decreasing the cost of energy on future turbines; this point can be achieved by both the new substructure concepts and the advanced sensing and control techniques.

The research and innovation aspects of the design have to be addressed in the multidisciplinary engineering domain of development of a future generation of lighter load-carrying structure offshore wind turbines. The focus is on issues that make significant impact on wind turbine operation and maintenance. The multidisciplinary nature of this work arises from the coupled hydro-aero-electro-mechanical dynamics of a wind turbine, so that advanced knowledge and experience of several engineering subjects is required and the work must be performed in a multidisciplinary framework. Indeed, to design, build and operate reliable and energy efficient wind turbines, knowledge from disciplines like mechanical engineering, metrology, fluid mechanics, condition and vibration monitoring, fault diagnosis, automatic control, signal processing, and computer simulation needs to be combined. It is the ambition of the sustainable control design to bring together specific capabilities and know-how to realize tailored solutions to the highlighted aims and objectives.

### 7.3.5 Sustainable Control Innovation Potentials

The main innovation in this sustainable control design is the development of radically new control and condition monitoring technologies and tools for offshore wind turbines and farm that allow for better control of structural loads based on information from an advanced wind turbine sensing system. This novel software sensing system provides further relevant information of the wind turbine states like the wind speeds at the blades, wind turbulence and direction, rotor and blade pitching speeds, rotor and blade pitching position, tower acceleration, tower and blade bending moments, etc.

The novel structural load control enables lighter load-carrying structures with larger rotors without decreasing the design bearing loads (on selected industrial wind turbines) and the nacelle efficacy, which decreases the cost of energy of the wind turbine. In this way, lighter load-carrying structures are possible only if the design bearing loads are decreased. Both the novel controller and novel sensing systems are made tolerant with respect to sensor and actuator faults. In this way, the sustainable control design should increase the availability of the wind turbine and thereby decrease the cost of energy. This issue represents a very important novel outcome of the proposed design strategy.

The control design is very challenging, and the disturbances acting on the system require proper estimation and compensation since the required control system bandwidths can be close to the wind disturbance frequencies for these powerful machines. The sensing system exploited by the controller uses standard industrial sensors and combines the acquired measurements exploiting models of the wind turbine to estimate the best obtainable values of the states, depending on available fault-free sensors.

The next expected design outcome is thus related to the design of novel control strategies that take into account disturbance, uncertainty, damage, and fatigue load models and load distribution mechanisms in the overall structure under complex loading conditions. The investigations involve a combination of simulations and practical testing with global optimization strategies. The innovations potential directly arises from the proposed novel control strategies being able to limit structural loads both in terms of fatigue and extreme loads.

It is worth noting that the development of condition monitoring, fault diagnosis, and fault tolerant control tools is another novel outcome of the proposed investigation. To this aim, different methodologies and technologies could be considered based on the high fidelity wind turbine and wind farm simulators and real operating installations, as well as test benches of wind turbines. Robustness and reliability of the developed modules have to be considered, since they may affect the performance of the overall sustainable control strategy. Further research issues that should be investigated will lead to novel scientific outcomes regarding the development of new observer, estimation and control schemes designed to optimize the general wind turbine working conditions and performances.

### 7.3.6 Sustainable Control Expected Impacts

The International Energy Agency and the European Commission estimates that EU electricity demand could double by 2030. Including the replacement of aging infrastructure, a global investment of some 10,000 billion euro will be necessary to meet this demand. It is a challenge to ensure that such an investment results in sustainable economic growth, high quality jobs, European technology and research leadership, and global competitiveness. Wind energy is perhaps the most advanced of the "new" renewable energy technologies, but there is still much work to be done, and this monograph aims to contribute to such a technologic area.

The main innovation in the sustainable control design is the development of radically new tools, based on advanced wind turbine sensing and control systems, that allow for a more effective control and shaping of the induced stresses on the supporting structure, such that its building, installation, and maintenance cost is reduced, specially in the case of four offshore wind turbines.

This novel sensing system can also provide further relevant information of the wind turbine states (e.g., the wind speeds at the blades, wind turbulence and direction, rotor and blade pitching speeds, rotor and blade pitching position, tower acceleration, tower and blade bending moments), which are very useful for all wind turbines, and especially for very large rotors (typically greater than 10 MW rating), since they allow for implementing advanced supervisory tools to improve availability, efficiency, and reliability of energy production by wind turbines.

An assessment of the research and technology impact is given briefly below, presenting the key benefits from both a scientific and an industrial point of view.

From a scientific point of view, the contribution of the sustainable control design will be in developing advanced sensing, diagnosis, and control tools, which can be integrated with the structural design tools used for wind turbines.

The control design is very challenging: the disturbances acting on the system require proper estimation and compensation since, for these powerful machines, the required control system bandwidths can be close to the wind disturbance frequencies. Because of such a not clear frequency separation between disturbances and system requirements, it has to take into account disturbance, damage and fatigue stresses, and load distribution in the overall system, e.g., on turbine rotor blades and on the lighter load-carrying structure, such that effective robust and fault tolerant control schemes can be developed.

The sensing system exploited by the controller uses standard industrial sensors to acquire the main system variables and combine them with the outputs of robust state observers, based on accurate models of the wind turbine, to get the best obtainable values of the states, identifying faults on sensors and actuators that then can be compensated for.

Therefore novel scientific outcomes regarding the development of new high fidelity wind turbine simulators as well as observer, estimation, and control schemes designed to optimize the general wind turbine working conditions are expected. To obtain the expected results, different methodologies and technologies have to be analyzed and applied, and the robustness and reliability of the developed modules must be carefully considered, since they may affect the performance of the overall sustainable control strategy.

The investigations should involve a combination of simulations and practical testing. From an industrial point of view (European wind turbine technology), the challenges addressed are intended to pave the way for higher technology levels and effective integration and validation of the associated technologies in future wind turbines. This aspect should be verified developing and using technological demonstrators and high-fidelity simulators of wind turbines.

The long-term impact of the sustainable control design is that it contributes to the solution to the global climate and energy challenges, by increasing the compatibility of wind turbines by lowering costs and increasing availability. Moreover, the proposed design should enhance innovation capacity; create new market opportunities, strengthen competitiveness and growth of the involved companies, thus bringing overall important benefits for society.

## 7.4 SUMMARY

This monograph summarized the most important modeling and control issues on wind turbines from a systems and control engineering point of view that were discussed in detail in the chapters of the book. A walk around the wind turbine control loops discussed the goals of the most common solutions and overviewed the typical actuation and sensing available on commercial turbines. The monograph intended to provide also an updated and broader perspective by covering not only the modeling and control of individual wind turbines, but also outlining a number of areas for further research, and anticipating new issues that can open up new paradigms for advanced control approaches. In fact, wind energy is a fast growing industry, and this growth has led to a large demand for better modeling and control of wind turbines. Uncertainty, disturbance and other deviations from normal working conditions of the wind turbines make the control challenging, thus motivating the need for advanced modeling and a number of so-called sustainable control approaches that should be explored to reduce the cost of wind energy. By enabling this clean renewable energy source to provide and reliably meet the world's electricity needs, the tremendous challenge of solving the world's energy requirements in the future are enhanced. The wind resource available worldwide is large, and much of the world's future electrical energy needs can be provided by wind energy alone if the technological barriers are overcome. The application of sustainable controls for wind energy systems is still in its infancy, and there are many fundamental and applied issues that can be addressed by the systems and control community to significantly improve the efficiency, operation, and lifetimes of wind turbines.

It is worth noting that further related researches can be carried out, aimed at optimizing the residual generators, in particular the neural network schemes, as different architectures, as well as different training algorithms, can be tested and compared, in order to optimize the modeling capability. Furthermore, also the AFTCs can be improved by investigating different solutions for the main controller, as the simple preexistent controller as been exploited in this work. Moreover, a noteworthy validation should be fulfilled by testing the proposed tools with real wind turbine installations. However, the overall achieved result induces future studies concerning the application to real wind turbine installations, where the currently adopted control strategies are often too conservative, as they involve the shutdown of the faulty turbine to wait for maintenance services.

Concluding, wind energy is a fast growing industry and this growth implies a large demand for better modeling and control. Possible faults, malfunctions, uncertainties, and disturbances make the control a challenging task to overcome. These considerations motivate the need for advanced modeling and further development of sustainable control strategy, with the main objective of reducing the wind energy cost. This clean and renewable energy source should match the global electricity needs, and it represents the proper candidate to comply with the future world's energy requirements, if the technological barriers are overcome. The industrial application of sustainable control is still in its prototyping phase, and there exist many opportunities to significantly improve the efficiency and the lifetime of wind turbines and wind farms.

# References

Adegas, F., Sloth, C., Stoustrup, J., 2012. Control of Linear Parameter Varying Systems with Applications. Chap. Structured linear parameter varying control of wind turbines. 1st ed. Springer-Verlag, London, pp. 303–337.

Adegas, F.D., Sonderby, I.B., Hansen, M.H., Stoustrup, J., 2013. Reduced-order LPV model of flexible wind turbines. In: Proceedings of the 2013 IEEE Multi-Conference on Systems and Control. IEEE, Hyderabad, India, pp. 424–429.

Adegas, F.D., Stoustrup, J., 2011. Robust structured control design via LMI optimization. In: Proc. of the 18th IFAC World Congress. IFAC, Milan, Italy. IFAC, pp. 7933–7938.

Adegas, F.D., Stoustrup, J., 2012. Structured control of LPV systems with application to wind turbines. In: Proc. of the 2012 American Control Conference. IEEE, Montreal, Canada. IEEE, pp. 756–761.

Amato, Francesco, Cosentino, Carlo, Mattei, Massimiliano, Paviglianiti, Gaetano, 2006. A direct/functional redundancy scheme for fault detection and isolation on an aircraft. Aerospace Science and Technology 10 (4), 338–345.

Amit Dixit, A., Suryanarayanan, S., 2005. Towards pitch-scheduled drivetrain damping in variable-speed, horizontal-axis large wind turbines. In: Proceedings of 44th IEEE Conference on Decision and Control. IEEE, pp. 1295–1300.

Appleby, B.D., Dowdle, J.R., Vander Velde, W., 1991. Robust estimator design using $\mu$ synthesis. In: Proc. of the 30th Conf. on Decision & Control. Brighton, UK, pp. 640–644.

Ayoubi, M., 1995. Neuro-fuzzy structure for rule generation and application in the fault diagnosis of technical processes. In: Proc. of the American Control Conference. ACC'95, Washington, USA, pp. 2757–2761.

Babuška, Robert, 1998. Fuzzy Modeling for Control. Kluwer Academic Publishers, Boston, USA.

Babuška, R., 2000. Fuzzy Modelling and Identification Toolbox. Version 3.1 ed. Control Engineering Laboratory, Faculty of Information Technology and Systems, Delft University of Technology, Delft, The Netherlands. Available at http://lcewww.et.tudelft.nl/~babuska.

Babuška, Robert, 2012. Fuzzy Modeling for Control, vol. 12. Springer Science & Business Media.

Basseville, M., 1988. Detecting changes in signals and systems: a survey. Automatica 24 (3), 309–326.

Basseville, M., 1997. Information criteria for residual generation and fault detection and isolation. Automatica 33, 783–803.

Basseville, M., Benveniste, A., 1986. Detection of Abrupt Changes in Signals and Dynamical Systems. Lecture Notes in Control and Information Sciences, vol. 77. Springer-Verlag, London.

Basseville, M., Nikiforov, I.V., 1993. Detection of Abrupt Changes: Theory and Application. Prentice Hall Inc.

Beard, R.V., 1971. Failure Accommodation in Linear Systems Through Self-Reorganisation. Technical Report MVT-71-1. Man Vehicle Lab., Cambridge, Mass.

Beghelli, S., Castaldi, P., Guidorzi, R.P., Soverini, U., 1994a. A comparison between different model selection criteria in Frisch scheme identification. Systems Science Journal 20 (1), 77–84. Wroclaw, Polonia.

Beghelli, S., Castaldi, P., Soverini, U., 1994b. Dynamic Frisch scheme identification: time and frequency domain approaches. In: 10th IFAC Symposium on System Identification. IFAC'94.

Beghelli, Sergio, Guidorzi, Roberto P., Soverini, Umberto, 1990. The Frisch scheme in dynamic system identification. Automatica 26 (1), 171–176.

Beghelli, S., Guidorzi, R.P., Soverini, U., 1997. A frequencal approach to the dynamic Frisch scheme identification. In: 4th European Control Conference. ECC'97, Brussels, Belgium.

Betz, A., Randall, D.G., 1966. Introduction to the Theory of Flow Machines. Pergamon Press, Oxford. ISBN 978-0080114330.

Bezdek, James C., 2013. Pattern Recognition with Fuzzy Objective Function Algorithms. Springer Science & Business Media.

Bianchi, F.D., Battista, H.D., Mantz, R.J., 2007. Wind Turbine Control Systems: Principles, Modelling and Gain Scheduling Design, 1st ed. Advances in Industrial Control. Springer. ISBN 1-84628-492-9.

Biegel, B., Madjidian, D., Spudic, V., Rantzer, A., Stoustrup, J., 2013. Distributed low-complexity controller for wind power plant in derated operation. In: Proc. of the 2013 IEEE Multi-conference on Systems and Control. IEEE, Hyderabad, India. IEEE, pp. 146–151.

Blanke, M., Kinnaert, M., Lunze, J., Staroswiecki, M., 2006. Diagnosis and Fault-Tolerant Control. Springer-Verlag, Berlin, Germany.

Blesa, J., Jimenez, P., Rotondo, D., Nejjari, F., Puig, V., 2014. Fault diagnosis of a wind farm using interval parity equations. In: Proceedings of the 19th IFAC World Congress, vol. 47. IFAC'14, IFAC, Elsevier, Cape Town, South Africa, pp. 4322–4327.

Bobál, V., Böhm, J., Fessl, J., Machácek, J., 2005. Digital Self-Tuning Controllers. Algorithms, Implementation and Applications, 1st ed. Advanced Textbooks Control and Signal Processing. Springer.

Bobál, V., Kubalcík, M., Ulehla, M., 1995. Auto-tuning of digital PID controllers using recursive identification. In: Proc. of the 5th IFAC Symposium on Adaptive Systems in Control and Signal Processing, vol. 1. Budapest, Hungary, IFAC ed., pp. 384–389.

Bonfè, Marcello, Castaldi, Paolo, Geri, Walter, Simani, Silvio, 2006. Fault detection and isolation for on-board sensors of a general aviation aircraft. International Journal of Adaptive Control and Signal Processing (ISSN 0890-6327) 20 (8), 381–408. Copyright 2006 John Wiley & Sons, Ltd.

Borcehrsen, A.B., Larsen, J.A., Stoustrup, J., 2014. Fault detection and load distribution for the wind farm challenge. In: Proceedings of the 19th IFAC World Congress, vol. 47. IFAC'14, IFAC, Elsevier, Cape Town, South Africa, pp. 4316–4321.

Bossanyi, E.A., 2003. Wind turbine control for load reduction. Wind Energy 6 (3), 229–244. http://dx.doi.org/10.1002/we.95.

Bossanyi, E.A., Hassan, G., 2000. The design of closed loop controllers for wind turbines. Wind Energy 3 (3), 149–164. http://dx.doi.org/10.1002/we.34. John Wiley & Sons, Ltd.

Bossanyi, E.A., Ramtharan, G., Savini, B., 2009. The importance of control in wind turbine design and loading. In: Proceedings of the 17th Mediterranean Conference on Control and Automation, 2009. MED'09, Thessaloniki, Greece, pp. 1269–1274.

Bottasso, C., Croce, A., Riboldi, C., 2013. Multi-layer control architecture for the reduction of deterministic and non-deterministic loads on wind turbines. Renewable Energy 51 (1), 159–169. http://dx.doi.org/10.1016/j.renene.2012.08.079.

Bottasso, C.L., Croce, A., Savini, B., 2007. Performance comparison of control schemes for variable-speed wind turbines. Journal of Physics. Conference Series 75, 012079. http://dx.doi.org/10.1088/1742-6596/75/1/012079. IOP Publishing.

Boyd, S., Ghaoui, L., Feron, E., Balakrishnan, V., 1994. Linear Matrix Inequalities in System and Control Theory. SIAM, Philadelphia.

Brown, M., Harris, C., 1994. Neurofuzzy Adaptive Modelling and Control. Prentice Hall.

Burton, T., Sharpe, D., Jenkins, N., Bossanyi, E., 2011. Wind Energy Handbook, 2nd ed. John Wiley & Sons, New York.

Calado, J.M.F., Korbicz, J., Patan, K., Patton, R.J., Sá da Costa, J.M.G., 2001. Soft computing approaches to fault diagnosis for dynamic systems. European Journal of Control 7 (2–3), 248–286.

Carpenter, G.A., Grossberg, S., 1987. A massively parallel architecture for a self-organizing neural pattern recognition machine. Computer Vision, Graphics, and Image Processing 37, 54–115.

Castaldi, P., Mimmo, N., Simani, S., 2014. Differential geometry based active fault tolerant control for aircraft. Control Engineering Practice 32, 227–235. http://dx.doi.org/10.1016/j.conengprac.2013.12.011. Invited Paper.

Chatzopoulos, A-P., Leithead, W.E., 2010. Reducing tower fatigue loads by a co-ordinated control of the supergen 2MW exemplar wind turbine. In: Proc. of the 3rd Torque 2010 Conference. Heraklion, Crete, Greece, pp. 667–674.

Chen, S., Billings, A.S., Cowan, C.F.N., Grant, P.M., 1990. Practical identification of NARMAX models using radial basis function. International Journal of Control 52, 1327–1350.

Chen, Wei, Ding, Steven X., Sari, A., Naik, Amol, Khan, Abdul Qayyum, Yin, Shen, 2011. Observer-based FDI schemes for wind turbine benchmark. In: Proceedings of IFAC World Congress, vol. 18, pp. 7073–7078.

Chen, Jie, Patton, R.J., 2000. Standard $H_\infty$ filter formulation of robust fault detection. In: Edelmayer, A.M. (Ed.), SAFEPROCESS 2000, 4th IFAC Symposium on Fault Detection, Supervision and Safety for Technical Processes, vol. 1. IFAC 2000, Budapest, Hungary, pp. 256–261.

Chen, J., Patton, R.J., 2001. Fault-tolerant control systems design using the linear matrix inequality method. In: European Control Conference. ECC'01, Porto, Portugal, pp. 1993–1998.

Chen, Jie, Patton, Ron J., 2012. Robust Model-Based Fault Diagnosis for Dynamic Systems, vol. 3. Springer Science & Business Media.

Chen, Z., Patton, R.J., Chen, J., 1997. Robust fault-tolerant system synthesis via LMI. In: Proc. of IFAC Symposium on Fault Detection, Supervision and Safety for Technical Processes, vol. 1. SAFEPROCESS'97, The University of Hull, UK, pp. 347–352.

Chen, J., Patton, R.J., Zhang, H.Y., 1993. A multi-criteria optimization approach to the design of robust fault detection algorithm. In: Proc. of Int. Conf. on Fault Diagnosis. TOOLDIAG'93, Toulouse, France.

Chen, J., Patton, R.J., Zhang, H.Y., 1996. Design of unknown input observer and robust fault detection filters. International Journal of Control 63 (1), 85–105.

Cheng, Chi-Tsong, 1984. Linear System Theory and Design. Holt-Saunders, Tokyo, Japan. ISBN 4-8337-0191-X.

Cheong, S.Y., Safonov, M.G., 2008. Bumpless transfer for adaptive switching controls. In: Proc. IFAC World Congress. Seoul, Korea, pp. 14415–14420.

Chiang, L.H., Russel, E.L., Braatz, R.D., 2001. Fault Detection Diagnosis in Industrial Systems. Advanced Textbooks Control and Signal Processing. Springer-Verlag London Limited, London, Great Britain.

Chow, E.Y., Willsky, A.S., 1980. Issue in the development of a general algorithm for reliable failure detection. In: Proc. of the 19th Conf. on Decision & Control. Albuquerque, NM.

Chow, E.Y., Willsky, A.S., 1984. Analytical redundancy and the design of robust detection systems. IEEE Transactions on Automatic Control 29 (7), 603–614.

Chui, N.L.C., Maciejowski, J.M., 1996. Realization of stable models with subspace methods. Automatica 32 (11), 1587–1595. http://dx.doi.org/10.1016/S0005-1098(96)00104-5.

Chung, Walter H., Speyer, Jason L., 1998. A game theoretic fault detection filter. IEEE Transactions on Automatic Control 43 (2), 143–161.

Clark, R.N., 1978. Instrument fault detection. IEEE Transactions on Aerospace and Electronic Systems AES-14 (3), 456–465.

Clark, R.N., 1989. Fault Diagnosis in Dynamic Systems: Theory and Application. Prentice Hall, pp. 21–45. Chap. 2.

Cutululis, N.A., Ceanga, E., Hansen, A.D., Sorensen, P., 2006. Robust multi-model control of an autonomous wind power system. Wind Energy 9 (5), 399–419. http://dx.doi.org/10.1002/we.194.

Daly, K.C., Gai, E., Harrison, J.V., 1979. Generalized likelihood test for FDI in redundancy sensor configurations. Journal of Guidance, Control, and Dynamics 2 (1), 9–17.

Darling, David J., 2008. The Encyclopedia of Alternative Energy and Sustainable Living. Wind Turbines. Worlds of David Darling.

Davis, L., 1991. Handbook of Genetic Algorithms. Van Nostrand Reinhold, New York.

De Persis, C., Isidori, A., 2000. On the observability codistributions of a nonlinear system. Systems & Control Letters 40 (5), 297–304.

De Persis, C., Isidori, A., 2001. A geometric approach to non-linear fault detection and isolation. IEEE Transactions on Automatic Control 45 (6), 853–865.

De Persis, C., Isidori, A., 2004a. Stabilizability by state feedback implies stabilizability by encoded state feedback. Systems & Control Letters 53 (3), 249–258. http://dx.doi.org/10.1016/j.sysconle.2004.05.003.

De Persis, C., Isidori, A., 2004b. On the design of fault detection filters with game-theoretic-optimal sensitivity. International Journal of Robust and Nonlinear Control 12 (8), 729–747.

Delmaire, G., Cassar, Ph., Staroswiecki, M., Christophe, C., 1999. Comparison of multivariable identification and parity space techniques for FDI purpose in M.I.M.O. systems. In: ECC'99. Karlsruhe, Germany.

Dexter, A.L., Benouarets, M., 1997. Model-based fault diagnosis using fuzzy matching. IEEE Transactions on Systems, Man and Cybernetics. Part A. Systems and Humans 27 (5), 673–682.

Diaz-Guerra, L., Adegas, F.D., Stoustrup, J., 2012. Adaptive control algorithm for improving power capture of wind turbines in turbulent winds. In: Proc. of the 2012 American Control Conference. IEEE, Montreal, Canada. IEEE, pp. 5807–5812.

Dietz, W.E., Kiech, E.L., Ali, M., 1989. Jet and rocket engine fault diagnosis in real time. Journal of Neural Network Computing 1, 5–18.

Ding, Steven X., 2008. Model-Based Fault Diagnosis Techniques: Design Schemes, Algorithms, and Tools, 1st ed. Springer, Berlin, Heidelberg. ISBN 978-3540763031.

Ding, X., Frank, P.M., 1990. Fault detection via factorization approach. Systems & Control Letters 14 (5), 431–436.

Ding, S.X., Jeinsch, T., Ding, E.L., Zhou, D., Wang, G., 1999. Application of observer-based FDI schemes to the three tank system. In: European Control Conference. ECC'99, Karlsruhe, Germany.

Ding, S.X., Jeinsch, T., Frank, P.M., Dind, E.L., 2000. A unified approach to the optimisation of fault detection systems. International Journal of Adaptive Control and Signal Processing 14 (7), 725–745.

Diversi, R., Guidorzi, R.P., 1998. Filtering-oriented identification of multivariable errors-in-variables models. In: Beghi, A., Finesso, L., Picci, G. (Eds.), Proc. of the MNST'98 Symposium. Il Poligrafo, Padova, Italy, pp. 775–778.

Dolan, Dale S.L., Lehn, Peter W., 2006. Simulation model of wind turbine 3p torque oscillations due to wind shear and tower shadow. In: Power Systems Conference and Exposition, 2006. PSCE'06, 2006 IEEE PES, pp. 2050–2057.

Doyle, J.C., et al., 1998. Essentials of Robust Control.

Duan, G.R., How, D., Patton, R.J., 2002. Robust fault detection in descriptor systems via generalised unknown input observers. International Journal of Systems Science 33 (5), 369–377.

Echavarria, E., van Bussel, G.J.W., Tomiyama, T., 2012. Finding functional redundancies in offshore wind turbine design. Wind Energy 15 (4), 609–626. http://dx.doi.org/10.1002/we.490.

Edelmayer, A., Bokor, J., Keviczky, L., 1997. A scaled $L_2$ optimisation approach for improving sensitivity of $H_\infty$ detection filters for LTV systems. In: Bányász, Cs. (Ed.), Preprints of the 2nd IFAC Symp. on Robust Control Design. RECOND 97, Budapest, Hungary, pp. 543–548.

Edwards, Christopher, 2004. A practical method for the design of sliding mode controllers using linear matrix inequalities. Automatica 40 (10), 1761–1769.

Edwards, C., Spurgeon, S., 1994. On the development of discontinuous observers. International Journal of Control 59 (1), 1211–1229.

Edwards, C., Spurgeon, S., 1998. Sliding Mode Control: Theory and Applications, 1st ed. Taylor & Francis, London. ISBN 978-0748406012.

Edwards, Christopher, Spurgeon, Sarah K., Patton, Ron J., 2000. Sliding mode observers for fault detection and isolation. Automatica 36 (1), 541–553.

Emami-Naeini, A.E., Akhter, M.M., Rock, M.M., 1988. Effect of model uncertainty on failure detection: the threshold selector. IEEE Transactions on Automatic Control 33 (12), 1105–1115.

Fantuzzi, C., Rovatti, R., 1996. On the approximation capabilities of the homogeneous Takagi–Sugeno model. In: Fuzzy Systems, 1996, Proceedings of the Fifth IEEE International Conference on, vol. 2. IEEE, pp. 1067–1072.

Fantuzzi, C., Simani, S., Beghelli, S., Rovatti, R., 2002. Identification of piecewise affine models in noisy environment. International Journal of Control 75 (18), 1472–1485.

Fantuzzi, C., Simani, S., Beghelli, S., Rovatti, R., 2003. Robust fault diagnosis of dynamic processes using parametric identification with eigenstructure assignment approach. IEEE Conference on Decision and Control 3 (1), 155–160. http://dx.doi.org/10.1109/CDC.2001.980090. Publisher: Taylor and Francis, Ltd.

Fischer, T., de Vries, W., Rainey, P., Schmidt, B., Argyriadis, K., Kühn, M., 2012. Offshore support structure optimization by means of integrated design and controls. Wind Energy Journal 15 (1), 99–117.

Frank, P.M., 1993. Advances in observer-based fault diagnosis. In: Proc. TOOLDIAG'93 Conference. CERT, Toulouse (F).

Frank, P.M., 1994. Enhancement of robustness on observer-based fault detection. International Journal of Control 59 (4), 955–983.

Frank, P.M., Ding, X., 1997. Survey of robust residual generation and evaluation methods in observer-based fault detection system. Journal of Process Control 7 (6), 403–424.

Frank, P.M., Ding, Steven X., Köpper-Seliger, Birgit, 2000. Current developments in the theory of FDI. In: SAFEPROCESS'00: Preprints of the IFAC Symposium on Fault Detection, Supervision and Safety for Technical Processes, vol. 1. Budapest, Hungary, pp. 16–27.

Franklin, G.F., Powell, J.D., Workman, M.L., 1998. Digital Control of Dynamic Systems. Addison Wesley Longman Control Engineering. 3rd ed. Addison Wesley, Menlo Park, CA, USA. ISBN 978-0-9791226-1-3.

Freire, N.M.A., Estima, J.O., Marques Cardoso, A.J., 2013. Open-circuit fault diagnosis in PMSG drives for wind turbine applications. IEEE Transactions on Industrial Electronics 60 (9), 3957–3967.

Friis, J., Nielsen, E., Bonding, J., Adegas, F.D., Stoustrup, J., Odgaard, P.F., 2011. Repetitive model predictive approach to individual pitch control of wind turbines. In: Proceedings of the IEEE CDC & ECC 2011. IEEE, Orlando, FL, USA. IEEE, pp. 3664–3670.

Füssel, D., Ballé, P., Isermann, R., 1997. Closed loop fault diagnosis based on a non-linear process model and automatic fuzzy rule generation. In: Proc. of IFAC Symposium on Fault Detection, Supervision and Safety for Technical Process. SAFEPROCESS'97, The University of Hull, UK.

Galdi, V., Piccolo, A., Siano, P., 2008. Designing an adaptive fuzzy controller for maximum wind energy extraction. IEEE Transactions on Energy Conversion 23 (2), 559–569.

Garcia-Sanz, M., Houpis, C.H., 2012. Wind Energy Systems: Control Engineering Design. CRC Press. ISBN 978-1439821794.

Gasch, Robert, Twele, Jochen, 2012. Wind Power Plants: Fundamentals, Design, Construction and Operation, 2nd ed. Springer. ISBN 978-3642229374.

Gertler, J., 1991. Generating directional residuals with dynamic parity equations. In: Proc. IFAC/IMACS Symp. SAFEPROCESS'91, Baden Baden (G).

Gertler, J., 1998. Fault Detection and Diagnosis in Engineering Systems. Marcel Dekker, New York.

Gertler, J., Monajemy, R., 1993. Generating directional residuals with dynamic parity equations. In: Proc. of the 12th IFAC World Congress 7. Sydney, pp. 505–510.

Gertler, J., Singer, D., 1990. A new structural framework for parity equation-based failure detection and isolation. Automatica 26 (2), 381–388.

Global Wind Energy Council, 2014. Wind Energy Statistics 2013. Report.

Gong, X., Qiao, W., 2013a. Bearing fault diagnosis for direct-drive wind turbines via current-demodulated signals. IEEE Transactions on Industrial Electronics 60 (8), 3419–3428. http://dx.doi.org/10.1109/TIE.2013.2238871.

Gong, X., Qiao, W., 2013b. Imbalance fault detection of direct-drive wind turbines using generator current signals. IEEE Transactions on Energy Conversion 27 (2), 468–476.

Graaff, Alexander Jakobus, Engelbrecht, Andries P., 2012. Clustering data in stationary environments with a local network neighborhood artificial immune system. International Journal of Machine Learning and Cybernetics 3 (1), 1–26.

Hadamard, J., 1964. La theorie des equations aus derivees partielles. Editions Scientifiques, Pekin.

Hagan, Martin T., Menhaj, Mohammad B., 1994. Training feedforward networks with the Marquardt algorithm. IEEE Transactions on Neural Networks 5 (6), 989–993.

Hameed, Z., Hong, Y.S., Cho, Y.M., Ahn, S.H., Song, C.K., 2009. Condition monitoring and fault detection of wind turbines and related algorithms: a review. Renewable & Sustainable Energy Reviews 13 (1), 1–39.

Hansen, M., 2011. Aeroelastic properties of backward swept blades. In: Proc. of the 49th AIAA Aerospace Sciences. AIAA, Orlando, Florida, USA. AIAA, pp. 1–19.

Haykin, Simon O., 2009. Neural Networks and Learning Machines, vol. 3. Pearson Education, Upper Saddle River.

Heier, S., 2014. Grid Integration of Wind Energy: Onshore and Offshore Conversion Systems. Engineering & Transportation. 3rd ed. John Wiley & Sons Ltd. ISBN 978-1119962946.

Himmelblau, D.M., 1978. Fault Diagnosis in Chemical and Petrochemical Processes. Elsevier, Amsterdam.

Himmelblau, D.M., Barker, R.W., Suewatanakul, W., 1991. Fault classification with the aid of artificial neural networks. In: IFAC/IMACS Symposium, vol. 2. SAFEPROCESS '91, Baden Baden, Germany, pp. 369–373.

Ho, W.K., Xu, W., 1998. PID tuning for unstable processes based on gain and phase-margin specifications. IEE Proceedings. Control Theory and Applications 145 (5), 392–396. http://dx.doi.org/10.1049/ip-cta:19982243.

Hoskins, J.C., Himmelblau, D.M., 1988. Artificial neural network models of knowledge representation in chemical engineering. Computers & Chemical Engineering 12, 881–890.

Hou, M., Patton, R.J., 1996. An LMI approach to $H_-/H_\infty$ fault detection observers. In: Control'96, UKACC International Conference on, Conf. Publ. No. 427, vol. 1. IET, pp. 305–310.

Hou, M., Patton, R.J., 1997. An $H_\infty/H_-$ approach to the design of robust fault diagnosis observers based upon LMI optimisation. In: Proceedings of the 4th European Control Conference. ECC'97, Brussels.

Intergovernmental Panel on Climate Change IPCC, 2011. Special Report on Renewable Energy Sources and Climate Change Mitigation. Report.

Ioannou, Petros A., Sun, Jing, 2012. Robust Adaptive Control. Courier Corporation.

Isermann, R., 1984. Process fault detection based on modeling and estimation methods: a survey. Automatica 20 (4), 387–404.

Isermann, R., 1992. Estimation of physical parameters for dynamic processes with application to an industrial robot. International Journal of Control 55, 1287–1298.

Isermann, R., 1994. Integration of fault detection and diagnosis methods. In: Proc. IFAC SAFEPROCESS Symposium '94. Espoo, Finland.

Isermann, R., 1997a. Supervision, fault detection and fault diagnosis methods: an introduction. Control Engineering Practice 5 (5), 639–652.

Isermann, Rolf, 1997b. Supervision, fault-detection and fault-diagnosis methods an introduction. Control Engineering Practice 5 (5), 639–652.

Isermann, R., 1998. On fuzzy logic applications for automatic control, supervision and fault diagnosis. IEEE Transactions on Systems, Man and Cybernetics. Part A. Systems and Humans 28 (2), 221–235.

Isermann, Rolf, 2005. Fault-Diagnosis Systems: An Introduction from Fault Detection to Fault Tolerance, 1st ed. Springer-Verlag, Weinheim, Germany. ISBN 3540241124.

Isermann, R., Ballé, P., 1997. Trends in the application of model-based fault detection and diagnosis of technical processes. Control Engineering Practice 5 (5), 709–719.

Isermann, R., Ballé, P., 1998. Process fault detection based on modelling and estimation methods – a survey. Automatica 20 (4), 709–719.

Jain, Anil K., Murty, M. Narasimha, Flynn, Patrick J., 1999. Data clustering: a review. ACM Computing Surveys (CSUR) 31 (3), 264–323.

Jang, J.S.R., 1993. ANFIS: adaptive-network-based fuzzy inference system. IEEE Transactions on Systems, Man and Cybernetics 23 (3), 665–684.

Jang, J.S.R., 1994. Structure determination in fuzzy modelling: a fuzzy CART approach. In: Proc. of IEEE International Conf. on Fuzzy Systems.

Jang, J.S.R., Sur, R., 1995. Neuro-fuzzy modeling and control. Proceedings of the IEEE 83 (3), 378–405.

Jazwinski, A.H., 1970. Stochastic Processes and Filtering Theory. Academic Press, New York.

Jensen, Niels Otto, 1983. A Note on Wind Generator Interaction.

Johnson, K.E., Fleming, P.A., 2011. Development, implementation, and testing of fault detection strategies on the National Wind Technology Center's controls advanced research turbines. Mechatronics 21 (4), 728–736. http://dx.doi.org/10.1016/j.mechatronics.2010.11.010.

Johnson, K.E., Pao, L.Y., Balas, M.J., Fingersh, L.J., 2006a. Control of variable-speed wind turbines: standard and adaptive techniques for maximizing energy capture. IEEE Control Systems Magazine 26 (3), 70–81. http://dx.doi.org/10.1109/MCS.2006.1636311.

Johnson, K.E., Pao, L.Y., Balas, M.J., Fingersh, L.J., 2006b. Methods for increasing region 2 power capture on a variable speed HAWT. ASME Journal of Solar Energy Engineering 126 (4), 1092–1100.

Johnson, K.E., Pao, L.Y., Balas, M.J., Kulkarni, V., Fingersh, L.J., 2004. Stability analysis of an adaptive torque controller for variable speed wind turbines. In: Proceedings of the 43rd IEEE Conference on Decision and Control, vol. 4. CDC'04, Paradise Island, Bahamas, pp. 4087–4094.

Jonkman, J.M., Buhl Jr., M.L., 2005. FAST User's Guide. Technical Report NREL/EL-500-38230. National Renewable Energy Laboratory, Golden, CO, USA.

Jonkman, J., Butterfield, S., Musial, W., Scott, G., 2009. Definition of a 5-MW Reference Wind Turbine for Offshore System Development. Technical Report NREL/TP-500-38060. National Renewable Energy Laboratory, Golden, CO, USA.

Jun, M., Safonov, M.G., 1999. Automatic PID tuning: an application of unfalsified control. In: Proc. IEEE Int. Symp. Computer Aided Control Syst. Design, pp. 328–333.

Jun, Wu, Shitong, Wang, Chung, Fu-lai, 2011. Positive and negative fuzzy rule system, extreme learning machine and image classification. International Journal of Machine Learning and Cybernetics 2 (4), 261–271.

Kabore, Pousga, Othman, Sami, McKenna, T.F., Hammouri, Hassan, 2000. Observer-based fault diagnosis for a class of non-linear systems application to a free radical copolymerization reaction. International Journal of Control 73 (9), 787–803.

Kaboré, Pousga, Wang, Hong, 2001. Design of fault diagnosis filters and fault-tolerant control for a class of nonlinear systems. IEEE Transactions on Automatic Control 46 (11), 1805–1810.

Kalman, R.E., 1982a. Identification from real data. In: Hazewinkel, M., Rinnoy Kan, A.H.G. (Eds.), Current Developments in the Interface: Economics, Econometrics, Mathematics. D. Reidel, Dordrecht, The Netherlands, pp. 161–196.

Kalman, R.E., 1982b. System Identification from Noisy Data. In: Bednarek, A.R., Cesari, L. (Eds.), Dynamical System II. Academic Press, New York, pp. 135–164.

Kalman, R.E., 1984. Identification of noisy systems. In: 50th Anniversary Symp. Steklov Institute of Mathematics. U.S.S.R. Academy of Sciences, Moscow.

Kalman, R.E., 1990. Nine Lectures on Identification. Lecture Notes in Economics and Mathematical Systems. Springer-Verlag, Berlin.

Khan, B., Valencia-Palomo, G., Rossiter, J.A., Jones, C., Gondhalekar, R., 2016. Long horizon input parameterisations to enlarge the region of attraction of MPC. Optimal Control Applications & Methods 37 (1), 139–153. http://dx.doi.org/10.1002/oca.2158.

Knudsen, T., Bak, T., Soltani, M., 2011. Prediction models for wind speed at turbine locations in a wind farm. Wind Energy 14 (7), 877–894. http://dx.doi.org/10.1002/we.491.

Korbicz, J., Koscielny, J.M., Kowalczuk, Z., Cholewa, W. (Eds.), 2004. Fault Diagnosis: Models, Artificial Intelligence, Applications, 1st ed. Springer-Verlag, London, UK. ISBN 3540407677.

Korbicz, J., Patan, K., Obuchowicz, A., 1999. Dynamic neural network for process modelling in fault detection and isolation systems. Applied Mathematics and Computer Science 9 (2), 519–546. Technical University of Zielona Gora, Poland.

Kuo, Benjamin C., 1995. Automatic Control Systems, seventh ed. Prentice Hall, Englewood Cliffs, New Jersey, 07632.

Kusiak, A., Li, W., 2011. The prediction and diagnosis of wind turbine faults. Renewable Energy 36 (1), 16–23. http://dx.doi.org/10.1016/j.renene.2010.05.014.

Laino, D.J., Hansen, A.C., 2002. User's Guide to the Wind Turbine Aerodynamics Computer Software AeroDyn. Technical Report TCX-9-29209-01. Windward Engineering, LC, Salt Lake City, UT, USA. Prepared for the National Renewable Energy Laboratory – NREL.

Laks, J.H., Pao, L.Y., Wright, A.D., 2009. Control of wind turbines: past, present, and future. In: Proceedings of the American Control Conference, 2009. ACC'09. ISSN 0743-1619. IEEE, St. Louis, MO, USA. ISBN 978-1-4244-4523-3, pp. 2096–2103.

Landau, L.D., Lifshitz, E.M., 1976. Mechanics, vol. 1 of Course of Theoretical Physics S, 3rd ed. Butterworth–Heinemann. ISBN 978-0750628969.

Laouti, Nassim, Sheibat-Othman, Nida, Othman, Sami, 2011. Support vector machines for fault detection in wind turbines. In: Proceedings of IFAC World Congress, vol. 2, pp. 7067–7072.

Leithead, W., Connor, B., 2000. Control of variable speed wind turbines: design task. International Journal of Control 73 (13), 1189–1212. http://dx.doi.org/10.1080/002071700417849.

Leontaritis, I.J., Billings, S.A., 1985a. Input–output parametric models for non-linear systems, part II: stochastic non-linear systems. International Journal of Control 41 (2), 329–344.

Leontaritis, I.J., Billings, S.A., 1985b. Input–output parametric models for non-linear systems, part I: deterministic non-linear systems. International Journal of Control 41 (2), 303–328.

Lewis, F.L., 1986. Optimal Control. John Wiley and Sons.

Linden, J.G., Larkowski, T., Burnham, K.J., 2012. Algorithms for recursive/semi-recursive biascompensating least squares system identification within the errors-in-variables framework. International Journal of Control 85 (11), 1625–1643. http://dx.doi.org/10.1080/00207179.2012.696145.

Lio, W.H., Jones, B.Ll., Lu, Q., Rossiter, J.A., 2015. Fundamental performance similarities between individual pitch control strategies for wind turbines. International Journal of Control 90 (1), 37–52. http://dx.doi.org/10.1080/00207179.2015.1078912.

Liu, Guo Ping, 2012. Nonlinear Identification and Control: A Neural Network Approach. Springer Science & Business Media.

Liu, G.P., Patton, R.J., 1998. Eigenstructure Assignment for Control System Design. John Wiley & Sons, England.

Ljung, L., 1999. System Identification: Theory for the User, second ed. Prentice Hall, Englewood Cliffs, N.J.

Lou, X., Willsky, A.S., Verghese, G.C., 1986. Optimal robust redundancy relations for failure detection in uncertainty systems. Automatica 22 (3), 333–344.

Mahmoud, M., Jiang, J., Zhang, Y., 2003. Active Fault Tolerant Control Systems: Stochastic Analysis and Synthesis. Lecture Notes in Control and Information Sciences. Springer-Verlag, Berlin, Germany. ISBN 3540003185.

Mahmoud, M., Jiang, J., Zhang, Y., 2004. Stochastic stability analysis of fault-tolerant control systems in the presence of noise. IEEE Transactions on Automatic Control 46 (11). IEEE Control Systems Society.

Mamdani, E., 1976. Advances in the linguistic synthesis of fuzzy controllers. International Journal of Man-Machine Studies 8, 669–678.

Mamdani, E., Assilian, S., 1995. An experiment in linguistic synthesis with fuzzy logic controller. International Journal of Man-Machine Studies 7 (1), 1–13.

Mangoubi, R., Appleby, B.D., Farrell, J.R., 1992. Robust estimation in fault detection. In: Proc. of the 31st Conf. on Decision & Control. Tucson, AZ, USA, pp. 2317–2322.

Manwell, J.F., McGowan, J.G., Rogers, A.L., 2002. Wind Energy Explained: Theory, Design, and Application. Wiley, West Sussex, England.

Marquardt, Donald W., 1963. An algorithm for least-squares estimation of nonlinear parameters. Journal of the Society for Industrial and Applied Mathematics 11 (2), 431–441.

Massoumnia, M.A., 1986. A Geometric Approach to Failure Detection and Identification in Linear Systems. PhD thesis. Massachusetts Institute of Technology, Massachusetts, USA.

Massoumnia, M., Verghese, G.C., Willsky, A.S., 1989. Failure detection and identification. IEEE Transactions on Automatic Control 34, 316–321.

Mattone, R., De Luca, A., 2006. Nonlinear fault detection and isolation in a three-tank heating system. IEEE Transactions on Control Systems Technology 14 (6), 1158–1166.

McDuff, R.J., Simpson, P.K., 1990. An adaptive resonance diagnostic system. Journal of Neural Network Computing 2, 19–29.

Medsker, Larry, Jain, Lakhmi C., 1999. Recurrent Neural Networks: Design and Applications. CRC Press.

Meneganti, M., Saviello, F.S., Tagliaferri, R., 1998. Fuzzy neural networks for classification and detection of anomalies. IEEE Transactions on Neural Networks 9 (5), 848–861.

Morozov, V.A., 1984. Methods for Solving Incorrectly Posed Problems. Springer, Berlin.

Muljadi, E., Butterfield, C., 1999. Pitch-controlled variable-speed wind turbine generation. In: 1999 IEEE Industry Applications Society Annual Meeting. Phoenix, Arizona, USA, pp. 470–474.

Munteanu, I., Bratcu, A.I., 2008. Optimal Control of Wind Energy Systems: Towards a Global Approach. Advances in Industrial Control. Springer. ISBN 978-1848000797.

Namik, H., Stol, K., 2010. Individual blade pitch control of floating offshore wind turbines. Wind Energy 13 (1), 74–85. http://dx.doi.org/10.1002/we.332.

Nelles, O., 2001. Nonlinear System Identification. Springer-Verlag, Berlin, Heidelberg, Germany.

Nelles, O., Isermann, R., 1996. Basis function networks for interpolation of local linear models. In: Proc. of the 35th IEEE Conference on Decision and Control, vol. 4. Kobe, Japan, pp. 470–475.

Niemann, H., Stoustrup, J., 1996. Filter design for failure detection and isolation in the presence of modelling errors and disturbances. In: Proc. of the 35th IEEE Conf. on Decision and Contr. Kobe, Japan, pp. 1155–1160.

Niemann, H., Stoustrup, J., 2005. An architecture for fault tolerant controllers. International Journal of Control 78 (14), 1091–1110.

Ningsu, Luo, Vidal, Y., Acho, L., 2014. Wind Turbine Control and Monitoring. Advances in Industrial Control. Springer. ISBN 978-3319084121.

Norgaard, M., Poulsen, N.K., Ravn, O., 2000. New developments in state estimation for nonlinear systems. Automatica 36, 1627–1638.

Odgaard, Peter Fogh, 2012. FDI/FTC wind turbine benchmark modelling. In: Patton, R.J. (Ed.), Workshop on Sustainable Control of Offshore Wind Turbines, vol. 1. Centre for Adaptive Science & Sustainability, University of Hull, Hull, UK.

Odgaard, P.F., Damgaard, C., Nielsen, R., 2008. On-line estimation of wind turbine power coefficients using unknown input observers. In: Proceedings of the 17th IFAC World Congress. The International Federation of Automatic Control. IFAC, Seoul, Korea, pp. 10646–10651.

Odgaard, Peter Fogh, Johnson, Kathryn, 2013. Wind turbine fault diagnosis and fault tolerant control – an enhanced benchmark challenge. In: Proc. of the 2013 American Control Conference – ACC. ISSN 0743-1619. IEEE Control Systems Society & American Automatic Control Council, Washington DC, USA. ISBN 978-1-4799-0177-7, pp. 4447–4452.

Odgaard, P.F., Patton, R.J., 2012. FDI/FTC wind turbine benchmark modelling. In: Patton, R.J. (Ed.), Workshop on Sustainable Control of Offshore Wind Turbines.

Odgaard, P.F., Shafiei, S.E., 2015a. Evaluation of wind farm controller based fault detection and isolation. In: Proceedings of the IFAC SAFEPROCESS Symposium 2015, vol. 48. IFAC, Paris, France, Elsevier ed., pp. 1084–1089.

Odgaard, Peter Fogh, Shafiei, Seyed Eshan, 2015b. Evaluation of wind farm controller based fault detection and isolation. IFAC-PapersOnLine 48 (21), 1084–1089.

Odgaard, P.F., Stoustrup, J., 2011. Orthogonal bases used for feed forward control of wind turbines. In: Proc. of the 18th IFAC World Congress. IFAC, Milan, Italy. IFAC, pp. 532–537.

Odgaard, P.F., Stoustrup, J., 2012a. Fault tolerant control of wind turbines using unknown input observers. In: Verde, C., Astorga Zaragoza, C.M., Molina, A. (Eds.), Proceedings of the 8th IFAC Symposium on Fault Detection, Supervision and Safety of Technical Processes, vol. 8. SAFEPROCESS 2012. National Autonomous University of Mexico, Mexico City, Mexico, pp. 313–319.

Odgaard, P.F., Stoustrup, J., 2012b. Karhunen Loeve basis used for detection of gearbox faults in a wind turbine. In: IFAC Proceedings, pp. 8891–8896.

Odgaard, Peter Fogh, Stoustrup, Jakob, 2012c. Results of a wind turbine FDI competition. In: 8th IFAC Symposium on Fault Detection, Supervision and Safety of Technical Processes, pp. 102–107.

Odgaard, Peter Fogh, Stoustrup, Jakob, 2012d. Results of a wind turbine FDI competition. In: Verde, C., Astorga Zaragoza, C.M., Molina, A. (Eds.), Proceedings of the 8th IFAC Symposium on Fault Detection, Supervision and Safety of Technical Processes, vol. 8. SAFEPROCESS 2012. National Autonomous University of Mexico, Mexico City, Mexico, pp. 102–107.

Odgaard, P.F., Stoustrup, J., 2013a. Fault tolerant wind farm control. A benchmark model. In: Control Applications (CCA), 2013 IEEE International Conference on. IEEE, pp. 412–417.

Odgaard, P.F., Stoustrup, J., 2013b. Fault tolerant wind farm control – a benchmark model. In: Proceedings of the IEEE Multiconference on Systems and Control. MSC 2013, Hyderabad, India, pp. 1–6.

Odgaard, Peter Fogh, Stoustrup, Jakob, 2015a. A benchmark evaluation of fault tolerant wind turbine control concepts. IEEE Transactions on Control Systems Technology 23 (3), 1221–1228.

Odgaard, Peter Fogh, Stoustrup, Jakob, 2015b. A benchmark evaluation of fault tolerant wind turbine control concepts. IEEE Transactions on Control Systems Technology 23 (3), 1221–1228.

Odgaard, Peter Fogh, Stoustrup, Jakob, Kinnaert, Michel, 2009a. Fault tolerant control of wind turbines – a benchmark model. In: Proceedings of the 7th IFAC Symposium on Fault Detection, Supervision and Safety of Technical Processes, vol. 1. Barcelona, Spain, pp. 155–160.

Odgaard, P.F., Stoustrup, J., Kinnaert, M., 2013. Fault-tolerant control of wind turbines: a benchmark model. IEEE Transactions on Control Systems Technology (ISSN 1063-6536) 21 (4), 1168–1182. http://dx.doi.org/10.1109/TCST.2013.2259235.

Odgaard, P.F., Stoustrup, J., Kinnaert, M., 2014. Frequency based fault detection in wind turbines. IFAC World Congress 19 (1), 5832–5837.

Odgaard, P.F., Stoustrup, J., Nielsen, R., Damgaard, C., 2009b. Observer based detection of sensor faults in wind turbines. In: Proceedings of European Wind Energy Conference – EWEA 2009. EWEA, Marseille, France, pp. 1–10.

Ostergaard, K.Z., Stoustrup, J., Brath, P., 2009. Linear parameter varying control of wind turbines covering both partial load and full load conditions. International Journal of Robust and Nonlinear Control 19 (1), 92–116. http://dx.doi.org/10.1002/rnc.1340.

Ozdemir, Ahmet Arda, Seiler, Peter, Balas, Gary J., 2011. Wind turbine fault detection using counter-based residual thresholding. In: Proceedings of IFAC World Congress, vol. 18, pp. 8289–8294.

Pao, L.Y., Johnson, K.E., 2009. A tutorial on the dynamics and control of wind turbines and wind farms. In: Proceedings of the American Control Conference, 2009. ACC'09, St. Louis, MO, USA. ISSN 0743-1619. IEEE. ISBN 978-1-4244-4523-3, pp. 2076–2089.

Pao, L.Y., Johnson, K.E., 2011. Control of wind turbines. IEEE Control Systems Magazine 31 (2), 44–62.

Parker, Max A., Ng, Chong, Ran, Li, 2011. Fault-tolerant control for a modular generator–converter scheme for direct-drive wind turbines. IEEE Transactions on Industrial Electronics 58 (1), 305–315.

Patton, Ron J., 2015. Fault-tolerant control. In: Encyclopedia of Systems and Control, pp. 422–428.

Patton, R.J., Chen, J., 1991. A review of parity space approaches to fault diagnosis. In: IFAC Symposium SAFEPROCESS '91. Baden-Baden.

Patton, R.J., Chen, J., 1993. Optimal selection of unknown input distribution matrix in the design of robust observers for fault diagnosis. Automatica 29 (4), 837–841.

Patton, R.J., Chen, J., 1994. A review of parity space approaches to fault diagnosis for aerospace systems. AIAA Journal of Guidance, Control, and Dynamics 17 (2), 278–285.

Patton, R.J., Chen, J., 1997. Observer-based fault detection and isolation: robustness and applications. Control Engineering Practice 5 (5), 671–682.

Patton, R.J., Chen, J., 2000. On eigenstructure assignment for robust fault diagnosis. International Journal of Robust and Nonlinear Control 10 (14), 1193–1208.

Patton, R.J., Chen, J., 2009. Design methods for robust fault diagnosis. Control Systems, Robotics & Automation 16 (1), 84–111.

Patton, R.J., Frank, P.M., Clark, R.N. (Eds.), 1989a. Fault Diagnosis in Dynamic Systems, Theory and Application. Control Engineering Series. Prentice Hall, London.

Patton, R.J., Frank, P.M., Clark, R.N. (Eds.), 2000. Issues of Fault Diagnosis for Dynamic Systems. Springer-Verlag, London Limited.

Patton, Ron J., Frank, Paul M., Clarke, Robert N., 1989b. Fault Diagnosis in Dynamic Systems: Theory and Application. Prentice Hall, Inc.

Patton, R.J., Hou, M., 1997. $H_\infty$ estimation and robust fault detection. In: Proc. of the 1997 European Control Conference. ECC'97, Brussels, Belgium (CD-ROM).

Patton, R.J., Lopez-Toribio, C.J., Uppal, F.I., 1999. Artificial intelligence approaches to fault diagnosis. Applied Mathematics and Computer Science 9 (3), 471–518.

Patton, R.J., Willcox, S.W., Winter, J.S., 1986. A parameter insensitive technique for aircraft sensor fault diagnosis using eigenstructure assignment and analytical redundancy. In: Proc. of the AIAA Conference on Guidance, Navigation & Control, No. 86-2029-CP. Williamsburg, VA.

Pau, L.F., 1981. Failure Diagnosis and Performance Control. Marcel Dekker, New York.

Pedersen, M.D., Fossen, T.I., 2012. Efficient nonlinear wind-turbine modeling for control applications. In: 7th Vienna International Conference on Mathematical Modelling (MATHMOD 2012), vol. 7. IFAC, Vienna, Austria, pp. 264–269.

Pintea, A., Popescu, D., Borne, P., 2010. In: Proc. of the 12th IFAC Symposium on Large Scale Systems: Theory and Applications, vol. 9. IFAC, Lille, France. IFAC, pp. 251–256.

Plumley, C.E., Leithead, B., Jamieson, P., Bossanyi, E., Graham, M., 2014. Comparison of individual pitch and smart rotor control strategies for load reduction. Journal of Physics. Conference Series 524, 012054. http://dx.doi.org/10.1088/1742-6596/524/1/012054.

Potter, I.E., Suman, M.C., 1977. Thresholdless Redundancy Management with Array of Skewed Instruments. Technical Report AGARDOGRAPH-224. AGARD, Integrity in Electronic Flight Control Systems.

Pramod, Jain, 2010. Wind Energy Engineering. McGraw-Hill. ISBN 978-0071714778.

Puig, V., 2010. Fault diagnosis and fault tolerant control using set-membership approaches: application to real case studies. International Journal of Applied Mathematics and Computer Science 20 (4), 619–635. http://dx.doi.org/10.2478/v10006-010-0046-y.

Rasmussen, F., Hansen, M., Thomsen, K., Larsen, T., Bertagnolio, F., Johansen, J., Madsen, H.A., Bak, C., Hansen, A., 2003. Present status of aeroelasticity of wind turbines. Wind Energy 6 (3), 213–228. http://dx.doi.org/10.1002/we.98.

Ray, A., Luck, R., 1991. An introduction to sensor signal validation in redundant measurement systems. IEEE Control Systems Magazine 11 (2), 44–49.

Rivkin, D., Anderson, L.D., Silk, L., 2012. Wind Turbine Control Systems, 1st ed. Jones & Bartlett Learning. ISBN 978-1449624538.

Rivkin, D., Silk, L., 2012. Wind Turbine Operations, Maintenance, Diagnosis, and Repair, 1st ed. Jones & Bartlett Learning. ISBN 978-1449624552.

Rodrigues, M., Theilliol, D., Aberkane, S., Sauter, D., 2007. Fault tolerant control design for polytopic LPV systems. International Journal of Applied Mathematics and Computer Science (ISSN 1641-876X) 17 (1), 27–37. http://dx.doi.org/10.2478/v10006-007-0004-5.

Rotondo, Damiano, Nejjari, Fatiha, Puig, Vicenc, Blesa, Joaquim, 2012. Fault tolerant control of the wind turbine benchmark using virtual sensors/actuators. In: Verde, C., Astorga Zaragoza, C.M., Molina, A. (Eds.), Proceedings of the 8th IFAC Symposium on Fault Detection, Supervision and Safety of Technical Processes, vol. 8. SAFEPROCESS 2012. National Autonomous University of Mexico, Mexico City, Mexico, pp. 114–119.

Rovatti, R., 1996. Takagi–Sugeno models as approximators in Sobolev norms: the SISO case. In: Fuzzy Systems, 1996, Proceedings of the Fifth IEEE International Conference on, vol. 2. IEEE, pp. 1060–1066.

Rovatti, R., Fantuzzi, C., Simani, S., 2000. High-speed DSP-based implementation of piecewise-affine and piecewise-quadratic fuzzy systems. In: Special Issue on Fuzzy Logic applied to Signal Processing. Signal Processing Journal 80 (6), 951–963. http://dx.doi.org/10.1016/S0165-1684(00)00013-X. Publisher: Elsevier.

Rovatti, R., Fantuzzi, C., Simani, S., Beghelli, S., 1998. Parameter identification for piecewise linear model with weakly varying noise. In: 1998 IEEE Conference on Decision and Control, vol. 4. CDC'98, Tampa, Florida, pp. 4488–4489.

Sadrnia, M.A., Chen, J., Patton, R.J., 1997. Robust $H_\infty/\mu$ observer-based residual generation for fault diagnosis. In: Proc. of the IFAC Symp. on Fault Detection, Supervision and Safety for Technical Processes. SAFEPROCESS'97, 1998 Pergamon ed. Univ. of Hull, UK, pp. 155–162.

Sakamoto, R., Senjyu, T., Kinjo, T., Urasaki, N., Funabashi, T., 2004. Output power leveling of wind turbine generator by pitch angle control using adaptive control method. In: Proceedings of 2004 International Conference on Power System Technology. POWERCON, pp. 834–839.

Sami, Montadher, Patton, Ron J., 2012. An FTC approach to wind turbine power maximisation via T–S fuzzy modelling and control. In: Verde, C., Astorga Zaragoza, C.M., Molina, A. (Eds.), Proceedings of the 8th IFAC Symposium on Fault Detection, Supervision and Safety of Technical Processes, vol. 8. SAFEPROCESS 2012. National Autonomous University of Mexico, Mexico City, Mexico, pp. 349–354.

Sauter, D., Rambeaux, F., Hamelin, F., 1997. Robust fault diagnosis in a $H_\infty$ setting. In: Proc. of the IFAC Symp. on Fault Detection, Supervision and Safety for Technical Processes. SAFEPROCESS'97, 1998 Pergamon ed. Univ. of Hull, UK, pp. 867–874.

Schilling, R.J., Carroll, J.J. Jr., Al-Ajlouni, A.F., 2001. Approximation of non-linear systems with radial basis function neural networks. IEEE Transactions on Neural Networks 12 (1), 1–15.

Shi, F., Patton, R.J., 2015. An active fault tolerant control approach to an offshore wind turbine model. Renewable Energy 75 (1), 788–798. http://dx.doi.org/10.1016/j.renene.2014.10.061.

Simani, Silvio, 2005. Identification and fault diagnosis of a simulated model of an industrial gas turbine. IEEE Transactions on Industrial Informatics 1 (3), 202–216. http://dx.doi.org/10.1109/TII.2005.844425.

Simani, S., 2012a. Application of a data-driven fuzzy control design to a wind turbine benchmark model. Advances in Fuzzy Systems (ISSN 1687-7101) 2012, 1–12. http://dx.doi.org/10.1155/2012/504368. e-ISSN 1687-711X. http://www.hindawi.com/journals/afs/2012/504368/. Invited Paper for the Special Issue: Fuzzy Logic Applications in Control Theory and Systems Biology (FLACE).

Simani, S., 2012b. Data-driven design of a PI fuzzy controller for a wind turbine simulated model. In: Vilanova, R., Visioli, A. (Eds.), Proceedings of the IFAC Conference on Advances in PID Control, vol. 2. PID'12, DII, Faculty of Engineering, University of Brescia, Italy. IFAC, University of Brescia, Brescia, Italy. ISBN 978-3-902823-18-2, pp. 667–672. Invited Paper.

Simani, S., Castaldi, P., 2011. Estimation of the power coefficient map for a wind turbine system. In: Proceedings of the 9th European Workshop on Advanced Control and Diagnosis, No. 13. ACD'11, pp. 1–7.

Simani, S., Castaldi, P., 2013. Data-driven and adaptive control applications to a wind turbine benchmark model. Control Engineering Practice (ISSN 0967-0661) 21 (12), 1678–1693. http://dx.doi.org/10.1016/j.conengprac.2013.08.009. Special Issue Invited Paper. PII: S0967-0661(13)00155-X.

Simani, Silvio, Castaldi, Paolo, 2014. Active actuator fault tolerant control of a wind turbine benchmark model. International Journal of Robust and Nonlinear Control 24 (8–9), 1283–1303. http://dx.doi.org/10.1002/rnc.2993. John Wiley.

Simani, S., Fantuzzi, C., 2002. Neural networks for fault diagnosis and identification of industrial processes. In: Proc. of the 10th European Symposium on Artificial Neural Networks, vol. 1. ESANN'02, Bruges, Belgium, ESANN ed. ISBN 2-930307-02-1, pp. 489–494. Invited Paper.

Simani, S., Fantuzzi, C., Patton, R.J., 2003a. Model-Based Fault Diagnosis in Dynamic Systems Using Identification Techniques, first ed. Advances in Industrial Control. Springer-Verlag, London, UK. ISBN 1852336854.

Simani, Silvio, Fantuzzi, Cesare, Patton, Ronald Jon, 2003b. Model-Based Fault Diagnosis Techniques. Springer.

Simani, S., Fantuzzi, C., Rovatti, R., Beghelli, S., 1999a. Non-linear algebraic system identification via piecewise affine models in stochastic environment. In: IEEE Conference on Decision and Control, vol. 1. CDC'99, Phoenix, AZ, USA, IEEE CSS ed., pp. 1083–1088.

Simani, S., Fantuzzi, C., Rovatti, R., Beghelli, S., 1999b. Parameter identification for piecewise linear fuzzy models in noisy environment. International Journal of Approximate Reasoning 1 (22), 149–167. Publisher: Elsevier.

Simani, Silvio, Fantuzzi, Cesare, Rovatti, Riccardo, Beghelli, Sergio, 1999c. Parameter identification for piecewise-affine fuzzy models in noisy environment. International Journal of Approximate Reasoning 22 (1), 149–167.

Simani, S., Fantuzzi, C., Spina, P.R., 1998. Application of a neural network in gas turbine control sensor fault detection. In: 1998 IEEE Conference on Control Applications, vol. 1. CCA'98, Trieste, Italy, pp. 182–186.

Simani, S., Farsoni, S., Castaldi, P., 2014. Residual generator fuzzy identification for wind farm fault diagnosis. In: Proceedings of the 19th World Congress of the International Federation of Automatic Control, vol. 19. IFAC'14, IFAC & South Africa Council for Automation and Control, IFAC, Cape Town, South Africa, pp. 4310–4315. Invited Paper for the Special Session "FDI and FTC of Wind Turbines in Wind Farms" organised by P.F. Odgaard and S. Simani.

Simani, S., Marangon, F., Fantuzzi, C., 1999d. Fault diagnosis in a power plant using artificial neural networks: analysis and comparison. In: European Control Conference. ECC'99, Karlsruhe, Germany, pp. 1–6.

Simani, S., Patton, R.J., 2002. Neural networks for fault diagnosis of industrial plants at different working points. In: Proc. of the 10th European Symposium on Artificial Neural Networks, vol. 1. ESANN'02, Bruges, Belgium, ESANN ed. ISBN 2-930307-02-1, pp. 495–500. Invited Paper.

Skogestad, S., 2003. Simple analytic rules for model reduction and PID controller tuning. Journal of Process Control 13 (4), 291–309. http://dx.doi.org/10.1016/S0959-1524(02)00062-8.

Sloth, C., Esbensen, T., Stoustrup, J., 2010. Active and passive fault-tolerant LPV control of wind turbines. In: Proceedings of the 2010 American Control Conference. ACC'10, Baltimore, Maryland, USA, pp. 4640–4646.

Sloth, C., Esbensen, T., Stoustrup, J., 2011. Robust and fault-tolerant linear parameter-varying control of wind turbines. Mechatronics 21, 645–659. http://dx.doi.org/10.1016/j.mechatronics.2011.02.001.

Sneider, H., Frank, P.M., 1996. Observer-based supervision and fault detection in robots using nonlinear and fuzzy logic residual evaluation. IEEE Transactions on Control Systems Technology 4 (3), 274–282.

Söderström, T., Stoica, P., 1987. System Identification. Prentice Hall, Englewood Cliffs, N.J.

Special Issue on Hybrid Control Systems, 1998. IEEE Transactions on Automatic Control 43 (4).

Speyer, Jason L., 1999. Residual sensitive fault detection filters. In: MED'99. Haifa, Israel, pp. 835–851.

Stamatis, D.H., 2003a. Failure Mode and Effect Analysis: FMEA from Theory to Execution, 2nd ed. ASQ Quality Press, Milwaukee, WI, USA. ISBN 0873895983.

Stamatis, D.H., 2003b. Guidelines for Failure Mode and Effects Analysis (FMEA). CRC Press.

Stoustrup, J., Grimble, M.J., Niemann, H., 1997. Design of integrated systems for the control and detection of actuator/sensor faults. Sensor Review 17 (2), 138–149.

Stoustrup, I., Niemann, H., 1998. Fault detection for nonlinear systems – a standard problem approach. In: Proc. of the 37th IEEE Conf. on Decision & Control. Tampa, Florida, USA, pp. 96–101.

Sugeno, M., Kang, G.T., 1988. Structure identification of fuzzy model. Fuzzy Sets and Systems 28, 15–33.

Svärd, Carl, Nyberg, Mattias, Stoustrup, Jakob, 2011. Automated design of an FDI system for the wind turbine benchmark. JCSE-Journal of Control Science and Engineering 2012, 19.

Tachibana, K., Furuhashi, T., 1994. A hierarchical fuzzy modelling method using genetic algorithm for identification of concise submodels. In: Proc. of 2nd Int. Conference on Knowledge-Based Intelligent Electronic Systems. Adelaide, Australia.

Takagi, T., Sugeno, M., 1985a. Fuzzy identification of systems and its application to modeling and control. IEEE Transactions on Systems, Man and Cybernetics SMC-15 (1), 116–132.

Takagi, Tomohiro, Sugeno, Michio, 1985b. Fuzzy identification of systems and its applications to modeling and control. IEEE Transactions on Systems, Man and Cybernetics 1, 116–132.

Tikhonov, A.N., Arsenin, V.Y., 1977. Solution of Ill-Posed Problems. Winston and Wiley, Washington.

Tou, J.T., Gonzalez, R.C., 1974. Pattern Recognition Principles. Addison–Wesley Publishing.

Towers, P.D., Jones, B.Ll., 2014. Real-time wind field reconstruction from LiDAR measurements using a dynamic wind model and state estimation. Wind Energy 19 (1), 133–150. http://dx.doi.org/10.1002/we.1824.

Valencia-Palomo, G., Rossiter, J.A., Lopez-Estrada, F.R., 2014. Improving the feed-forward compensator in predictive control for setpoint tracking. ISA Transactions 53 (1), 755–766. http://dx.doi.org/10.1016/j.isatra.2014.02.009.

Valencia-Palomo, G., Rossiter, J.A., Lopez-Estrada, F.R., 2015. Programmable logic controller implementation of an auto-tuned predictive control based on minimal plant information. ISA Transactions 50 (1), 755–766. http://dx.doi.org/10.1016/j.isatra.2010.10.002.

van Engelen, T., Schuurmans, J., Kanev, S., Dong, J., Verhaegen, M., Hayashi, Y., 2011. Fault tolerant wind turbine production operation and shutdown (sustainable control). In: Proceedings of EWEA 2011. Brussels, Belgium.

Venkatasubramanian, V., Chan, K., 1989. A neural network methodology for process fault diagnosis. AIChE Journal 35, 1993–2002.

Visioli, A., 2003. Modified anti-windup scheme for PID controllers. IEE Proceedings. Control Theory and Applications 150 (1), 49–54.

Willecke, A., Fischer, T., 2013. Large Monopiles for Offshore Wind Farms in the German North Sea. Technical report. COME, Hamburg, Germany. Available from https://www.eoi.es/.

Willsky, A.S., 1976. A survey of design methods for failure detection in dynamic systems. Automatica 12 (6), 601–611.

Wu, Z.Q., Harris, C.J., 1996. Neuro-fuzzy modelling and state estimation. In: IEEE Medit. Symp. on Control and Automation: Circuits, Systems and Computers '96. Hellenic Naval Academy, Piraeus, Greece, pp. 603–610.

Wünnenberg, J., 1990. Observer-Based Fault Detection in Dynamic Systems. PhD thesis. University of Duisburg, Duisburg, Germany.

Wünnenberg, J., Frank, P.M., 1987. Sensor fault detection via robust observer. In: Tzafestas, S., et al. (Eds.), System Fault Diagnosis, Reliability, and Related Knowledge-Based Approaches, vol. 1, pp. 147–160.

Ying, H., 1994. Sufficient conditions on general fuzzy systems as function approximators. Automatica 30, 521–525.

Zhang, Y., Jiang, J., 2008. Bibliographical review on reconfigurable fault-tolerant control systems. Annual Reviews in Control 32, 229–252.

Zhang, Xin-fang, Xu, Da-ping, Liu, Yi-bing, 2004. Adaptive optimal fuzzy control for variable speed fixed pitch wind turbines. In: Intelligent Control and Automation, 2004, Fifth World Congress on, vol. 3. WCICA 2004. IEEE, pp. 2481–2485.

Zhang, Xiaodong, Zhang, Qi, Zhao, Songling, Ferrari, Riccardo M.G., Polycarpou, Marios M., Parisini, Thomas, 2011. Fault detection and isolation of the wind turbine benchmark: an estimation-based approach. In: Proceedings of IFAC World Congress, vol. 2, pp. 8295–8300.

Zhao, W., Stol, K., 2007. Individual blade pitch for active yaw control of a horizontal-axis wind turbine. In: Proceedings of the 45th AIAA Aerospace Sciences Meeting and Exhibit. AIAA, Reno, NV, USA. AIAA.

Zhou, K., Doyle, J.C., Glover, K., 1996a. Robust and Optimal Control. Prentice Hall, New Jersey.

Zhou, Kemin, Doyle, John Comstock, Glover, Keith, et al., 1996b. Robust and Optimal Control, vol. 40. Prentice Hall, New Jersey.

Ziegler, J.G., Nichols, N.B., 1942. Optimum settings for automatic controllers. Transactions of the American Society of Mechanical Engineers 64, 759–768.

Zolghadri, A., 1996. An algorithm for real-time failure detection in Kalman filters. IEEE Transactions on Automatic Control 41 (10), 1537–1539.

# Index

## Symbols

$H_\infty$ filter
  residual generation, 77
$H_\infty$ optimization, 77
  residual generation, 78
$\mathcal{H}_-/\mathcal{H}_\infty$, 81
  fault estimation, 81
  performance index, 84
$\mu$ synthesis
  residual generation, 78

## A

Active fault tolerance, 98
Active fault tolerant control, 97
Active FTC, 124
Actuator
  model, 23, 154
Adaptive algorithm, 90
Adaptive control, 88, 90, 91
Adaptive controller, 87, 92, 118
Adaptive filter, 85
Adaptive Frisch scheme, 92
Adaptive FTC, 126, 127
Adaptive NLGA, 131
Adaptive nonlinear filter, 131
Adaptive parametric model, 125
Adaptive PI controller, 93
Adaptive PI regulator, 129
Adaptive residual generation, 78
Advanced control, 117, 120, 121, 189
Aerodynamic
  map, 17
  rotor torque, 17
Aerodynamic model, 16, 131, 148
Aerodynamics, 6
  map, 16, 150
AFTC, 8, 99, 187
AFTCS, 124
Analytical redundancy, 7, 186
Anemometer, 29, 185
ANFIS, 69
Anti-windup, 120
Approach
  model-based, 7, 186
Availability, 3

## B

Baseline controller, 94, 130
Betz limit, 5, 186
Bias compensation, 90

Blade, 185
  bending, 18
  Element Momentum, 16, 150
Blades, 4
Brake, 4
Braking system, 19, 151
Bumpless control, 120
Bumpless transfer, 27, 96

## C

Capacity factor, 20
Centrifugal forces, 24
Comparative analysis, 142
Control
  adaptive, 11, 189
  advanced, 13, 24, 147, 154
  feedback, 27
  hierarchical, 11, 188
  overall extremum seeking, 10, 188
  PI, 26, 156
  power, 26
  reconfiguration, 3
  self-tuning, 11, 189
  speed, 26, 156
  sustainable, 9, 14, 148, 187
  torque, 24, 154
  variable-speed, 26, 156
Control algorithm, 89
Control reconfiguration, 98
Converter, 185
Converter system, 6
Cumulative SUM, 136
CUSUM, 136

## D

Data-driven approach
  control design, 87
Data-driven controller, 118
Data-driven fault diagnosis, 107, 139
Data-driven FDI, 113
Data-driven FTC, 125
Disturbance, 2, 11
Disturbance decoupling, 189
Drive-train, 16, 150
Dynamics
  Lagrangian, 17, 151

## E

Effect analysis, 38
Efficiency, 13, 147

Eigenstructure assignment, 8, 186
EIV
  fuzzy systems, 63
EKF, 54
Energy capture, 191
Energy cost, 8, 187
Error, 2
Errors-In-Variables, 63, 88
Extended Kalman Filter, 54

## F

Failure, 2
Failure mode, 38
False alarm rate, 115
FAST, 10, 21, 188
Fatigue damage, 189
Fault, 2
  abrupt, 3
  accommodation, 3
  actuator, 6, 186
  additive, 3
  controller, 6, 186
  detection, 3, 7, 186
  diagnosis, 3, 15, 148
  identification, 3
  incipient, 3
  intermittent, 3
  isolation, 3
  multiplicative, 3
  process, 6, 186
  propagation analysis, 36
  requirements, 39
  sensor, 6, 186
  specification, 39
  tolerance, 13, 147
Fault analysis, 36
Fault compensation, 98, 132
Fault detection, 140
  $H_\infty$ methods, 72, 73, 77
  active robustness, 73
  disturbance decoupling, 71
  neural networks, 68
  passive robustness, 73
  performance index, 73
Fault detection estimator, 115
Fault detection thresholds, 109, 111
Fault diagnosis, 97
  fuzzy systems, 62
  model-based, 43
  neural networks, 63, 67
  nonlinear system, 79

Printed in the United States
By Bookmasters